Lecture Notes in Computer Science 4853

Commenced Publication in 1973
Founding and Former Series Editors:
Gerhard Goos, Juris Hartmanis, and Jan van Leeuwen

Frederico Fonseca   M. Andrea Rodríguez
Sergei Levashkin (Eds.)

GeoSpatial Semantics

Second International Conference, GeoS 2007
Mexico City, Mexico, November 29-30, 2007
Proceedings

 Springer

Volume Editors

Frederico Fonseca
Pennsylvania State University
School of Information Sciences and Technology
USA
E-mail: fredfonseca@ist.psu.edu

M. Andrea Rodríguez
Universidad de Concepcíon de Chile
Chile
E-mail: andrea@udec.cl

Sergei Levashkin
National Polytechnic Institute
Centro de Investigacion en Computacion
Mexico
E-mail: sergei@levashkin.com

Library of Congress Control Number: Applied for

CR Subject Classification (1998): H.2, H.2.8, H.3, I.2, H.4, C.2

LNCS Sublibrary: SL 3 – Information Systems and Application, incl. Internet/Web and HCI

ISSN   0302-9743
ISBN-10   3-540-76875-0 Springer Berlin Heidelberg New York
ISBN-13   978-3-540-76875-3 Springer Berlin Heidelberg New York

Springer is a part of Springer Science+Business Media

springer.com

© Springer-Verlag Berlin Heidelberg 2007
Printed in Germany

Typesetting: Camera-ready by author, data conversion by Scientific Publishing Services, Chennai, India
Printed on acid-free paper   SPIN: 12193790   06/3180   5 4 3 2 1 0

Preface

GeoS 2007 was the second edition of the International Conference on Geospatial Semantics. It was held in Mexico City, November 29–30, 2007.

Semantics has become one of the most important topics of research in computer and information sciences. After many of the basic problems in information sharing and use were solved, it became time to face the most challenging one of all: how to make sense of all the information available even when it was collected and organized by different people under different premises?

Geographic information science is no different in this aspect. Although we start with more or less the same spatio-temporal cognitive abilities, when it comes to splitting the world into objects and fields, fiat and bona fide boundaries, and events and processes, we have a hard time understanding each other's information and its underlying assumptions. This is where GeoS 2007 tries to fit in and to bring some of the of state-of-the-art research results related to the many facets of modeling and processing geospatial semantics.

This volume contains 19 papers, which were selected from among 35 submissions received in response to the Call for Papers. Each submission was reviewed by three or four Program Committee members and 15 full and 4 short papers were chosen for presentation. The papers focused on geo-ontologies ranging from alignment and integration aspects to how to create and use them in information retrieval. Formal representations for geospatial data and the integration of semantics into spatial query processing were also favorite topics among the researchers who presented at Geos 2007. Overall, a very diverse body of research was presented coming from institutions in Brazil, France, Germany, India, Italy, Mexico, Spain, UK and the USA

We are in debt to many people who made this event happen. The members of the Program Committee offered their help with reviewing submissions. Our thanks go also to Marco Moreno, Miguel Torres, Rolando Quintero, and Giovanni Guzmán, who formed the Local Organizing Committee and took care of all the logistics. The Centro de Investigación en Computación, Mexico City, Mexico, was the local host and co-sponsored GeoS 2007. Finally, we would like to thank all the authors who submitted papers to GeoS 2007.

November 2007

Frederico Fonseca
M. Andrea Rodríguez
Sergei Levashkin

Organization

Organizing Committee

General Chair   Sergei Levashkin
Centro de Investigación en Computación,
Mexico City, Mexico

Program Chairs   Frederico Fonseca
The Pennsylvania State University, USA
M. Andrea Rodríguez
Universidad de Concepción, Chile

Local Organization   Centro de Investigación en Computación,
Mexico City, Mexico
Marco Moreno-Ibarra (Chair)
Miguel Torres
Rolando Quintero
Giovanni Guzmán

GeoS 2007 Program Committee

Ioan Marius Bilasco	Laboratoire d'Informatique de Grenoble (LIG), France
Stefano Borgo	LOA, ISTC-CNR, Italy
Boyan Brodaric	Geological Survey of Canada, Canada
Gilberto Câmara	INPE, Brazil
Isabel Cruz	University of Illinois at Chicago, USA
Clodoveu Davis Jr.	Pontificia Universidade Católica de Minas Gerais, Brazil
Denis Dean	Colorado State University, USA
Andrew U. Frank	Technical University Vienna, Austria
Christian Freksa	University of Bremen, Germany
Mark Gahegan	The Pennsylvania State University, USA
Marinos Kavouras	National Technical University of Athens, Greece
Alexander Klippel	The Pennsylvania State University, USA
Craig Knoblock	University of Southern California, USA
Margarita Kokla	National Technical University of Athens, Greece
Dave Kolas	BBN Technologies, USA
Werner Kuhn	University of Münster, Germany

Christopher Jones   Cardiff University, UK
Michael Lutz   European Commission - DG Joint Research
                              Centre, Italy
Marco Painho   ISEGI, Universidade Nova de Lisboa, Portugal
Matthew Perry   Wright State University, USA
Christoph Schlieder   Universtity of Bamberg, Germany
Shashi Shekhar   University of Minnesota, USA
Stefano Spaccapietra   Swiss Federal Institute of
                              Technology - Lausanne, Switzerland
Kathleen Stewart Hornsby   University of Iowa, USA
Nancy Wiegand   University of Wisconsin-Madison, USA
Stephan Winter   University of Melbourne, Australia
Mike Worboys   University of Maine, USA

GeoS 2007 Additional Reviewers

Lina Al-Jadir   Ecole Polytechnique Fédérale de Lausanne,
                             Switzerland
Fabiano Beppler   Universidade Federal de Santa Catarina, Brazil
Jose Antonio Fernandes   Ecole Polytechnique Fédérale de Lausanne,
  de Macedo   Switzerland
Krzysztof Janowicz   University of Münster, Germany
Celine Lopez-Velasco   Laboratoire d'Informatique de Grenoble,
                             France
Roberto Lucchi   European Commission - Joint Research Centre,
                             Ispra, Italy
Matthew Michelson   University of Southern California, USA
Martin Michalowski   University of Southern California, USA
Florian Probst   University of Münster, Germany
Andrey Soares   The Pennsylvania State University, USA
Snehal Thakkar   University of Southern California, USA
Rattapoom Tuchinda   University of Southern California, USA

Sponsoring Institutions

Instituto Politécnico Nacional (IPN), COFAA-IPN, SIP - IPN, Mexico
Centro de Investigación en Computación (CIC), Mexico
Consejo Nacional de Ciencia y Tecnología (CONACyT), Mexico

Table of Contents

Formal Representation for GeoSpatial Data

Integration of Semantics into Spatial Query Processing

Short Papers

Two Types of Hierarchies in Geospatial Ontologies

Sumit Sen

Dept. of Computer Science and Engineering, IIT Bombay,
Powai, Mumbai 400 076 India
sumitsen@uni-muenster.de

Abstract. Geospatial ontologies contain hierarchical structures, which are either based on the taxonomy of entity classes or functions and roles these entities can take. While the taxonomic hierarchies can be extracted from noun phrases contained in the formal texts that describe the geospatial domain, the hierarchies of action concepts can be traced from the verb phrases. This paper reports a simple case study of extracting the two types of such hierarchies from formal texts of traffic code. Problems of concurrent use of both hierarchies for ontology reasoning are dis-cussed, particularly, in context of the different views on geospatial ontologies. An approach based on separation of action and entity concepts. Use of probabilistic linkages between entities and actions is discussed as a way for integration of the two views. The initial results of this approach provide a first step towards an integration of the two existing views in the geospatial domain.

1 Introduction

Geographic Information (GI) is being increasingly used across domains and more recently across geographies as well. Semantics of geographic information has received focused attention in the recent years. Geospatial ontologies, which can be defined as *explicit specification of concepts*[1]. in the geospatial domain, have been suggested as the tool for semantic interoperability and translations [2], [3]. The content and also the nature of such geospatial ontologies have been different. This is evident in the divergent approaches that have evolved over the years. An existing challenge of ontology engineering in the geospatial domain is an integration of knowledge that are (or will be in future) specified in ontologies of different kinds and different approaches [4].

One major difference in the approaches, which is by no means trivial, is the relative significance given to functions of geospatial entities (or many times the lack of it) in the ontologies[5]. Geospatial ontologies have traditionally focused on taxonomic hierarchies of geospatial entities with little or no attention to the *occurents* in which they participate. This can be seen in the broader context the theory or SNAP and SPAN[6], where *occurents* can be described independently based on their own hierarchies. Functional properties can be seen as placeholders for such *occurents* and thus a way to specify knowledge about "possibilities" of occurrences of functions. For example, it is possible to say that a human can walk on a pedestrian road, which point to a function of *walking* and an affordance of *walkability*[1]. Such functions could

---

[1] For the function *walking* the agent is human and the environment consists of pedestrian roads.

F. Fonseca, M.A. Rodríguez, and S. Levashkin (Eds.): GeoS 2007, LNCS 4853, pp. 1–19, 2007.
© Springer-Verlag Berlin Heidelberg 2007

themselves have hierarchies based on entailment relationships that hold between two functions. Examples of such relationships are abundant among ontologies from the CAD domain [7],[8] . It is also important to note how such functions are related to roles in the broader sense and that function-based hierarchies are only a part of the bigger, complex hierarchy of roles. Given the significance of functions and roles in geospatial ontologies it is imperative to discuss strategies to integrate the two views and hence integrate function-based hierarchies with the taxonomic hierarchies of geospatial entities.

The need of such integration is evident in the transportation domain where semantics of symbols for GI artifacts such as roads, bridges and ferry-lines are often defined by the roles that they play and the possibility of entailment of other roles. Thus whether a *motorway* in the United Kingdom is semantically equivalent to a *freeway* in California can only be determined based on the function it has, with respect to agents such as motorcars, trucks, humans or bicyclists. Ontological reasoning, naturally, provides different results for equivalent concepts in the function-based hierarchies as compared to the taxonomic hierarchies of entities.

The function-based approach also differs from the taxonomy-based approach in terms of ontology extraction techniques as well. While nouns and noun phrases provide the taxonomic hierarchies of geospatial entities the role-based hierarchies is rather focused on the verb and verb phrases. This paper briefly describes both the approaches based on a case study of three different formal texts of traffic code, namely the highway code of the UK (HWC[2]) California state motor driving instructions (CSMDI[3]) NY state driver's manual (NYDM[4]). Two hierarchies each are extracted from these texts and are used for testing the hypothesis that an integrated approach gives better equivalence matching. This approach can be seen as an extension of existing work in the CAD domain, which segregates role concepts from relations [9]. The next subsection provides further background into the use of taxonomic hierarchies and role based hierarchies in geospatial ontologies.

The remaining of the paper is arranged as follows. The present section provides an introduction to the background motivation of this work. Previous work on extracting ontologies from texts is also summarized. Section 2 begins with the case study. The noun based taxonomic hierarchy and the verb based role hierarchy is presented therein. Section 3 discusses the differences between the two approaches and problems faced in integration of the two. Particular relevance in context of the two views among geographers about geospatial knowledge is shown. Some possible approaches to resolve this issue are discussed. Section four discusses the use of role placeholders for an integrated approach and presents some initial results based on equivalence among top concepts occurring in the case study. Finally we provide conclusions along with directions for future work in this area in section 5.

1.1 Motivation

Geospatial space consists of both geospatial entities and geospatial actions in relation to such entities [10]. The hierarchies of entities, which can be see in most

[2] http://www.highwaycode.gov.uk/
[3] http://www.dmv.ca.gov/pubs/hdbk/driver_handbook_toc.htm
[4] http://www.nydmv.state.ny.us/dmanual/default.html

conventional geospatial ontologies [11], [12] are often the backbone of geospatial ontologies. The view on ontologies as a categorization of things in the world based on their structural properties is simple and mostly effective. For example, we tend to classify wider ways as roads and narrow ways as streets. Such approaches are evident in classification of water bodies into rivers and streams [13]. However in most cases it is easy to identify that the semantics of symbols used to represent geospatial categories are strongly related to the geospatial activities of humans. Thus rivers would be those water bodies where activities such as boat navigation are possible.

What is important to bear in mind is that artifacts (such as roads, footpaths, crossings, etc) as opposed to naturally occurring entities (such as mountains, rivers, etc) are more likely to be clearly described by the human actions associated with them [5]. Such artifacts in the transportation domain, namely road network entities are the focus of our work. The view of actions or functions of entities that we adopt in this paper is rather broad and encompasses the viewpoint of the designed functionality as well as that of the affordance notion [14]. We argue that the complex notion of functions [15] is important to characterize the geospatial entities in question.

It has been argued that "increasing complexities of human actions leads to increased complexities of the concepts of the environment around them and not the other way around." [5] It is also mentioned that it is most likely that there are fewer concepts of such actions as compared to the entities. Under such a hypothesis it is important to analyze the hierarchies resulting from the original source with a similar methodology but with differing approaches as outlined above. The important aspects of this investigation are

i) Deriving hierarchies of concepts based on geospatial entity concepts: In this paper we restrict ourselves to the road network subspace of entities in the world. The hierarchies are taxonomies of such entities based on the principle that the properties of the parent entity are inherited by the child entities. The norm of inheritance in this case is that of specialization [16]. The inherited properties are rather structural in nature than role based.

ii) Function or action hierarchies are also extracted from the same subspace. Here the notion of entailment of verbs and hence the generalization-specialization relation between different functions are captured in a hierarchy. To say that a function Y is parent of another function W is distinctly different from saying (i) Y is a precondition for W (presupposition) or (ii) Y is a part of W (part of). These distinctions are similar to those experienced in forming hierarchies of entities. The additional clarification required in case of hierarchies of functions (and hence the corresponding verbs) is that of specialization of manner and specialization of goals. Sumigawa et al [9] points out the problems of using entities concepts and role concepts in a single hierarchy.

It is important to note that instances in the real world could also be categorized on the basis of such action-based categories. Thus it is possible to attach instances of so called Footpaths to the action concept $walk \cap (\neg drive)$. Although we do not deal with instances in our paper it is imperative to understand that function based hierarchies have equal if not higher capability to characterize things in the real world.

1.2 Background

In this section we describe the background work already done in the area of (i) extraction of ontologies from formal texts (ii) natural language processing concepts and tools used for such extraction.

Ontology sources: Creation of ontologies from existing knowledge bases has been a topic of research for many researchers and such efforts have included techniques of Part-Of-Speech (POS) tagging [17], concept clustering and formal concept analysis [18]. Documents on the web have also been explored for machine-generated ontologies. Kuhn [5] suggests that analyses of formal texts (such as traffic codes), as opposed to informal texts form a good source for ontology creation in a particular domain. This approach avoids logical inconsistencies and incompleteness that is often seen among informal documents. For our purpose it is important to avoid cross-linguistic issues and hence we choose traffic code texts from different countries but with the same language.

Text preparation: A next step in the analysis after choosing the text is that of preparing it for analysis. This includes identification of the words and phrasals, parts of speech tagging besides role and sense of the word or phrase. These individual tasks are research areas by themselves but we stay within practical limits of available tools.

Part of speech tagging: Machine based part of speech tagging includes identification of the part-of-speech based on the role played by each individual word in a sentence and tagging them with symbols such as the *Penn treebank tagset [19]*. By analyzing the word sense (for example WordNet senses) and the role of the given word in the sentence (based on chunking) it is possible to build efficient part of speech. GAMBL[17] is reported to be an efficient engine and utilizes a genetic algorithm taggers to assign part-of-speech along with Word Sense Disambiguation (WSD).

Frequency analysis: The concepts that occur in the texts have different frequencies of occurrence. If we assume that the weightage of each sentence in the text is equal to any other, we can obtain the significance of information available about a certain entity based on its frequency of occurrence. This frequency needs careful consideration because several forms of the conceptualization can occur. For example, the concept of walking could occur in several of its synonyms. Also in the absence of anaphora resolution [20] the frequency numbers could be significantly low (arguably more so for the very frequent noun terms).

Hypernym identification: Hypernyms are good sources for extracting taxonomic relations. However, hypernyms in linguistic sources (such as WordNet [21]) can be misleading due to misinterpretations of the inheritance relation with part of relations [22]. When used carefully, such relations serve as good starting point for building Directed Acyclic Graphs (DAGs) of concept hierarchies as shown by Kuhn [23].

We have used GAMBL and WordNet to achieve automation of our text analysis. Concept hierarchies based on analysis of text have been attempted by Madache, *et al* [18]. Some common principles and processes employed include POS and Word Clustering. Besides Kuhn [5], and Sen and Janowicz [24] extraction of action hierarchies can be found in the work of Kitamura, *et al* [7] and Sasijama, *et al* [25] .

However limited automation is available in all these approaches. Kuhn [5] has also noted that automation of text analysis is problematic and manual intervention is rather effective.

2 Case Study

Given the motivation and the background information about available methodologies and tools we now proceed towards our case study of extracting hierarchies of concepts from the three formal texts in question.

This section is divided into (i) the general analysis of the text, (ii) the extraction of noun-based entity hierarchies and (iii) extraction of verb based action hierarchies. Some assumptions made for our case study are summarized as below.

(1) It is assumed that the formal texts are both consistent and complete. This is rather idealistic and means that (i) meanings of words used in the text remain unchanged throughout the text and (ii) there is no extra information about the entities described in the text than what is available within the text.
(2) The text-analysis tools such as the POS tagger and the WordNet lexicon have their own limitations and for our case study we assume that the performance of the tools is consistent for both texts and caters to both US and UK versions of English.
(3) A noun phrase-verb phrase linkage is sufficient to establish the linkage between the corresponding entity concept and the action concept. Each evidence of such a linkage in the text accounts for a stronger belief that a certain entity is linked to a certain action. A negative linkage is established when a negation of the linkage is found in the text. (e.g. "You must not cycle on pavements")
(4) Pronouns are currently ignored in our case study but since this is done for all entities occurring in the text, the effect on the relative values of the entity-action linkages can be ignored initially. It can be speculated that the effect of analyzing pronouns for the entity-action linkages is rather pronounced in the case of more frequently occurring noun phrases. This is an area for future work.

2.1 Analysis of Formal Texts

The online texts of the UK Highway Code (HWC), California State Motor Driving Instructions (CSMDI) and the NY Driver's Manual (NYDM) were downloaded and analysed for Parts of Speech using GAMBL POS tagger. The output of the tagged text included the POS as well as the WordNet Sense definition (sample shown in table 1 below). The outputs are imported as a worksheet to help the further analysis.

Other steps involved in the text analysis include:

Frequency analysis: The worksheets containing the tagged texts are analysed for the occurrence of each word along with the particular WordNet sense (which includes the POS and the role of that word in the sentence).

Hypernym relation extraction: The word list generated from the frequency analysis is analysed for occurrences of hypernyms using WordNet. It is important to note that there are several senses of every word and it is necessary to use the particular sense of the word, which actually occurs in the word list generated previously.

Table 1. Part of tagged output text from GAMBL for NYDM (token 179-189)

#	Token	Lemma	POS	Chunk	Relation	Sense	Sense Definition
179	You	you	PRP	NP-B	NPSBJ-B	no-sense	
180	must	must	MD	VP-B	VP-B	no-sense	
181	come	come	VB	VP-I	VP-I	come%2:38:04::	reach a "destination " arrive by movement or by making "progress " "She arrived home at 7 o'clock " "He got into college " She didn't get to Chicago until after midnight
182	to	to	TO	PP-B	PNP-B	no-sense	
183	a	a	DT	NP-B	PNP-I	no-sense	
184	stop	stop	NN	NP-I	PNP-I	stop%1:11:00::	the event of something "ending " it came to a stop at the bottom of the hill
185	before	before	IN	PP-B	PNP-B	no-sense	
186	the	the	DT	NP-B	PNP-I	no-sense	
187	stop-line	stop-lin	NN	NP-I	PNP-I	stop-line%1:11:0	line indicating end of road;

2.2 Noun Based Hierarchies

For generating the noun-based hierarchies, steps described in 2.1 are carried out on words that occur as noun forms (NN, NNS, NNP, NNPS). Frequency lists of such

Table 2. Most frequent Road network entities from the New York Driver's manual

Term	Sense token	Sense definition
Driveway	Driveway%1:06:00	a road leading up to a private "house "; they parked in the driveway
Road	road%1:06:00	an open way (generally public) for travel or transportation
Lane	lane%1:06:00-(default)	a narrow way or road
Way	way%1:04:01	how a result is obtained or an end is "achieved " "a means of control " "an example is the best agency of instruction " the true way to success
Crosswalk	Crosswalk%1:06:00	a path (often marked) where a street or railroad can be crossed
Two-way(road)	two-way%5:00:00:bidirectional:00-(default)	operating or permitting operation in either of two opposite "directions " "a two-way valve " "two-way traffic " two-way streets
Street	street%1:06:00	a thoroughfare (usually including sidewalks) that is lined with buildings; they walked the streets of the small town; he lives on Nassau Street
U-turn	u-turn%1:04:00	complete reversal of direction of travel
Path	path%1:04:00	a course of conduct; the path of virtue; we went our separate ways; our paths in life led us apart; genius usually follows a revolutionary path
Route	route%1:15:00	an established line of travel or access
Incline	incline%1:06:00-(default)	an inclined surface or roadway that moves traffic from one level to another or axle (as in vehicles or other machines)
Expressway	expressway%1:06:00	a broad highway designed for high-speed traffic
Sidewalk	Sidewalk%1:06:00	walk consisting of a paved area for "pedestrians " usually beside a street or roadway

noun-based words are generated. To keep our efforts focused on the road network and for simplicity, we skip (i) non-spatial and other nouns, which do not form a part of a road network (ii) less frequent words (cut-off frequency). Table 2 shows the list of words obtained from the NYDM and HWC along with their sense definition. Similar table for CSDMI was obtained.

The terms from the three texts do overlap to a certain extent (e.g. Lane, Street, etc) but some words in the HWC (e.g. Motorway, Carriageway, etc) and some in the NYDM (e.g. Expressway, Crosswalk, etc) or the CSDMI (e.g. Freeway, Roundabout) do not find any corresponding matches in the other texts.

The hypernym relation analysis helps to generate the hierarchies of the concepts in the two given texts. Firstly the nouns are written as classes in an OWL ontology. The hypernym relations are modelled as subclass relations. Such an OWL-DL representation of the concepts is a primitive ontology of the conventional nature. Ontologies use relations and axioms to represent the knowledge in a particular domain. For the case of simplicity we continue with simple hierarchies of concepts and represent them as a Directed Acyclic Graph (DAG). The hierarchies shown in figure 1,2 and 3 below are the results of T-Box reasoning carried out on the OWL ontologies of NYDM, CSDMI and the HWC (we use Racer and RICE [Haarslev, 2003]). The DAGs are mainly concerned with network elements, which are represented as edges and assuming the term *Way* as the top concept.

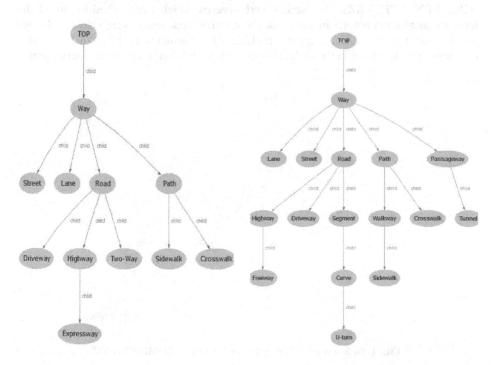

Fig. 1. Entity concept hierarchy for CSDMI (arcs denote *is-a* relations)

Fig. 2. Entity concept hierarchy for NYDM (arcs denote *is-a* relations)

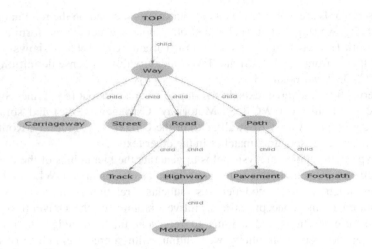

Fig. 3. Entity concept hierarchy for HWC

2.3 Verb Based Hierarchies

Steps similar to those described in 2.2 were repeated for verb occurrences (VB, VBD, VBG, VPN, VPP, VBZ). Only spatial verbs (verbs which were linked to any of the most frequent noun entities in any of the three traffic code texts) were selected and are listed in table 3 below. The frequency ranking for the terms were different for the two different texts as shown in the table. However all verb based terms were overlapping.

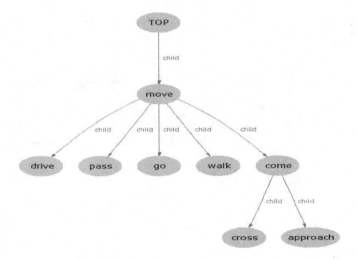

Fig. 4. Action concept hierarchy for NYDM, CSDMI and HWC

The single concept hierarchy (considering that frequency of occurrence has no effect on the concept hierarchy) is shown below in figure 4 as a DAG. It is important to note that the notion of verb entailment used by Kuhn [5] is broader than the notion used here. As discussed earlier, the hypernym notion is applied carefully to obtain the specialisation-generalisation type inheritance relations only. This relation is reflected by the troponymy relations in verbs (see [26]).

Table 3. Frequent Action concepts from both HWC and NYDM

Term	NYDM freq rank	CSDMI freq rank	HWC freq rank	Sense definition
go	1	4	1	move away from a place into another "direction " "Go away before I start to cry " The train departs at noon
cross	2	1	3	travel across or pass "over " The caravan covered almost 100 miles each day
drive	3	3	2	operate or control a vehicle; drive a car or bus; Can you drive this four-wheel truck?
pass	4	2	5	go across or "through " "We passed the point where the police car had parked "
approach	5	5	7	move "towards " "We were approaching our destination "
come	6	6	6	reach a "destination " arrive by movement or by making "progress "
turn	7	8	8	change orientation or "direction," also in the abstract sense; Turn towards me;
walk	8	7	4	use one's feet to "advance " advance by "steps " "Walk, don't run! "

3   Problems of Linking Entity and Their Functions

The previous section has described the extraction of hierarchies of entities and their functions independent of each other. Such a process is different from the processes of ontology extraction described by Maedche et al. [18]or Buitelaar et al[27]. Independent hierarchies allow working with the two hierarchies and this is particularly important in the different context of usage. For example noun-based ontology is usually helpful for databases of landforms (mountains, rivers and other physical structures). In other cases such as landuse databases or road network databases it is important to employ the action based concepts.

At the same time it is important to provide a link between these two hierarchies. However the linkages between the entity concepts and the function concepts are problematic because

1. There exist many to many linkages between entities and functions. Thus every entity has multiple functions and also the other way around [28].
2. The linkages are probabilistic and it is only possible to say that an entity has a certain function in certain cases. This probability can be used to express the relative strength of a linkage between an entity and a function in comparison to the linkage to another function. Thus a linkage of *walking* to *Footpaths* can be said to be higher than that of *driving* to *Footpaths*.

3. These probabilistic linkages undergo revision, as new knowledge about the entities is made available. Such a belief revision, which updates current linkages can be seen in the context of ontology evolution and is a continuous process.

Using functions as just another property of an entity has been a frequently used way of integrating functions inside entity hierarchies. However there are several issues in doing so. Evidently this approach subjugates the importance of action or function concepts in ontologies. This is not helpful, if we want to provide equal, or higher weightage to the knowledge about functions. We now examine some critical issues of attaching the function or action based properties to entity concepts.

3.1 Inheritance of Roles

Role concepts are defined by Sunagawa *et al*[9] as the concept of the role that an entity plays in a certain context. Thus *Road*s play a role of moving *Cars* from the starting node to the end node. It is important to note that there are many times when the *Roads* do not play that role, namely when it is under repair or it there is traffic congestion. In summary we can state that role based properties do not hold true for instances of entity concepts for some time during their lifetime. This observation is held strongly against inclusion of role-based properties in Subsumption hierarchies of ontologies[29] .

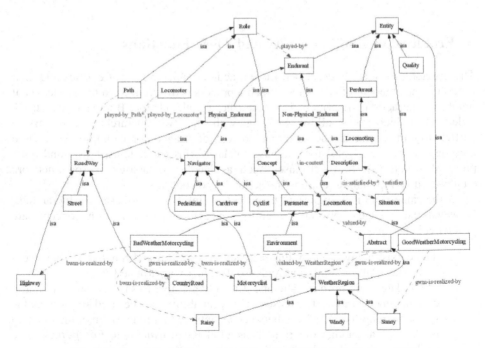

Fig. 5. Role and Entity hierarchies linked by "played by" relations [31]

Fan *et al*[30]. have also advocated the distinction between entities and roles. The view that role hierarchies can exist independent of the entities can also be seen in the work of Loos and Porzel [31] reproduced in figure 4 below.

The overlap between concept of roles and that of actions or functions is quite large. Actions or functions give rise to different roles and roles can be seen as bundles of such functions or actions. At the same time, an individual function could also serve as a role (for example *walk* can lead to a role of *walking areas* and all the entities such as *Footpaths* and *Pavements* are related to it). We now inspect issues in mixing such roles and entities.

3.2 Rigidity of Roles and Its Implication in Ontologies

Rigidity is an important tool from philosophy deployed in ontology engineering [32] and help to do away with unnecessary and problematic concepts in formal ontologies. A property is said to be rigid if it holds true for all instances of the entity class throughout its lifetime. Properties, which are not true for certain time periods in the lifetime of instances of a class, are classified as anti-rigid. Thus role-based properties are anti-rigid by nature. According to ontology engineering principles, anti-rigid concepts can only be subsumed by anti-rigid ones. (see Guarino and Welty [29]).

As discussed earlier, entities like *Footpaths* do not always serve as *walking areas* at all times because sometimes the footpaths are under construction or under repair (or even encroached upon by street vendors). Clearly, being a *Footpath* is rigid while being a walking area is anti-rigid. Thus it is not appropriate to include *areas of walking* as a subclass of *Footpath*.

3.3 Significance of Human Activities in Geography

While entities and their structures form a core part of the study of physical geography, human geography has traditionally relied on phenomenological account in which identity of entities emerge from social action. This important distinction has formed a part of the discussion on issues to integrate the two views of geography (see [33], [34] and [35]). In physical geography, identity is established using classical methods of ontology based on meaningful collections of attributes or on 'essences' of identity. The underlying principles of realism [36] and efforts to characterize geospatial objects as 'mesoscopic entities' are distinctly different from the cognitive view of the human geographer. The debate is similar to the realist vs. cognitive principles in ontologies (see[37]).

Massey [33] has discussed the need for physical geographers to remain within the claims of arguments of physics while stating their ontologies. She argues that human geography tends to look at these arguments in the context of human sciences and social interactions. She mentions arguements of the human geographer, that (social) spatiality and entities such as 'places' are products of our (social) interactions. This is similar to the view stated earlier in this paper, which gives primary status to actions and functions as opposed to the entities themselves. Although ontologies based on approaches of human geography is rather ambitious and untried, ontologies that treat actions and functions as first class citizens have been reported [5], [38].

Any approach to integrate the two different views of geography has to adapt one of the systems to the other along with the knowledge to do this. The issue in this case is that such knowledge is usually learnt during the lifetime of humans. Any claim of such knowledge is only a snapshot in time. Formal texts such as traffic codes are also such snapshots and it is possible to extract the linkage in a particular version of the text. Arguably the text itself undergoes revisions over time.

We now embark on extraction of the knowledge to adapt one of the views to the other. This is done by introducing probabilistic linkages. We recollect that such linkages are probabilistic in nature and undergo revision over time.

3.4 Probabilistic Linkages

Knowledge about linkages between the entities and the actions are available in the formal texts. Kuhn [Kuhn, 01] has reported that knowledge about the action concepts

Table 4. Linkages between Entities and Functions based on co-occurrence in NYDM and HWC

HWC	Street	Road	Footpath	Motorway	Lane	Way	Path	Crosswalk	Expresswa
move	0.015	0.049		0.012	0.107	0.035	-	-	-
walk	-	0.026	0.056	0.000	-	-	-	-	-
drive	0.057	0.062	0.000	0.069	0.000	-	-	-	-
enter	-	0.025	-	-	0.000	0.020	-	-	-
stop	0.010	0.075	-	0.000	0.000	0.051	-	-	-
be	0.014	0.215	0.006	0.028	0.061	0.033	0.014	-	-
cross	0.029	0.135	-	0.000	0.024	0.067	0.020	-	-
turn	0.038	0.059	-		0.042	0.041	-	-	-
wait	-	0.040	-	0.000	0.009	0.031	-	-	-
approach	0.022	0.052	-	0.016	0.065	0.045	0.023	-	-
go	-	0.021	-	-	0.063	-	-	-	-
pass	-	0.038	-	-	0.032	0.012	0.017	-	-
CSDMI									
move	-	0.025	-	-	0.036	-	-	-	-
walk	0.037	-	-	-	-	-	-	-	-
drive	0.006	0.117	-	-	0.100	0.008	0.009	0.000	-
enter	0.013	0.042	-	-	0.043	0.019	0.030	0.022	-
stop	0.019	0.041	-	-	0.005	0.015	0.018	0.069	-
be	0.049	0.119	-	-	0.111	0.035	0.011	0.030	-
cross	0.149	0.000	-	-	0.042	-	0.083	0.020	-
turn	0.094	0.034	-	-	0.191	0.036	0.013	-	-
wait	0.033	-	-	-	0.016	-	-	0.034	-
approach	-	-	-	-	-	0.024	-	-	-
go	0.027	0.043	-	-	0.029	0.083	-	0.023	-
pass	-	0.060	-	-	0.065	0.011	-	0.012	-
NYDM									
move	0.026	0.032	-	-	0.107	-	0.032	-	-
walk	-	0.010	-	-	-	-	-	-	-
drive	0.020	0.061	-	-	0.056	-	-	-	0.047
enter	0.025	0.048	-	-	0.077	0.041	-	0.053	0.064
stop	0.019	0.048	-	-	0.038	0.026	-	0.059	0.026
be	0.011	0.068	-	-	0.089	0.026	0.004	0.009	0.024
cross	0.061	0.033	-	-	0.017	0.071	-	0.030	
turn	0.037	0.080	-	-	0.094	0.051	0.029	0.018	0.008
wait	0.040	-	-	-	0.009	0.059	-	-	0.029
approach	0.015	0.060	-	-	0.034	-	-	-	0.026

afforded by different entities can be analysed based on a cross tabulation of entities and actions. We advance this piece of work by allowing probabilistic information (rather than deterministic information used in [5]) and use machine processing by employing a verb-noun co-occurrence model similar to the approaches of Cyre[39].

Table 4 above depicts the results of the analysis for each of the HWC, CSDMI and NYDM. The values represent the ratio of co-occurrence of the verb-noun to occurrence of the noun alone. In this table an occurrence of a negation (e.g. "You must not cycle on pavements") results in a zero value for the linkage. It is important to note that Motorway and Footpath have no occurrences in the NYDM whereas Crosswalk and Expressway have no occurrences in the HWC. We treat the cases of non-occurrences of the verb-noun pairs and the non-occurrence of noun similarly and state that no knowledge is available about their linkage (Note that all the verbs have occurrences in both the texts).

4 Linking Entities and Entity Functions

We have obtained quantitative values for the linkages between the entity concepts and function concepts in the previous section. However a clear formalism of using them in an ontology framework is difficult for the following reasons.

1. The problem of inheritance of roles and inferences across hierarchies is difficult to resolve
2. There is a need to link multiple function concepts to multiple action concepts. Such as requirement is quite messy in the context of undesirable multiple inheritances within an ontology.
3. Probabilistic information is difficult to use in formal ontologies and although formalisms exist to model uncertainty in ontology frameworks they are not easy to translate into popular formalisms such as OWL.

We discuss these aspects as below.

4.1 Using Roles and Role-Holder Concepts

As discussed earlier in section 3.2 roles and entities have been reported to be distinctly different in the context of ontologies. In particular, it is necessary to avoid mixing roles in a hierarchy of entities and vice versa. The role-holder concept used by Hozo [9] is one way of dealing with roles and entities separately. Role-holders (as shown in the figure 5 below) help to model the entities.

Role holder concepts are clearly an advantage because they serve as a link between both roles and entities that can take up such roles. However the knowledge used is deterministic and similar to the tabulation shown by Kuhn [5],as opposed to the (richer) probabilistic information available to us.

In the context of Hozo we also need to point out that reasoning capabilities on such ontologies is also restricted and the use of role-holders in Subsumption reasoning is unresolved.

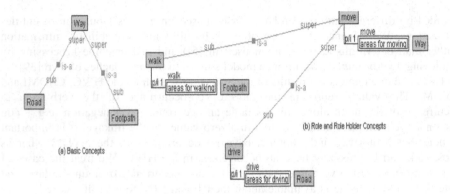

Fig. 6. Basic Concepts, Roles and Role holder concepts in Hozo

4.2 Requirements of Multiple Inheritances

Frank [28] has shown the importance of multiple inheritances in modelling spatial databases in view of the multiple applications, which use the same data. The underlying principle is that there are many to many correspondences between the database concepts and the application concepts. Parameterised inheritance is often used to cater to such requirements[16].

In the context of geospatial ontologies, which serve as a specification of the conceptualisation of different domains that interact with each other about the same entities (or otherwise), the need to allow users to construct classes based on multiple concepts is clear. However, use of multiple inheritances is a complicated issue and ontological engineering principles advocate avoiding them totally [29].

Nevertheless it is evident in many cases that entities and actions in geospatial space have many-many relation between themselves. Thus it is possible to show the concept *'Road'* related to *'walking'* or *'driving'* or even *'cycling'*. Such information is available from table 4.

4.3 Probabilistic Knowledge in Ontologies

The information contained in table 4 is probabilistic rather than deterministic and states that the link between an entity concept and a function concept is a given value between 0 and 1. We have stated already that such information is amenable to revision based on updated knowledge or revision of the formal text. It is therefore imperative to handle the links between entities and functions in a probabilistic framework.

Some probabilistic frameworks available in ontologies include that of Ding *et al.* [40]and that of Holi and Hyvönen [41]. Both the approaches are strikingly similar in their approaches of using Bayesian network to represent and ontology, BayesOWL provides not only a framework to build the probabilistic ontologies but also build a reasoning framework around them. These approaches are distinctively different from other probabilistic extensions proposed such as P-CLASSIC [42], Fuzzy DL [43] besides others [44]. The Bayesian network based ontologies do not propose new logic formalism for ontology representation but allows translation of existing DL based ontologies along with probabilistic knowledge [40]. The main steps in the construction of such ontologies as shown by us [45] include

1) Construction of two DAGs (as already discussed in section 2) for actions and entity concepts. Thereafter, use degree of similarity from the WordNet lexicon (See [46] to construct the Conditional Probability Tables (CPTs).
2) Linking enitity concepts to action concepts using the values from table 4 as conditional probabilities.

The results of such BayesOWL based geospatial ontologies are positive based on human subjects testing [47].

5 Conclusions and Future Work

In this section we discuss the conclusions of the work presented in this paper related to extraction of ontologies from text. We also present some ideas for future work in the context of the work reported in this paper. Some of these are being explored and probabilistic ontologies based on BayesOWL provide good intitial results.

5.1 Conclusions

In the paper we have presented an approach to extract ontologies from formal texts based on three different aspects of geographic space, viz, entities and actions. This approach assumes that ontologies in the form of hierarchies of entity concepts and hierarchies of action concepts can be specified independently. We have also seen that

1. In both cases, techniques in text analysis permitted the automation of ontology extraction from formal text of a particular domain, to a large extent. However such techniques require careful examination.
2. Several ontological principles govern the use of role-based concepts in ontologies and action concepts cannot be directly used in the ontologies.
3. Information about the linkages between these two types of concepts are available as well and were extracted from the traffic code texts. This information is relative in nature and requires a probabilistic framework for its representation.

We have employed simple and freely available tools for our case study and have extracted hierarchies for two different traffic code specifications.

The most important aspect of this case study is that it shows that the two different views of geographic space can exist independent of each other but also that there are linkages between them. Despite technical difficulties in linking the two types of ontologies a framework of ontology specification and reasoning is required which can combine both types of specifications (in the form of entity and action hierarchies) within a probabilistic framework.

5.2 Future Work

The case study reported in this paper is only the groundwork for a major step towards a new paradigm of ontology specifications in the geospatial domain. The integration of the two approaches of ontology specification requires a probabilistic framework for geospatial ontologies. It also requires techniques to link and translate between concepts of ontologies such that we are able to translate between *Motorway* concept of the HWC and the *Expressway* concept of the NYDM as discussed earlier.

Some of the areas that need to be furhter investigated are as follows

1. Since our tool involved use of available tools for text analysis, the results were limited to a certain extent by the limitations of the tools themselves. For example the GAMBL POS tagger shows an efficiency of about 80%. The WordNet lexicon has its own limitations. Improvement in the automation process would help to reduce the manual interventions required.
2. The analysis of verb-noun linkages currently ignores pronouns as discussed in section 3.4. It is important to evaluate the effect of considering pronouns in such analysis by using anaphora resolution tools [20].
3. A natural step forward from this case study would be to further develop the probabilistic framework for linking the two hierarchies (which in-turn represents the two different views of the geospatial domain) [45].
4. Finally we need to develop a mechanism to translate entity concepts from one ontology to another, based on their linkage to common action concepts. Such a mechanism would enable semantic reference translation and projection necessary to devise semantic reference systems as proposed by Kuhn [5].

Acknowledgements

The author acknowledges help from members of Institute of GeoInformatics, especially Werner Kuhn, for ideas presented in this paper. This work was completed with financial support from Ordnance Survey, UK.

References

1. Gruber, T.R.: Toward Principles for the Design of Ontologies Used for Knowledge Sharing. In: Guarino, N., Poly, R. (eds.) Formal Ontology in Conceptual Analysis and Knowledge Representation, Kluwer Academic Publishers, Dordrecht (1993)
2. Fonseca, F., Egenhofer, M.: Ontology-Driven Geographic Information Systems. In: 7th ACM Symposium on Advances in Geographic Information Systems, Kansas, ACM Press, New York (1999)
3. Kuhn, W.: Semantic Reference Systems. International Journal of Geographical Information Science 17(5), 405–409 (2003)
4. Agarwal, P.: Ontological Considerations in GIScience. International Journal of Geographical Information Science 19(5), 501–536 (2005)
5. Kuhn, W.: Ontologies in Support of Activities in Geographic Space. International Journal of Geographical Information Science 15(7), 613–631 (2001)
6. Grenon, P., Smith, B.: SNAP: and SPAN: Towards Dynamic Spatial Ontology. Spatial Cognition and Computation 4(1), 69–104 (2004)
7. Kitamura, Y., Koji, Y., Mizoguchi, R.: An Ontological Model of Device Function and Its Deployment for Engineering Knowledge Sharing. In: First Workshop FOMI 2005 - Formal Ontologies Meet Industry, Castelnuovo del Garda (VR), Italy (2005)

8. Chandrasekaran, B., Goel, A., Iwasaki, Y.: Functional Representation as Design Rationale. IEEE Computer 26, 48–56 (1993)
9. Sunagawa, E., Kozaki, K., Kitamura, Y., Mizogucho, R.: A Framework for Organizing Role Concepts in Ontology Development Tool: Hozo. In: Roles, an Interdisciplinary Perspective: Ontologies, Programming Languages, and Multiagent Systems (AAAI Fall Symposium) (2005)
10. Camara, G., Monteiro, A.M., Paiva, J., Souza, R.C.: Action-Driven Ontologies of the Geographical Space. In: GIScience 2000, Savannah, GA (2000)
11. ESRI: ArcGIS Transportation Data Model. ESRI (2001)
12. Sen, S., Somavarapu, S., Sarda, N.L.: Class structures and Lexical similarities of class names for ontology matching. In: ODBIS (VLDB Workshop on Ontologies-based techniques for DataBases and Information Systems), Seoul, South Korea (2006)
13. Kavouras, M., Kokla, M.: A Method for the Formalization and Integration of Geographic Categorizations. International Journal of Geographical Information Science 16(5), 439–453 (2002)
14. Gibson, J.: The Theory of Affordances. In: Shaw, R., Bransford, J. (eds.) Perceiving, Acting, and Knowing - Toward an Ecological Psychology, pp. 67–82. Lawrence Erlbaum Ass, Hillsdale, New Jersey (1997)
15. Barsalou, L., Sloman, S., Chaigneau, S.: The HIPE Theory of Function. In: Carlson, L., van der Zee, E. (eds.) Representing Functional Features for Language and Space: Insights from Perception, Categorization and Development, pp. 131–147. Oxford University Press, New York (2005)
16. Brady, A.F.: A Taxonomy of Inheritance Semantics. In: 7th international Workshop on Software Specification and Design, Redondo Beach, California, IEEE Computer Society Press, Los Alamitos (1993)
17. Decadt, B., Hoste, V., Daelemans, W.: GAMBL, Genetic Algorithm Optimization of Memory-Based WSD. In: Third International Workshop on the Evaluation of Systems for the Semantic Analysis of Text (Senseval-3), Barcelona, Spain (2004)
18. Maedche, A., Staab, S.: Semi-automatic Engineering of Ontologies from Text. In: 12th International Conference on Software Engineering and Knowledge Engineering (SEKE2000), Chicago, IL, USA (2000)
19. Marcus, M., Kim, G., Marcinkiewicz, M.A., MacIntyre, R., Bies, A., Ferguson, M., Katz, K., Schasberger, B.: The Penn Treebank: A Revised Corpus Design for Extracting Predicate Argument Structure. In: ARPA Human Language Technology Workshop, Morgan-Kaufman, Princeton (1994)
20. Mitkov, R., Evans, R., Orasan, C.: A New, Fully Automatic Version of Mitkov's Knowledge-Poor Pronoun Resolution Method. In: Third International Conference on Intelligent Text Processing and Computational Linguistics, Mexico (2002)
21. Fellbaum, C.: WordNet. An Electronic Lexical Database. MIT Press, Cambridge (1998)
22. Gangemi, A., Guarino, N., Mosolo, C., Oltramari, A.: Sweeting WordNet with DOLCE. AI Magazine 24(3) (2003)
23. Kuhn, W.: Modeling the Semantics of Geographic Categories through Conceptual Integration, in Geographic Information Science. In: Egenhofer, M.J., Mark, D.M. (eds.) GIScience 2002. LNCS, vol. 2478, pp. 108–118. Springer, Berlin (2002)
24. Sen, S., Janowicz, K.: Semantics of Motion verbs. In: Workshop on Spatial Language and Dialogue (WOSLAD-05), Delmenhorst, Germany (2005)
25. Sasijima, M., Kitamura, Y., Ikeda, M., Mizoguchi, I.: FBRL: A Function and Behavior Representation Language. In: IJCAI (1995)

26. Fellbaum, C.: On the Semantics of Troponymy. In: Green, R., Bean, C., Myaeng, S. (eds.) The Semantics of Relationships: An Interdisciplinary Perspective, pp. 23–24. Kluwer, Dordrecht, Holland (2002)
27. Buitelaar, P., Olejnik, D., Sintek, M.: A Protégé Plug-In for Ontology Extraction from Text Based on Linguistic Analysis. In: 1st European Semantic Web Symposium (ESWS), Heraklion, Greece (2003)
28. Frank, A.: Multiple Inheritance and Genericity for the Integration of a Database Management System in an Object-Oriented Approach. In: 2nd International Workshop on Object-Oriented Database Systems, Springer, Heidelberg (1987)
29. Guarino, N., Welty, C.: An Overview of OntoClean. In: Staab, S., Struder, R. (eds.) Handbook on Ontologies, Springer, Heidelberg (2004)
30. Fan, J., Barker, K., Porter, B., Clark, P.: Representing Roles and Purpose. In: International Conference on Knowledge Capture (K-Cap2001), Victoria, B.C., Canada, ACM Press, New York (2001)
31. Loos, B., Porzel, R.: Towards Ontology-based Pragmatic Analysis. In: DIALOR 2005. Ninth Workshop on the Semantics and Pragmatics of Dialogue (SEMDIAL), Nancy, France (2005)
32. Guarino, N., Welty, C.: Identity, Unity, and Individuality: Towards a Formal Toolkit for Ontological Analysis. In: Proceedings of the 14th European Conference on Artificial Intelligence, ECAI-2000, IOS Press, Amsterdam (2000)
33. Massey, D.: Space-Time, 'Science' and the Relationship between Physical Geography and Human Geography. Transactions of the Institute of British Geographers 24(3), 261–276 (1999)
34. Lane, S.N.: Constructive Comments on D Massey Space-time, "Science" and the Relationship Between Physical Geography and Human Geography. Transactions of the Institute of British Geographers 26(2:243) (2001)
35. Raper, J., Livingstone, D.: Let's Get Real: Spatio-Temporal Identity and Geographic Entities. Transactions of the Institute of British Geographers 26 (2001)
36. Smith, B.: Formal Ontology, Commonsense and Cognitive Science. International Journal of Human and Computer Studies 43 (1995)
37. Smith, B.: Beyond concepts: Ontology as Reality Representation. In: FOIS 2004. International Conference on Formal Ontology and Information Systems, Turin, Italy, IOS Press, Amsterdam (2004)
38. Raubal, M., Kuhn, W.: Ontology-Based Task Simulation. Spatial Cognition and Computation 4(1), 15–37 (2004)
39. Cyre, W.R.: Knowledge Extractor: A Tool for Extracting Knowledge from Text. In: Delugach, H.S., Keeler, M.A., Searle, L., Lukose, D., Sowa, J.F. (eds.) ICCS 1997. LNCS, vol. 1257, Springer, Heidelberg (1997)
40. Ding, Z., Peng, Y., Pan, R.: BayesOWL: Uncertainty Modelling in Semantic Web Ontologies. In: Soft Computing in Ontologies and Semantic Web, Springer, Heidelberg (2005)
41. Holi, M., Hyvöonen, E.: Probabilistic Information Retrieval based on Conceptual Overlap in SemanticWeb Ontologies. In: 11th Finnish AI Conference, Web Intelligence. Finnish AI Society, Finland (2004)
42. Koller, D., Levy, A., Pfeffer, A.: P-CLASSIC: A Tractable Probabilistic Description Logic. In: AAAI-97 (1997)
43. Straccia, U.: A fuzzy description logic. In: 15th Nat.Conf. on Artificial Intelligence (AAAI-98), Madison, USA (1998)

44. Stuckenschmidt, H.,Visser, U.: Semantic Translation Based on Approximate Re-Classifcation. In: KR workshop on semantic approximation granularity and vagueness, Breckenridge, (CO US) (2000)
45. Sen, S.: Linking Hierarchies of Entities and their Functions in Geospatial Ontologies. Journal of Geomatics 1(1) (2007)
46. Patwardhan, S., Pedersen, T.: Using WordNet Based Context Vectors to Estimate the Semantic Relatedness of Concepts. In: EACL 2006 Workshop Making Sense of Sense - Bringing Computational Linguistics and Psycholinguistics Together, Trento, Italy (2006)
47. Sen, S.: Human Perspectives in Semantic Translations - Role of Entity Functions. In: GI_Forum, Salzburg (2007)

Semantic Annotation of Maps Through Knowledge Provenance

Nicholas Del Rio, Paulo Pinheiro da Silva,
Ann Q. Gates, and Leonardo Salayandia

The University of Texas at El Paso, Computer Science,
500 W. University Ave. El Paso TX 79968 USA
ndel2@miners.utep.edu, {paulo,agates,leonardo}@utep.edu

Abstract. Maps are artifacts often derived from multiple sources of
data, e.g., sensors, and processed by multiple methods, e.g., gridding and
smoothing algorithms. As a result, complex metadata may be required to
describe maps semantically. This paper presents an approach to describe
maps by annotating associated provenance. Knowledge provenance can
represent a semantic annotation mechanism that is more scalable than
direct annotation of map. Semantic annotation of maps through knowledge provenance provides several benefits to end users. For example, a
user study is presented showing that scientists with different levels of
expertise and background are able to evaluate the quality of maps by
analyzing their knowledge provenance information.

1 Introduction

Maps are expected to be generated, understood, accepted, shared, and reused
by scientists like many other scientific products, e.g., reports and graphs. Semantic annotation of maps is often necessary to assure that scientists are able
to understand and evaluate information represented by maps. For example, map
annotation can be used by scientists not involved in a map generation process
to understand the properties of the map, e.g., recency, geospatial coverage, and
data sources used, and evaluate the map against some established criteria, e.g.,
that the data used in the map generation of the map came from a reliable source.
Once a scientist understands and accepts a given map, the scientist can confi-
dently reuse and share the map to save time and resources of other collaborators
that would otherwise be required to be regenerated.

There are different methods for annotating maps and images in general, each
with their respective benefits. For instance, semantic annotation of maps may be
achieved by defining map artifacts as instances of semantic concepts comprising
an ontology and may involve the annotation of the resources used to generate maps (e.g., source data types, intermediate data types, and transformation
methods). However, a small variation in the generation process of a map, e.g.,
the use of a different filtering algorithm, would require the introduction of at
least a new class in the ontology, along with new semantic annotations. Another
challenge of this approach is that it becomes difficult to reuse existing domain

F. Fonseca, M.A. Rodríguez, and S. Levashkin (Eds.): GeoS 2007, LNCS 4853, pp. 20–35, 2007.

ontologies to annotate semantic information. For example, suppose there existed an ontology developed by a third party that contained semantic annotations for general-purpose filtering algorithms; the annotations provided by such ontology might not be rich enough to capture the relationship between a filtering algorithm and its particular application to generate a map artifact.

Provenance information in general is meta-information that can be used to document how products such as maps are generated. Provenance often includes meta-information about the following: original datasets used to derive products; executions of processes, i.e., traces of workflow executions and composite services execution; methods called by workflows and composite services, i.e., services, tools, and applications; intermediate datasets generated during process executions; and any other information sources used. This paper refers to the term *Knowledge provenance* (KP) [1], to account for the above meta-information that includes *provenance meta-information*, which is a description of the origin of a piece of knowledge, and *process meta-information*, which is a description of the reasoning process used to generate the answer, which may include intermediate datasets referred to as *intermediate results*. We have used the phrase "knowledge provenance" instead of data provenance intentionally. Data provenance [2,3] may be viewed as the analog to knowledge provenance aimed at the database community. That community's definition typically includes both a description of the origin of the information and the process by which it arrived in the database. Knowledge provenance is essentially the same except that it includes proof-like information about the process by which knowledge arrives in the knowledge base. In this sense, knowledge provenance broadens the notion of data derivation that can be performed before data is inserted into a database or after data is retrieved from a database. Nevertheless, data provenance and knowledge provenance have the same concerns and motivations. In this paper we describe how KP can be used to semantically annotate maps and how this semantic information can help scientists to understand and evaluate map products.

The rest of this paper is organized as follows. Section 2 introduces a scenario where a map is generated through a workflow executing over cyberinfrastructure services. Section 3 describes how these services are instrumented to log KP about the workflow execution. Section 4 describes how KP annotation can be used by scientists to better understand how maps are generated. Section 5 describes a user study that demonstrates the need of scientists to have access to KP associated with maps. Section 6 discusses the pros and cons of annotating maps while Section 7 concludes the paper

2   Gravity Map Annotation: An Example

2.1   Gravity Map Scenario

Contour maps generated from gravity data readings serve as models from which geophysicists can identify subterranean features. In particular, geophysicists are often concerned with data anomalies, e.g., spikes and dips, because these are usually indicative of the presence of some subterranean resource such as a water table or an oil reserve. The Gravity Map scenario described in this section is based on a cyberinfrastructure application that generates such gravity contour maps from the Gravity and Magnetic Dataset Repository[1] hosted at the Regional Geospatial Service Center at the University of Texas at El Paso. In this scenario, scientists request the generation of contour maps by providing a footprint defined by a pair latitude and longitude coordinates; this footprint specifies the 2D spatial region of the map to be created. The following sequence of tasks generate gravity data contour maps in this scenario:

1. *Gather Task*: Gather the raw gravity dataset readings for the specified region of interest
2. *Filter Task*: Filter the raw gravity dataset readings (remove unlikely point values)
3. *Grid Task*: Create a uniformly distributed dataset by applying a gridding algorithm
4. *Contour Task*: Create a contoured rendering of the uniformly distributed dataset

Each of the tasks involved in this scenario are realized by a web service, thus emphasizing the use of a loosely coupled, distributed environment comparable to that of a cyberinfrastructure, where semantic annotation information is particularly critical. Furthermore, this particular scenario can be viewed as a pipeline, where the output of a task is used as input in the subsequent task. The specification stating that these tasks must be sequentially executed in the order described above can be viewed as an executable workflow and it is further described in Section 3.2. Of course it is possible to implement the required functionalities as a single autonomous application, however, the availability of these services over the Web as smaller cohesive tasks allows for greater possibility of reuse especially in other domains; tasks 3 and 4 are not specific to gravity data.

2.2   Gravity WDO

Services, datasets, and workflow specifications in the scenario need to be semantically described by an ontology if one wants to understand contour maps about gravity data. In this paper, we rely on the Gravity Workflow Driven Ontology as a source of gravity map concepts and relationships.

Dr. Randy Keller, a leading expert on gravity data, worked with Flor Salcedo to encode his knowledge in the gravity field as an ontology. The development of the ontology was part of the NSF-funded GEON Cyberinfrastructure project [4], and it is part of a concentrated effort to capture essential knowledge about the Gravity domain as it is applied to Geophysical studies. The initial motivation for the effort was to document and share gravity terminology and resources within the GEON community. At the time of this writing, the Gravity ontology contains more than 90 classes fully documented.

---

[1] http://irpsrvgis00.utep.edu/repositorywebsite/

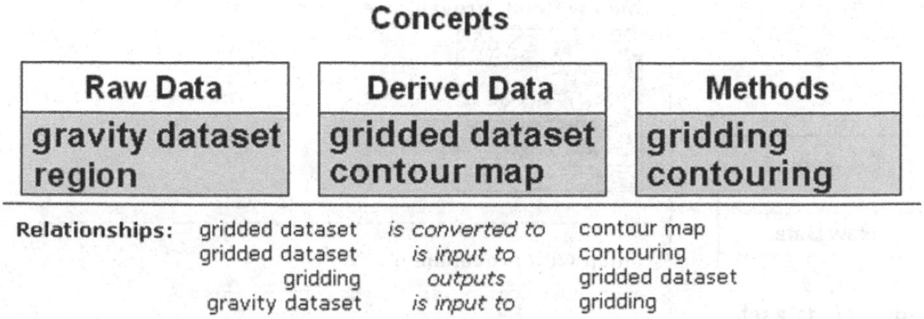

Concepts

Raw Data	Derived Data	Methods
gravity dataset region	gridded dataset contour map	gridding contouring

Relationships:			
	gridded dataset	*is converted to*	contour map
	gridded dataset	*is input to*	contouring
	gridding	*outputs*	gridded dataset
	gravity dataset	*is input to*	gridding

Fig. 1. Gravity Ontology

Figure 1 presents a visual representation of three upper level classes in the ontology class hierarchy from the Gravity ontology and some of the subclasses related to producing a gravity contour map, e.g., *region* and *gridding*. The Gravity ontology specifies multiple relationships between classes across the three hierarchies; for clarity the relations that are associated with the classes are listed in the sidebar of the figure rather than shown graphically. In the case of workflow-driven ontologies, it is expected that the different types of services published on the cyberinfrastructure are represented as classes defined in an ontology used to create workflows. Consequently, services that correspond to classes under the *Raw Data* and *Derived Data* hierarchy of the ontology are services that provide access to data repositories; services that correspond to classes under the Method hierarchy are services that take data as input, provide some functionality that can transform the data, and outputs the transformed data; and services that correspond to classes under the Product hierarchy are services that provide access to an artifact library.

The relationships between classes provide the basic roadmap to specify complex functionality through composition of services. As an example, consider the second row of the relationship sidebar in Figure 1 that shows the *outputs* between the classes *gridding* and *gridded dataset*. This relationship suggests that, given a service that corresponds to the *gridding* class, a service composition is viable that would result in *derived data* corresponding to a *gridded dataset* class.

2.3   Semantic Annotation of the Gravity Map and Related Work

Semantics are associated with artifacts, such as maps, through appended meta-information known as annotations. Annotations serve as the link between concepts defined in ontologies and artifacts; annotations are simply tags that refer to some concept. For instance, the gravity contour map resulting from the gravity map scenario, can be associated with the *contour map* concept defined in the gravity ontology as shown in Figure 2 without provenance. Scientists or agents would be able to unambiguously identify this artifact as a *contour map*.

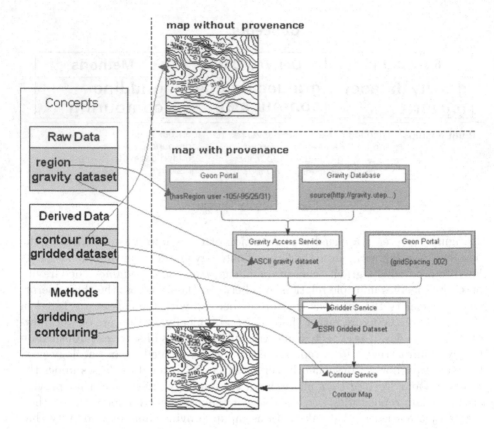

Fig. 2. Annotations with and without provenance

The current practice is to associate each object in some domain with only one concept in an ontology. Usually, only a single artifact itself (i.e., a map) is annotated by either concepts in some ontology or with arbitrary terms or captions as in Google Maps [5], ArcGis [6], and XML for Image and Map Annotations (XIMA) [7]. In Google Maps, annotations are limited to the map as a whole or for particular latitudes/longitudes (i.e., single points) in the map. In contrast, ArcGis and XIMA allow users to annotate whole maps, points on a map and sub-regions (i.e., subsections defined by polygons) of a map using text based captions. In all of these cases however, only the final map or image is annotated, where as the approach presented in this paper aims to annotate both the map and associated knowledge provenance.

For many cases however, a single concept or annotation may provide adequate semantic information for both human and software agents to correctly manage the artifact. In the gravity map example, the geospatial region provided by the scientist is associated with a *region* concept in the gravity ontology. This provides enough information to the service represented by task 1 to know that the input contains both upper and lower latitudes and longitudes in some particular format and thus facilitate correct parsing; the format of *region* would also be defined in the gravity ontology. A single concept annotation however may not always be enough to define a complex artifact such as a map. Referring to the gravity scenario, the gravity ontology defines a concept *contour map*, which can be used to semantically define the resultant gravity contour map. However, this concept, by itself, says nothing about what kind of map it is (i.e. what kind of data was used to generate this map). If an ontology is very rich, then perhaps complex reasoning might provide answers to the questions posed. Even so, in the case of the gravity ontology, there are many methods defined which generate contour maps (i.e. they all have an *outputs* relationship with the *contour* concept). Reasoning alone could not indicate which methods were used to generate a particular instance of a contour map. A more explicit way to semantically define the map may be to associate both the map and its knowledge provenance to concepts in the ontology as shown in Figure 2 with provenance. In this case, most of the KP can be associated with some concept in an ontology providing better utilization of the knowledge and a much richer description of the artifact. KP already contains the process by which the map was generated including all intermediate data such as the raw gravity dataset. If the gravity dataset, contained in the KP, was defined as an instance of *gravity data*, then any scientist or software agent could quickly realize that the contour map was generated by gravity data and is thus a gravity contour map. In this sense, KP is the medium through which additional semantics, that might otherwise have to be deduced by reasoning, can be appended to the artifact. Adding semantics to KP associated with some artifact in turn adds richer descriptions of the artifact itself. A few systems including PSW, described in Section 3, and MyGrid [8], from the e-science initiative, provide provenance associated with complex artifacts while leveraging ontologies to further enrich the provenance descriptions.

Once KP has been annotated with concepts in the ontology, tools can be used to view this semantically defined provenance. Section 4 further explores such a tool and potential uses.

3 Capturing Gravity Map Knowledge Provenance

3.1 The Inference Web and the Proof Markup Language (PML)

The Inference Web [9,10] is a knowledge provenance infrastructure for conclusions derived from inference engines which supports interoperable explanations of sources (i.e. sources published on the Web), assumptions, learned information, and answers associated with derived conclusions, that can provide users with a level of trust regarding those conclusions. The goal of the Inference Web is the same as the goal of this work which is to provide users with an understanding of how results are derived by providing them with an accurate account of the derivation process (i.e. knowledge provenance), except that this work deals with workflows rather than inference engines; workflow knowledge provenance encompasses a range of complex artifacts such as datasets and corresponding visualizations while inference Web provenance always consists of logical statements leading to some final conclusions and can thus be regarded as a justification.

Inference Web provides the Proof Markup Language (PML) to encode KP. PML is an RDF based language defined by a rich ontology of provenance and justification concepts which describe the various elements of automatically generated proofs. The main concept defined in PML is *node set*, which contains both a conclusion (i.e., a logical expression) and a collection of inference steps each of which provide a different justification of the conclusion; in its simplest composition, a single PML node set simply represents a single proof step. Inference steps themselves contain a number of elements including antecedents, rule, and inference engine, which correspond to the rule antecedents, the name of the rule applied to the antecedents, and the name inference engine responsible for the derivation respectively. In PML, antecedents are simply references to other node sets comprising the rest of a justification. Thus PML justifications are graphs with node sets as nodes and antecedents acting as edges. This graph is directed and acyclic, with the edges always pointing towards the direction of root, the conclusion of the entire proof. In this sense, node sets always contribute to the final conclusion.

PML justifications can also be used to store KP information associated with scientific workflow execution. From this perspective, node sets represent the execution of a particular web service; the node set conclusion serves as the output of the service (i.e., and intermediate result) while the inference step represents provenance associated with the service's function. For example, elements antecedent, rule, and the inference engine can be used to describe the service's inputs, function, and name or hosting organization respectively. Additionally, the links between nodesets can be viewed as an execution sequence of a workflow.

PML itself is defined in OWL [11,12] thus supporting the distribution of proofs throughout the Web. Each PML node set comprising a particular justification can reside in a uniquely identified document published on the Web separately from the others. The workflows considered in this research are service oriented and thus distributed. The support provided by PML is so well suited for scientific workflows that it is used as the provenance interlingua for out KP browser Probe-It! briefly described in Section 4. It is also relevant to mention that PML addresses only the encoding issues related to provenance but prescribes no specific method for collecting it.

3.2   Workflows and the PML Service Wrapper (PSW)

The gravity map scenario is realized by a service-oriented workflow composed of four Simple Object Access Protocol (SOAP) services, which gather, filter, grid and contour gravity datasets respectively. These Web services are piped or chained together; the output of one service is forwarded as the input to the next service specified in the workflow. A workflow director is responsible for managing the inputs/outputs of each service as well as coordinating their execution. KP associated with scientific workflows of this nature might include the services execution sequence as well as each of their respective outputs, which we refer to as *intermediate results*.

PML Service Wrapper (PSW) is a general-purpose Web service wrapper that logs knowledge provenance associated with workflow execution as PML documents. In order to capture knowledge provenance associated with workflows execution, each service composing the workflow has an associated PSW wrapper that is configured to accept and generate PML documents specific to it. Since PML node sets include the *conclusion* element, which is used to store the result of an inference step or Web service, the provenance returned by the wrappers also includes the service output thus workflows can be composed only of these PSWs; this configuration introduces a level of indirection between service consumers (i.e. workflow engine) and the target services that performs the required function. In this sense, PSW can be seen as a server side provenance logger.

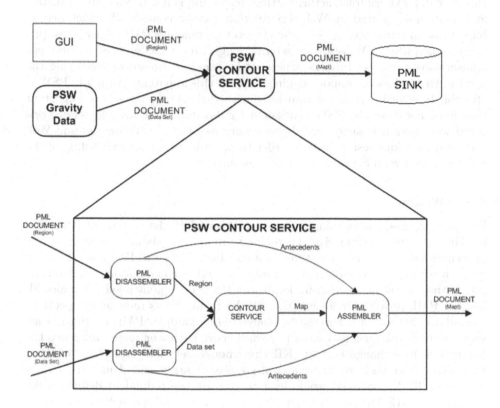

Fig. 3. Example of PSW configured for a contouring service

The logging capability provided by PSW can be decomposed into three basic tasks: decompose, consume, and compose as illustrated in Figure 3. Upon invocation, the wrapper decomposes the conclusion of an incoming PML document, i.e., extracts the data resident in the PML conclusion using Inference Web's PML API. PSW then consumes the target service, forwarding the extracted data as an input to the target service. The result and associated provenance of the target service is then composed to produce the resultant PML document, the PSW output. For example, a contouring service requires 2D spatial data to map and the region to consider in the mapping therefore a PSW wrapper for this contouring service would require two PML documents, one containing 2D spatial data, coming from some data retrieval service, and the other containing a region, (e.g. specified by latitude and longitude) specified by some user. The output of the contour service is a map, from which a new PML document is created, referencing the two input PML node sets as antecedents.

PSW has been developed in support of scientific workflows able to execute in a distributed environment such as the cyberinfrastructure. In traditional Inference Web applications [13,10], inference engines are instrumented to generate PML. However in a cyberinfrastructure setting, reasoning is not necessarily deductive and is often supported by Web services that can be considered "black boxes" hard to be instrumented at source-code level to generate PML. This is the primary reason why PSW, a sort of external logger, must be deployed to intercept transactions and record events generated by services instead of modifying the services themselves to support logging. Despite this apparent limitation, PSW is still able to record provenance associated with various target systems' important functions. For example, PSW configured for database systems and service oriented workflows can easily record provenance associated with queries and Web service invocations respectively in order to provide a thorough recording of the KP associated with cyberinfrastructure applications.

3.3   IW-Base

For querying and maintaining large quantities of KP, the parsing of PML files has shown to be too expensive. Therefore, to increase scalability, certain generic provenance elements are also stored in a database known as IW- Base [14]. The result are PML documents that can reference KP elements stored in IW-Base rather than including their defintion in the PML document itself. This also alleviates PML provenance loggers (i.e., PSW) from always re-generating certain meta-data that could otherwise be shared. For example, PML documents associated with conclusions from the Java Theorem Prover might reference the Knowledge Interchange Format (KIF) provnenace element stored in IW-Base, to indicate that their resulting logical statement are encoded in KIF. Otherwise, each PML document would have to contain the redundant definition describing the KIF format. Additionally, having a centralized defintion of some elements supports interoperability when sharing KP among Inference Web tools and between Inference Web tools and other Semantic Web tools in general. Thus, IW-Base can serve as standard of defintions, for provenance elements that are commonly used. This paper proposes that an ontology can supplement the information contained in IW-Base, by providing additional semantic defintions of certain PML elements. For example, traditional PML documents associated with services that retrieve gravity data might reference the *ASCII dataset* definition in IW-Base to indicate that the dataset is in ASCII tabular format. This paper proposes that PML documents should also reference concepts in an ontology, such as *gravity data*, in order to provide a richer description of the services' outputs. In an inference Web scenario, inference engines mainly output logical statements, which semantics are provided within the statement itself, thus only the format of the statement is an issue. In a cyberinfrastructure scenario, conclusions range from datasets and reports to complex visualizations, thus associated semantic defintions of these different data becomes more necessary.

IW-Base critically depends on the IW-Base registry and IW-Base registrar. An IW-Base registrar is a collection of applications used for maintaining an IW-Base registry. From a human user point of view, the registrar is an interactive application where the user can add, update, and browse the registry contents. From a software agent point of view, the registrar is a collection of services for querying and updating the registry. The registrar is also responsible for keeping the synchronization between the registry database or provenance elements and the OWL files representing those elements.

4   Using Annotated Gravity Map

Users who store their provenance as a collection of PML documents can use Probe-It!, a KP visualization tool, to view their information. Probe-It! is capable of graphically rendering every aspect of KP associated with map generation on the cyberinfrastructure. Figure 4 illustrates the renderings provided by Probe-It! in visualizing the KP associated with a gravity contour map. The left side of the

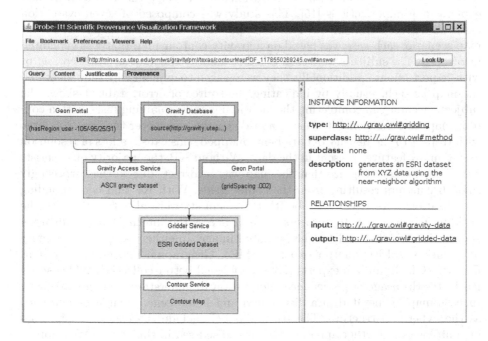

Fig. 4. Probe-It! Provenance Viewer

screen presents the KP associated with the execution trace visualized as a DAG. In this representation, data flow is represented by edges; the representation is such that data flows from the leaf nodes towards the root node of the DAG, which represents the final service invoked in the workflow. The DAG essentially contains two types of nodes, workflow inputs and information transformation services corresponding to the workflow inputs and invoked Web services respectively. Upon clicking on the nodes (i.e. KP elements) comprising the execution DAG, the associated semantic information is displayed on the right pane. For example, a highlighted border surrounding the gridding service node denotes that this KP item is selected, and thus the semantic information is presented. According to the gravity ontology, this service is an instance of type *gridding* inheriting from the *method* concept. Additionally, this service requires *gravity-data* as input and outputs *gridded-data*. Scientists can use this rich information to get a very good understanding of the how the map was generated in their own terminology.

5   Evaluation

The premise of our work is that KP is a valuable resource that will soon become an integral aspect of all cyberinfrastructure applications. The use of ontologies is becoming more pervasive in the sciences, however the use of KP is still being researched and its various applications are still being explored, thus a widespread adoption of KP has yet to take place. A previous study has indicated however that providing scientists with visualizations of KP helps them to identify and explain map imperfections [15]. This study was composed of seven evaluation cases all derived from the different possible errors that can arise in the gravity map scenario; each case was based on a gravity contour map that was generated incorrectly. The subjects were each asked to identify the map as either correct or with imperfections. Additionally, they were asked to explain why they identified the map as such, usually by indicating the source of error. Table 1 shows the subjects accuracy in completing the identifying and explaining tasks with a contour map that was generated using a grid spacing parameter that was too large with respect to the density of data being mapped; this causes a loss of resolution hiding many features present in the data. Without KP, the majority of scientists were not able to recognize that the map was incorrect, due to the surprisingly smooth contours resulting from the course grids. With KP and corresponding visualizations provided by Probe-It!, the scientists were able to either see the gridding parameter in the process trace or access the intermediate result associated with gridding and see the pixelated image. In either case, every category of scientists: subject matter experts (SME), Geographic Information Systems Experts (GISE), an non experts (NE), performed better collectively. This study motivates the usage of provenance information to understand complex artifacts, such as maps, generated in a distributed and heterogeneous environment such as the cyberinfrastructure. The study did not include the concept of leveraging ontologies to further annotate KP, as is discussed in this paper. We strongly

Table 1. Percentage of correct identifications and explanations of map imperfections introduced by the inappropriate gridding parameter [No Provenance (NP), Provenance (P)]

	(%) Correct Identifications		(%) Correct Explanations	
Experience	NP	P	NP	P
SME	50	100	25	100
GISE	11	78	11	78
NE	0	75	0	75
all users	13	80	6	80

believe however that adding formal semantics to the provenance will only increase the accuracy of the scientist in understanding scientific results.

6 Discussion

6.1 PML Support for Semantic Annotations

PML node sets contain the conclusion derived as a result of applying a particular inference step. Additionally, the node set contains an element *language*, which is used to indicate the language the conclusion is encoded in; this makes more sense in a theorem proving scenario, where the result of each proof step is some logical statement encoded in a first order language such as Knowledge Interchange Format (KIF). The possible entries for this particular element are any of the languages registered in IW-Base. Similarly, the elements comprising the inference step: rule and inference engine, can be annotated with any registered entries for rules and theorem provers.

In the same way PML documents reference entries in IW-Base, they could also be adapted to reference concepts defined in an ontology, as suggested in this work. For example, the third task in the gravity map scenario outputs an ESRI gridded-dataset, thus a PML node set associated with this task would contain a dataset as it's conclusion. The corresponding language element of this PML node set could be annotated with the URI of the *gridded-data* concept contained in the gravity ontology, instead of a language registered in IW-Base; similarly, the corresponding inference engine element could be annotated with the *gridding* concept. The result is a PML document describing the gridding service of the gravity map scenario as outputting a *gridded-dataset* generated from a *gridding* service using only standard PML elements and the gravity ontology.

6.2 Pre vs. Post Processing Annotation

Knowledge provenance can be annotated with semantic information during workflow execution or after as a post-processing step. KP annotation during workflow execution implies that the PML service wrappers be equipped with the capability to semantically annotate PML node sets, prior to execution. As the wrappers generate the PML provenance, it can incorporate the semantic annotations. This entails that PSW be coupled with a particular ontology of some domain. At the cost of a more complex wrapper, this configuration may be the most straight forward way to annotate KP.

On the other hand, annotating the PML documents after execution of work-flow provides greater flexibility; instead of PSW annotating the KP with concepts of some fixed domain, the KP can be semantically annotated by concepts of any domain, provided an ontology. Of course it would be up to a scientist to correctly associate the KP elements to concepts of some ontology. In order to automate the post annotation process, the program would require a mapping of provenance elements stored in IW-Base and instances of a some ontology. This is because standard PML only references provenance elements stored in IW-Base. In order to compliment the IW-Base entries with concepts of some ontology, a mechanism is needed to ensure that the IW-Base entries and concepts are congruent.

6.3   Annotation Granularity

PSW is capable of logging most aspects of KP associated with scientific workflows including the execution trace (i.e. the sequence of services that were invoked), in-termediate results, and information describing the functionality provided by each service composing the workflow. Because the gravity ontology is very detailed, ev-ery aspect of KP associated with the gravity map scenario can be semantically annotated. The gravity ontology defines concepts for all the inputs/outputs and services that comprise the gravity map workflow. Additionally, the gravity ontol-ogy defines the relationships between the data and different methods that operate on that data. Semantically annotating KP elements would not be possible if the gravity ontology were not defined with scientific processes in mind. Therefore, the level of KP that can be annotated depends upon the granularity of the ontology. If an ontology is defined at such a high level, that relationships between data and methods are not explicit, then annotation of KP elements regarding the output of each service may not be possible.

6.4   Distributed Provenance (PML) vs. Workflow-Level Provenance

Service oriented workflows, such as the gravity map workflow can be segmented into two parts: the workflow engine or director and the services comprising the workflow activities. The workflow director is responsible for forwarding the out-put of each service to the next service specified in the sequence, therefore the director must know details about the services such as where they are located (i.e. what are the services endpoint URI) and what the data type of there respective input/output parameters. The services, on the other hand are not aware of the workflows they belong to; they simply execute upon request and return their results to the calling application, which may or may not be a workflow.

Just as there are two main segments composing a service-oriented workflow, there are two points from which to collect knowledge provenance. Knowledge provenance can be collected from either the workflow engine side or on the service side such as is done with PSW. Typically, systems that record KP on the workflow engine side are tightly coupled to the workflow engine itself, thus only aspects of KP visible to the workflow engine can be recorded. Kepler [16], a workflow engine, records KP on the engine side, thus information regarding the input/outputs of each service and the sequence of their execution can be logged. However, from the workflow engine side, the services composing the workflow are simply "black boxes", only their location, input and outputs are known. On the other hand, PSW and other service side KP loggers have the benefit of being closely coupled with the service they are logging and can usually provide more detailed KP regarding their functionality. Additionally, with these types of configurations, the responsibility of logging KP is removed from the workflow engine and placed on the service side.

A side-effect of service side logging however is that a layer of indirection is added between the workflow engine and the target service that performs the desired function. This overhead may be a small price to pay in order to obtain rich KP associated with a service's functionality. If PSW is wrapping a service from the "black box" perspective then the wrapper can only log very basic provenance, such as the services end-point URI. Despite this limitation, the wrapper is still able to log process meta-information and intermediate results, which at the level of single service correspond to name of the service and its output data respectively. If PSW or other service side loggers have intimate details about the services they are wrapping (i.e., the source code of the services is available) then the wrapper may be configured to capture richer provenance such as the employed algorithm or the hosting organization. In contrast, provenance captured by Kepler does not include any description or indication of the organization hosting the invoked services or their supporting algorithm because provenance is captured on the workflow side; from the point of view of the Kepler workflow engine, services are "black boxes" located at some end-point address.

Without provenance related to a service's function however, scientists may not be able to identify what algorithm was employed leading to a weaker understanding of what function the service provides and thus a weaker understanding of the quality of the final result. Although from a computer science perspective, the "black box" nature of service-oriented architecture is very beneficial, especially in terms of designing highly scalable systems, it makes it difficult to analyze the output of systems designed as such. From the study discussed in Section 5, it was determined that scientists need rich KP associated with all aspects of the workflow execution, including the algorithm supported by each service in order to fully understand complex results. Additionally, measurements such as trust that are derived from provenance can not easily be obtained through the use of workflow-side captured provenance such as provided by Kepler. For the provenance use cases outlined by Kepler developers however, the detail of provenance recorded is more than adequate. Additionally, tracing provenance in Kepler only inflicts minimal processing time penalties, because there is no level of indirection introduced between workflows and target services, as is the case in PSW.

7   Conclusions

Ontologies provide a formal definition of concepts in some domain, essentially establishing a standard vocabulary, from which both scientists and software agents can use to better understand artifacts. Knowledge provenance provides a detailed description of the origins of some artifact generated by complex processes such as scientific workflows. When used in conjunction as described in this paper, scientists are provided with very rich knowledge about some artifact, including a description of its origins defined by an ontology. This paper demonstrates how knowledge provenance is leveraged as a medium, from which rich semantics can be associated with complex artifacts such as a maps. Semantically annotating KP associated with maps, such as gravity contour maps, provides a richer description than is available when annotating only the artifact itself. Scientists need detailed information regarding the generation of artifacts in order to accurately reuse them. From the positive results achieved in the user study evaluating the need of KP, we believe that further annotating KP with semantics will only further aid scientists in better understanding and thus better utilizing complex artifacts.

Acknowledgements

This work has been partially supported by the University of Texas at El Paso GIS Center.

References

1. Pinheiro da Silva, P., McGuinness, D.L., McCool, R.: Knowledge Provenance Infrastructure. IEEE Data Engineering Bulletin 25(2), 179–227 (2003)
2. Buneman, P., Khanna, S., Tan, W.C.: Why and Where: A Characterization of Data Provenance. In: Proceedings of 8th International Conference on Database Theory, pp. 316–330 (January 2001)
3. Cui, Y., Widom, J., Wiener, J.L.: Tracing the Lineage of View Data in a Warehousing Environment. ACM Trans. on Database Systems 25(2), 179–227 (2000)
4. Aldouri, R., Keller, G., Gates, A., Rasillo, J., Salayandia, L., Kreinovich, V., Seeley, J., Taylor, P., Holloway, S.: GEON: Geophysical data add the 3rd dimension in geospatial studies. In: Proceedings of the ESRI International User Conference 2004, San Diego, CA, 1898 (2004)
5. Google: Google Map Features, http://maps.google.com/
6. GIS, E., Software, M.: Annotation Features. http://www.esri.com
7. Evans, J.: Discussion Paper: XML for Image and Map Annotations (XIMA) Draft Candidate Inferface Specification, http://portal.opengeospatial.org
8. Zhao, J., Wroe, C., Goble, C., Quan, D., R.S., Greenweed, M.: Using Semantic Web Technologies for Representing E-science Provenance. In: Proceedings of the 3rd International Semantic Web Conference, pp. 92–106 (November 2004)
9. McGuinness, D.L., Pinheiro da Silva, P.: Infrastructure for Web Explanations. In: Fensel, D., Sycara, K.P., Mylopoulos, J. (eds.) ISWC 2003. LNCS, vol. 2870, pp. 113–129. Springer, Heidelberg (2003)

10. McGuinness, D.L., Pinheiro da Silva, P.: Explaining Answers from the Semantic Web. Journal of Web Semantics 1(4), 397–413 (2004)
11. Dean, M., Schreiber, G.: OWL web ontology language reference. Technical report, W3C (2004)
12. McGuinness, D.L., van Harmelen, F.: OWL Web Ontology Language Overview. Technical report, World Wide Web Consortium (W3C) Recommendation (February 10 2004)
13. Murdock, J.W., McGuinness, D.L., Pinheiro da Silva, P., Welty, C., Ferrucci, D.: Explaining Conclusions from Diverse Knowledge Sources. In: Cruz, I., Decker, S., Allemang, D., Preist, C., Schwabe, D., Mika, P., Uschold, M., Aroyo, L. (eds.) ISWC 2006. LNCS, vol. 4273, pp. 861–872. Springer, Heidelberg (2006)
14. McGuinness, D.L., Pinheiro da Silva, P., Chang, C.: IW-Base: Provenance Metadata Infrastructure for Explaining and Trusting Answers from the Web. Technical Report KSL-04-07, Knowledge Systems Laboratory, Stanford University (2004)
15. Rio, N.D., da Silva, P.P.: Identifying and Explaining Map Imperfections Through Knowledge Provenance Visualization. Technical report, The University of Texas at El Paso (June 2007)
16. Bowers, S., McPhillips, T., Ludascher, B., Cohen, S., Davidson, S.B.: A Model for User-Oriented Data Provenance in Pipelined Scientific Workflows. In: Moreau, L., Foster, I. (eds.) IPAW 2006. LNCS, vol. 4145, Springer, Heidelberg (2006)

Architecture for a Grounded Ontology of Geographic Information

Allan Third, Brandon Bennett, and David Mallenby*

School of Computing, University of Leeds, Leeds, LS2 9JT, UK
{thirda,brandon,davidm}@comp.leeds.ac.uk

Abstract. A major problem with encoding an ontology of geographic information in a formal language is how to cope with the issues of vagueness, ambiguity and multiple, possibly conflicting, perspectives on the same concepts. We present a means of structuring such an ontology which allows these issues to be handled in a controlled and principled manner, with reference to an example ontology of the domain of naive hydrography, and discuss some of the issues which arise when grounding such a theory in real data — that is to say, when relating qualitative geographic description to quantitative geographic data.

1 Introduction

A major problem with encoding an ontology of geographic information in a formal language is how to cope with the issues of vagueness, ambiguity and multiple, possibly conflicting, perspectives on the same concepts. We present a means of structuring such an ontology which allows these issues to be handled in a controlled and principled manner, with reference to an example ontology of the domain of naive hydrography, and discuss some of the issues which arise when grounding such a theory in real data — that is to say, when relating qualitative geographic description to quantitative geographic data.

We take an encoding of the "ontology" of a particular domain to be a collection of sentences in some formal language defining the terms of that domain and constraining their interpretation by means of axioms. We refer to such a collection as an *ontology* of that domain. One of the purposes of encoding an ontology is to assist the integration of heterogenous data sources and to enable the automatic handling of queries and reasoning tasks with regard to the natural high-level concepts associated with the domain in question. Such tasks may involve the relationships between the concepts themselves, or the application of those concepts to actual data gathered by domain experts.

In order to integrate different data sources, it is necessary to relate the terms defined in an ontology to data objects and their attributes. In terms of an ontology in a formal language such as first-order logic, a specific data set ideally provides a *model* for that ontology — that is to say, the formulae in the ontology

---

* The authors gratefully acknowledge the support of EPSRC grant no. EP/D002834/1.

F. Fonseca, M.A. Rodríguez, and S. Levashkin (Eds.): GeoS 2007, LNCS 4853, pp. 36–50, 2007.

should all be *true* in the data set. The process of computing the relationship between terms and data — that is, providing concrete interpretations of predicates in terms of (sets of) data objects — we refer to as *grounding*.

However, as we noted above, geographic information is not straightforward. In particular, many natural geographic terms are *vague* (what is the difference between a hill and a mountain?) and ambiguous ("stream" can refer either to any channel containing flowing water, or to a small such channel such as a brook). The problem of ambiguity is exacerbated by the wide range of both the physical phenomena relevant to geography, and the variety of different human activities to which geographic information is relevant. A hydrographer may define a term such as "estuary" in terms of the relative salinity of different regions of water ([1]), whereas the cartographer, or the navigator of a boat, may each have quite different definitions. Such agents may disagree over which regions are considered "estuary", even if there exists a general commonly-understood meaning of the term to which all agree, and of which the particular meaning used by each is a specialisation. Similarly, two hydrographers, or the same hydrographer on different occasions, may vary in their interpretations of a single term, depending on the context. Thus different perspectives on reality can lead to ambiguity in the interpretation of common terms. In the context of information systems, different perspectives such as these can be reflected in the different kinds of information recorded in data sets: the hydrographer may collect data of no interest to the cartographer or navigator, and vice versa, and yet the same high-level geographic terms can be interpreted over the data gathered by each. Such ambiguity cannot be idealised away, nor, we believe, is it desirable to try to do so.

A further difficulty, which we believe is likely to apply to many situations in which abstract qualitative descriptions are related to quantitative data, is that humans tend to ignore "insignificant" deviations in reality from the abstract description. For example, small tree-less regions on the edge of a wooded area may nonetheless be included as part of a forest, and, as we discuss later, a river can still be classed as being vaguely linear overall, even if there are sections of it which are definitely non-linear, provided those sections are small enough. We show by example a way of handling such irregularities as part of the grounding process.

In light of these issues, we believe it is more useful to try to handle the ontology of geography in such a way as to accommodate vagueness, ambiguity and the existence of different perspectives, rather than attempt to anticipate and accommodate every possibilty. In this matter, we are in agreement with [2], who outlines a semantic framework incorporating an explicit notion of *context* which allows contextual variation for vague and ambiguous terms. The work we present here concerns the internal structure of an ontology, and its relation to data; we believe that any ontology of the kind we discuss could be slotted quite straightforwardly into the framework of [2].

We argue here for the use of a layered architecture for an ontology of geographic information which allows the vagueness and ambiguity of the general terms of that domain to be handled in a straightforward way. The structure we propose allows a principled approach to the problem of grounding the same ontology in different kinds of data. We illustrate this architecture by means of a simple ontology of inland water features, grounded in this case in two-dimensional "map" data.

2   The Semantics of Vagueness

A predicate p is said to be *vague* if there are elements of the relevant real-world domain which are neither clearly p nor clearly not p. The natural language term "river" is vague, for example, because there exist flowing water features about which it is unclear whether they are small rivers or large streams (or even elongated lakes). It is important to remember that the phenomenon of vagueness is distinct from that of ambiguity. A word can have more than one meaning, each of which is perfectly precise, and a word with a single meaning can have unclear boundaries of application. Many words, of course, exhibit both phenomena.

There are a variety of approaches in the philosophical and knowledge representation literature to the semantics of vague terms, from fuzzy logic [3], in which statements about borderline cases of vague terms are treated as partially true, to epistemic models [4], in which the lack of clarity about borderline cases is treated as a kind of ignorance, to supervaluation semantics [5]. In [6] and [7], it is argued that many vague geographic terms are such that, given a partial denotation of a term — for example, the set of clear-cut cases of river — there remain many "acceptable" ways of making that term precise. That is to say, one interpretation may include certain borderline cases of river as genuine rivers, and another may not, without either interpretation contradicting our intuitive understanding of the term. This argument suggests, then, that vague geographic terms can be interpreted using *supervaluation semantics*.

According to the standard account of supervaluationism, vague terms are interpreted relative to a set of *admissible* interpretations, each of which is a classical interpretation of those terms. A single admissible interpretation corresponds to one way of making all vague terms precise. A sentence containing a vague term is *supertrue (superfalse)* if it is true (false) on all admissible interpretations, and is neither true nor false otherwise.

To illustrate this idea, consider Figure 1, in which a range of different sources have shaded the region each considers to have some particular (vague) property p. The property p might be, for example, the property of being an estuary. All of our different sources have agreed that region A does *not* lie within the p-region, and all agree that region B *does* lie within it. We can thus identify a *core* region, considered to be p by *every* source, and a *fringe* region, the largest region which *any* source considers to be p. In Figure 1, the core region is that shaded by source 1, which is a subregion of those shaded by the other sources, and the fringe region is that shaded by source 3 — the regions shaded by both other sources are subregions of it. Core and fringe regions can be identified even if some sources provide fuzzy boundaries, by considering which regions are definitely (non-fuzzily) p and definitely not p.

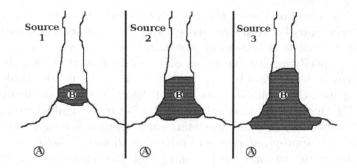

Fig. 1. Alternative interpretations of p

In terms of supervaluation semantics, the different sources correspond to different ways of making the term p precise, and the core and fringe regions identify the range of admissible interpretations. Every admissible interpretation must include the core region as p, and no admissible interpretation includes any region outside the fringe as being p. Supertrue sentences — those which are true in *all* admissible interpretations — turn out to be those to which every source would agree, and superfalse sentences those to which no source would agree. It is possible, of course, for some sentences to be true in some interpretations and not in others: these are the sentences to which some sources would agree, and some would disagree. Supervaluation semantics, then, models the situation where multiple agents, who can each have their own internally-consistent theory governing the use of a term nonetheless share a common understanding of it.

So a supervaluationist semantics of, say, vague hydrographic terms would contain both admissible interpretations in which a borderline stream/river would be classified as a stream, and other interpretations in which the same object would be classified as a river. A clear case of a river — the Amazon, say — would be interpreted as a river on every admissible interpretation, and so sentences referring to it as a river would be supertrue.

We believe that a supervaluationist approach is more appropriate than the use of a multi-valued or fuzzy logic for a variety of reasons. It is not clear, given a set of terms representing data objects, and a vague predicate, how exactly to assign truth values to formulae involving them. The notion of entailment in fuzzy logic is also not as strong as classical or supervaluationist entailment, and can weaken logical relationships between concepts.

In much of the philosophical literature on supervaluationism, the idea of the "admissibility" of an interpretation is often left worryingly undefined. We analyse admissibility, by means of the following observation. The applicability of the vague terms in which we are interested turns out to be dependent on certain precise properties which can take a range of values. For example, it seems clear that a river should be wider than a stream, even if it is unclear at precisely which specific value of width the boundary between them lies. We can model this phenomenon in terms of *threshold values* on the relevant properties — for example, a threshold on the average width of a channel of flowing water determining the boundary between river and stream. A specification of values for each threshold corresponds to a classical interpretation of the vague terms, making each precise. Specifying that the values of a threshold must lie in a given *range* therefore fixes a *set* of ways in which the relevant terms can be made precise. By building such thresholds into the formal definitions of vague terms, we acquire a straightforward means of controlling the set of admissible interpretations of those definitions, and, when grounding such definitions in real data, we can experiment with appropriate ranges of values for those thresholds, and quantify over them to be able to carry out reasoning and draw supertrue conclusions. In this paper, we leave the choice of constraints on threshold values open; however, we show that we can still give a semantically rich logical representation which works modulo the setting of thresholds. Detailed discussion of this particular approach to the logic of vagueness can be found in [8].

Throughout this paper, we represent an n-ary vague predicate p whose interpretation depends on m thresholds $t_1, \ldots, t_m$ by the notation

$$\mathsf{p}[t_1, \ldots, t_m](x_1, \ldots, x_n)$$

Some vague predicates, such as "small", for example, depend for their interpretation on a *comparison class*: what counts as small for a small man, say, is different to what counts as small for a small mouse. We indicate that c is the relevant comparison class for a predicate p (always unary in this paper) with the notation

$$\mathsf{p{:}c}[t_1, \ldots, t_m](x)$$

Any such p thus in fact represents a *family* of predicates, not dependent on a comparison class, one for each c.

3   Ontological Architecture

We divide an ontology into three separate layers, or modules: the *general, grounding* and *data* layers.

The general layer is a high-level theory of the structure of the domain, defining symbols corresponding to natural language terms. Where these terms are vague, we model the vagueness by means of parameters, in accordance with the preceding discussion. So, for example, in an ontology of geographic information, this layer defines basic notions such as types and classes of matter – water, oxygen, solid, fluid, and so on – basic spatial predicates, such as the languages of the Region Connection Calculus [9] or Region-Based Geometry [10] and temporal structure [11]. Such basic notions can then be used to define and axiomatise the high-level terms of the domain, such as planet, latitude, longitude, two-dimensional projections, and so on. It can also define the general, commonly-understood meanings of vague or ambiguous terms such as "river" and "lake".

The grounding layer takes predicates which are treated as primitive in the general layer, and provides definitions for those predicates in terms of precise predicates of the kind found in collections of data. For example, a grounding layer for a geographic ontology may take the high-level, vague definition of a river as, say, a large narrow stretch of water, and flesh out the idea of *stretch* with reference to the "linear" features of the two-dimensional geometry of a water network viewed from above. Different grounding layers can be given for the same general layer, depending on the kind of data one has in mind. Clearly, the detailed definition of a stretch of water in terms of two-dimensional data will not be sufficient to ground the definition of a river in a set of data incorporating three-dimensional topographic and bathymetric information. Similarly, different grounding layers can be used to accommodate different perspectives on terms in the general layer — for example, to enable the grounding of the same high-level concept — that of river estuary, say — in completely different data, relating to geometry and salinity, respectively, for example.

We believe that varying the grounding layer in this way can provide the infrastructure for dealing with some of the ambiguity of natural language terms mentioned in the general layer. Different senses of a given term can be encoded as different ways of grounding that term in reality, while those aspects of meaning which are common to all senses of a term can be encoded in the general layer. A full discussion of issues relating to ambiguity, vagueness and multiple perspectives on meaning can be found in [12].

The particular choice of grounding layer depends very heavily, therefore, on the data in which one wishes to ground the ontology, and is not constrained, as the general layer is, to contain commonly-understood terms. Rather, it provides the means of relating such common terms to the specifics of a particular perspective or set of data. It is thus a good place also to define technical terms which are not necessarily widely shared.

Finally, the data layer provides a concrete ground interpretation of the relevant grounding layer, and, by extension, an interpretation of the general layer. From a data set of, say, two-dimensional spatial regions with attributes such as "water", "land", and so on, it is possible to extract a set of ground atomic formulae in which each region in the data is represented by a constant and each attribute as a predicate. Such a set of formulae containing only predicates which are considered primitive in the grounding and general layers can represent a model of those higher-level layers based on actual data.

It is not necessarily straightforward to map the predicates of a high-level theory onto the attributes and relations found in data-sets such as Geographic Information Systems (GISs). Consider, for example, a high-level concept such as river, and suppose that we stipulate in its definition that a river should be vaguely geometrically linear. Suppose further that the actual data in which we want to ground our theory consists, not unusually, of a set of spatial regions and a flag stating whether each represents an actual region of water or land. The problem remains of identifying which subsets of these data can be identified as linear or not, subject to a vague parametrisation of linear. This problem is distinct from the issues of giving both context-independent, and specific context-dependent, definitions of high-level terms, and depends very much on specific data. This dependence is an advantage: in a discussion of rivers, say, the interpretation of terms such as long is very heavily context-dependent. We thus locate such segmentation in the data layer, which therefore consists of a set of data which has been analysed and marked up with the denotations of derived, but low-level, predicates such as linear, long or deep. Hence, the data layer is the most specific yet.

To summarise, then, in order to handle issues such as vagueness, ambiguity, and the grounding problem, we divide an ontology of a particular domain, such as geographic information, into three layers: the general layer, consisting of context-independent definitions of high-level predicates and including, for example, a general description of the structure of the planet, among other prerequisites for any geographical discussion; the data layer, corresponding to a specific data-set and consisting of a set of individual objects and the denotations of a range of "basic" predicates over that domain, often of an observational, quantitative nature, such as land, water, and so on, and more complex, but still low-level predicates which can be derived from the data, such as linear. Between these layers, we have what we call the grounding layer, which varies with context and relates the high-level terms of the general layer to the low-level terms of the data layer.

Figure 2 illustrates this structure, showing how a sample general layer for an ontology of geography can be related to two different grounding layers, one intended to ground high-level general predicates to two-dimensional topographic data, and one intended to ground those same predicates in three-dimensional topographic and bathymetric data. The two-dimensional grounding layer is then related to two different data layers, which share a definition of linear, but have different definitions of the highly context-sensitive term small. The formulae in Figure 2 are intended solely to be illustrative: clearly, a genuine attempt at an appropriate ontology requires much more detail. Note, however, how the threshold parameters for vague predicates are passed down through the layers.

This division into three layers is, we claim, a natural one. As we noted above, the applicability of certain high-level concepts may depend on the context or perspective in which they are interpreted, and it may be possible, or common, to interpret the same concepts as applying to different kinds of data. The separation between the general and the grounding layers is thus motivated. The role of the data layer we take to be more evident still: there is no general interest in a theory of any domain which applies only to one specific set of data, or to no data at all.

4   Example Ontology: Naive Hydrography

In order to illustrate the architecture proposed above, we present an example ontology of common water features, and ground its general layer in two-dimensional, map-like data. We define terms such as "river" and "lake" in a way which we believe to represent a formal encoding of the intuitions of the average native speaker of English confronted with an unlabelled map; no doubt a trained hydrographer would take issue with some or all of our definitions, but,

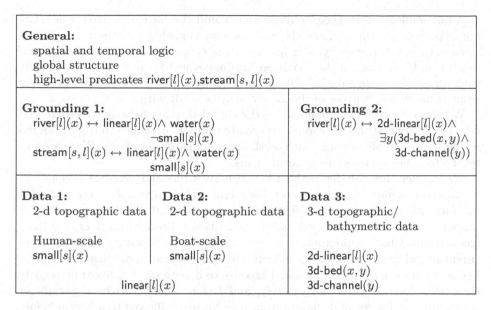

Fig. 2. Example of a layered structure relating the same general layer to multiple grounding and data layers

after all, the main aim of our proposed structure is precisely to accommodate such disagreement.

4.1   General Layer

We work in a first-order language with equality interpreted over regions of space, and assume an axiomatisation of a suitable set of spatial relations — for example, the binary relations of RCC-8 [9]. We assume henceforth that the reader is familiar with RCC-8.

We also assume a metric function d such that for any pair p_1, p_2 of points, $d(p_1, p_2)$ is the (shortest) distance between p_1 and p_2.

Although a fully general theory of geography must of course be able to represent time, for the moment we ignore issues regarding time, and possible changes in the nature of geographic entities over time, for simplicity. We anticipate that extension to include time in our theory can be carried out along the lines of the proposal in [13]. This paper is also the source of our interpretation of matter types. Briefly, mass nouns such as water are interpreted as referring to the sum of all spatial regions which contain only water. Thus, the interpretation WATER of the term water is itself a region, and so to express that a given region r contains only water, we simply need to write $\mathrm{P}(r, \mathrm{WATER})$, where P is the RCC part relation. Terms referring to families of matter types (classes), such as solid, can be interpreted as the sum of the interpretations of all matter types in that class, and so again are interpreted as regions.

A full ontology of the geographic domain would also have to define terms relating to planetary structure, and the various ways of projecting from three dimensions onto two. Since everything henceforth is concerned with two-dimensional regions on the surface of the Earth, we assume, solely for reasons of space, that such an ontology can satisfactorily be constructed, and issues such as precisely what is meant by "surface of the Earth" can be dealt with.

We concentrate for the moment on the the distinction between river-like and lake-like water features. Obviously, there are many more issues to be considered in the hydrographic domain, such as the precise definition of sea. We intend to return to issues such as this in future work.

What, then, are the distinctions between river-like and lake-like regions? In the absence of data regarding water flow, temporal change, and so on, and with the intuition that many water features can be classified simply by considering shapes on a map, the obvious geometric condition is linearity. A river or stream has a course which approximates a line, albeit often one with a high degree of curvature, whereas a lake or a pond exhibits a non-linear, more disc-like, shape. Let us, then, take *linear-channel* and *expanse* to denote vaguely linear or vaguely circular regions of water, respectively, and l (for "linear") to be a parameter controlling the degree of deviation from true linearity allowed to a region before it is considered definitely not to be linear. The conditions of application of these predicates will be supplied by the grounding layer (as described below).

The more natural terms can then be defined in terms of *linear-channel* and *expanse*.

$$\mathsf{river}[l, n](x) \leftrightarrow \mathsf{linear\text{-}channel}[l](x) \wedge \neg\mathsf{narrow\text{:}linear\text{-}channel}[n](x)$$

$$\mathsf{stream}[l, n](x) \leftrightarrow \mathsf{linear\text{-}channel}[l](x) \wedge \mathsf{narrow\text{:}linear\text{-}channel}[n](x)$$

$$\mathsf{lake}[l, s](x) \leftrightarrow \mathsf{expanse}[l](x) \wedge \neg\mathsf{small\text{:}expanse}[s](x)$$

$$\mathsf{pond}[l, s](x) \leftrightarrow \mathsf{expanse}[l](x) \wedge \mathsf{small\text{:}expanse}[s](x)$$

where *narrow:linear-channel*$[n](x)$ and *small:expanse*$[s](x)$ are dependent on the width and size of x and the comparison classes of *linear-channel* and *expanse*, respectively, and parameters n, s (for "narrow", "small", respectively) modelling the vagueness of *narrow* and *small*.

Obviously, in general there are other issues relevant to the lake/river distinction such as speed of flow, geometry of the lake/river bed, and so on, but these involve temporal considerations and three-dimensional properties of space, which may not always be recorded in the data. Another temporal issue is that of how to characterise a "hydrographic feature" which does not always contain water, but, for example, only flows seasonally. These are issues to be dealt with in future work.

4.2   Grounding Layer

The preceding discussion outlines the general layer of our case study of an ontology of naive hydrography. We now continue by defining the predicates needed to ground the above definitions in actual data automatically.

In the context of the grounding layer ontology, we assume that the data available consist of a polygonal representation two-dimensional regions of water and land, as might be found in a GIS, and that analysis yields a segmentation of water regions into polygons classified as either linear or non-linear, according to some supplied threshold l of linearity. We also assume that each region in the segmentation is *maximal* with respect to linearity, by which we mean that no (linear/non-linear) region is a proper part of any other (linear/non-linear) region. We have implemented a geometric analysis algorithm [14], so our approach is already applicable to real-world data.

The purpose of the grounding layer is to relate the properties of the information in the data layer to the relevant primitive predicates of the general layer. The requirements of this layer have thus been informed by observations of the output at the data layer. The main issue which has been observed is that regardless of the threshold value for linearity, there are sections of regions of water representing real-world rivers which are not classified as linear. These sections seem to correspond to bends in the river course, junctions in which one river flows into another or irregularities in the shape of the river banks. Each of these phenomena can make a local section of water appear to be closer in shape to a circle than to a straight line, which leads to its classification as non-linear. Figure 3 illustrates this phenomenon, being a graphical representation of our sample data set, the river Humber on the east coast of the UK, and showing, by shading, which regions of that data set are classified as linear by our analysis tool. Unshaded areas in what intuitively one would consider to be linear stretches of water can be observed with any particular linearity threshold. However, casual observation of these "gaps" showed them generally to be insignificant in size compared to the surrounding linear regions. We believe this phenomenon arises from the fact that linear is an abstraction from the actual shape of a river or stream, as indeed any general geometric term applied to these features must be, and there are always likely to be irregularities such as these when describing natural, qualitative features in abstract terms. What is important, however, is to be able to deal with them in a principled way, which motivates the following definitions.

$$\text{stretch}[l](x) \quad\leftrightarrow\quad \text{P}(x, \text{WATER}) \wedge \text{linear}[l](x) \wedge$$
$$\forall y(\text{P}(y, \text{WATER}) \wedge \text{linear}[l](y) \wedge$$
$$\text{P}(x, y) \rightarrow \text{EQ}(x, y))$$

$$\text{interstretch}[c, l](x) \quad\leftrightarrow\quad \neg\text{linear}[l](x) \wedge$$
$$\forall y(\text{EC}(x, y) \rightarrow (\text{land}(y) \vee$$
$$(\text{water}(y) \wedge \text{linear}[l](y) \wedge \text{close-to}[c](x, y))))$$

where EQ is RCC equality of regions, and EC is the "external connection" relation. So a *stretch* is a maximal linear region of water, and an *interstretch* is a region of water externally connected only to land, or to regions of water which are *close-to* it, where c, for "close", parametrises the vague predicate *close-to*. Thus an interstretch is an "insignificant" region of water between stretches.

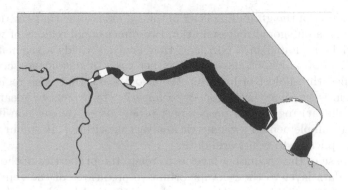

Fig. 3. Linear sections of the Humber estuary

The *linear-channel* predicate required by the definition of *river* and *stream* is then defined to apply to any region that is equal to the sum of a maximal sequence of regions $r_1, \ldots, r_n$ such that for each i, $1 < i < n$, r_i is either a stretch or an interstretch between r_{i-1} and r_{i+1}, and for $i = 1$ or $i = n$, r_i is either a stretch or a *stretch-source*, and

$$\text{stretch-source}[l](x) \leftrightarrow \begin{aligned} &\mathrm{P}(x, \text{WATER}) \wedge \\ &\exists y(\text{stretch}[l](y) \wedge \mathrm{EC}(x, y) \wedge \\ &\forall z(\mathrm{EC}(x, z) \wedge \neg \mathrm{P}(z, y) \rightarrow \\ &\neg \mathrm{P}(z, \text{WATER}))) \end{aligned}$$

That is, a stretch source is a region of water externally connected to a stretch but otherwise entirely surrounded by non-water. Stretch-source is intended to capture the situation where a water channel appears to come out of the ground, at a spring, say. Such a region, it is easy to check, will always be classified as non-linear, but is not an interstretch, being connected to only one stretch.

An *expanse* plays a similar role for lake-like regions as stretch does for river-like regions, and is defined as follows.

$$\begin{aligned} \text{expanse}[c, l](x) \leftrightarrow\ &\mathrm{P}(x, \text{WATER}) \wedge \neg \text{linear}[l](x) \wedge \\ &(\neg \text{interstretch}[c, l](x) \wedge \neg \text{stretch-source}(x)) \wedge \\ &\forall y(\mathrm{P}(y, \text{WATER}) \wedge \neg \text{linear}[l](y) \wedge \mathrm{P}(x, y) \rightarrow \mathrm{EQ}(x, y)) \end{aligned}$$

The discussion so far has ignored the fact that linear channels can be, and often are, connected to one another, and that these connections occur in different kinds. Specifically, it is possible for two rivers, say, to merge to form a larger river (a "confluence"), for one smaller river to flow into a larger river (a "tributary"), and for a single river to divide into two separate channels, which may rejoin each other further downstream, as happens, for example, when an island occurs. We give an outline of a naive way of handling these different kinds of junction in our sample theory. We consider only the case of junctions between two channels, for simplicity. It is hoped that more complex junctions can be decomposed in terms of simpler cases.

Suppose, then, that we have two linear channels, c_1 and c_2, each of which is composed of a connected sequence of stretches, interstretches, stretch sources and stretch inlets according to the constraints given above. Let c_1 consist of the sequence $r_1, \ldots, r_n$ of such regions, and let c_2 consist of $s_1, \ldots, s_m$, such that r_i is connected to r_{i+1} and s_j is connected to s_{j+1} for all i, j, $1 \leq i < n, 1 \leq j < m$. There is a junction between c_1 and c_2 if $r_1, \ldots, r_n$ and $s_1, \ldots, s_n$ have either a common initial subsequence, a common final subsequence, or both (provided in this final case that both channels also contain distinct subsequences of regions, otherwise $c_1 = c_2$). That is to say, either, for some i, $1 \leq i < n, i < m$, $r_j = s_j$ for all $j \leq i$ and $r_{i+1} \neq s_{i+1}$, or, for some i, $1 < i \leq n, i \leq m$, $r_j = s_j$ for all j, $i \leq j \leq n, j \leq m$ and $r_{i-1} \neq s_{i-1}$, or both of these hold simultaneously. For ease of exposition, let us assume that c_1 and c_2 have a common *final* subsequence, representing the case where two linear channels merge to form a new channel. We refer to that common final sequence as c_3, and let c'_1, c'_2 be the initial sequences of c_1 and c_2, respectively, so that c_1 is the concatenation of c'_1 and c_3 and c_2 is the concatenation of c'_2 and c_3. We identify the junction of c_1 and c_2 with the triple (r, s, t) consisting of the final regions r, s from each of the sequences c'_1, c'_2, respectively, and the first region t in the sequence c_3.

The idea of junction is thus precise: given a segmentation of water regions into linear channels as described above, the interpretation of junction is fixed. One source of vagueness, however, lies in the notions of tributary and confluence. We assume that there are two possible ways to characterise the merging of two linear channels: either two similarly sized channels flow together to form a new, "large" channel, or one "small" channel flows into a "large" channel. We refer to these cases as confluence and tributary, respectively.

In order to interpret these terms in our theory, we need a vague notion of "similar size". Let us say, then, than for any two members r_i, s_j of the sequences of regions making up linear channels c_1 and c_2, similar-size-to$[c](r_i, s_j)$ holds if the difference between the average widths w_i, w_j of r_i, s_j are less than c, using the same vagueness parameter c we used for close-to to represent a "small" distance in the relevant context.

We can now define confluence and tributary. We say that the junction (r, s, t) of c_1 and c_2 is a confluence if similar-size-to$[c](r, s)$; thus neither c'_1 nor c'_2 can be identified as the "main" channel into which the other is flowing. We say that c'_2 (which, it is easy to check, will be a linear-channel) is a tributary of c_1 if similar-size-to$[c](r, t)$, and the average width of s is less than, and *not* close-to the widths of r, t. We believe that physical constraints rule out the possibility that none of r, s, t are a similar-size-to either of the others, and the possibility that the widths of any of r, s, t are not appropriately representative of the "widths" of c'_1, c'_2 and c'_3. Note that since confluence and tributary depend on similar-size-to, both of these predicates depend on the vagueness parameter c of close-to.

Naively, we interpret similar-size-to to refer to the physical size of the channels at the junction. A more sophisticated approach can always, of course, interpret these with respect to more hydrographically relevant considerations, such as size of catchment for each channel, supposing that such information is available in, or can be deduced from, the data.

The case where two channels diverge, and then rejoin further downstream, is more straightforward, and can be used, relatively easily, to identify regions of land which can be described as *islands*.

These definitions provide enough information to ground the high-level terms given earlier in actual data consisting of regions of ground and water classified as linear or otherwise.

4.3   Data Layer

The grounding layer we have outlined above is intended to relate predicates of the general layer to actual data in the form of two-dimensional regions of both water and ground. We assume that such data is relatively common, and that it is relatively straightforward to compute which spatial (RCC-8) relations hold between regions. What remains is to compute the denotations of the remaining predicates of the grounding layer. In the theory we have given above, those predicates are close-to, small:expanse, narrow:linear-channel and linear.

All of the predicates that we must interpret are parametrised in order to model their vagueness, and it is at the level of the data layer that values, or ranges of values, for those parameters must be set. The predicate close-to, which relates two regions, is relatively easy to deal with, by computing the largest distance between any two points in those regions, and stipulating that close-to holds when that distance is less than the value of the parameter c. Such a definition of close-to corresponds to intuition and reflects, through a suitable choice of value for c, the highly context-sensitive aspect of its meaning. The related terms small:expanse and narrow:linear-channel can be handled similarly, with the relevant threshold parameters being compared to, say, area and average width, respectively.

The more difficult spatial predicate to interpret is, as might be expected, linear, where by "linear region", we mean, loosely speaking, a region whose width is small, and relatively constant, with respect to its length. A detailed discussion of how we compute linearity can be found in [14]. Roughly speaking, though, we wish to classify a region as *linear* if it does not exhibit "too much" variation in width along its length, with the notion of "too much" being controlled by the parameter l.

Figure 3 earlier shows which regions are considered linear by the algorithm of [14] in our sample set of input data, which, as stated above, represents the Humber estuary on the east coast of the UK. This classification depends on a particular choice of linearity threshold l. The shaded regions are linear. Note the presence of unshaded regions which lie within the area one might intuitively wish to classify as river. These are locally non-linear regions, which with the interpretation of close-to, can be classified in terms of the higher-level theory as either interstretch or stretch-source.

5   Conclusions

We have presented a layered architecture for ontology concerning vague, ambiguous and context-dependent terms, and illustrated this architecture with reference to a simple ontology of water features. We have discussed the grounding of this ontology in two-dimensional data. Our architecture is designed to assist the grounding of high-level definitions in actual data, without having to sacrifice the vagueness and ambiguity inherent to many geographic terms, for example.

We have implemented a system which is able to take appropriate sets of data, and an encoding of a high-level geographic ontology, and by means of suitable grounding definitions, carry out various tasks relating the high-level terms to the data, such as identifying to which data objects those terms apply, and evaluating complex formulae over those objects.

This system enables us to test different definitions and explore the effect on the resulting classification of the data; it was such testing which identified the need for the grounding term interstretch, the gap between the abstract description of rivers in two dimensions as vaguely linear water features, and the irregularities and "insignificant" deviations from this abstraction which occurs in actual data. This resulting accommodation is carried out in a principled manner which we believe reflects the approach humans take to the interpretation of such terms.

It should be noted that the specific ontology we have presented is by no means intended to be prescriptive, but merely to demonstrate the features both of our theoretical framework and its practical applications. An obvious application is to use a system such as ours automatically to label low-level geographic data in terms of high-level concepts, particularly in cases where the specific data was not originally collected with those particular high-level concepts in mind. It is also possible to use such a system to *test* different proposed definitions of terms, and compare the results of grounding to the expectations of domain experts. Other directions for future work include the extension of the classification to a larger and more finely discriminating set of hydrographic features, and the incorporation of temporal aspects into the theoretical framework.

Although we have focused here on the domain of geography, and more specifically still, on hydrographic features, the approach we have taken can, we believe, extend to much more general domains. The ability of our framework to accommodate vague and ambiguous natural language terms in a flexible fashion makes it suitable for application to a wide range of fields which have so far been difficult to handle using standard modelling techniques.

References

1. Cameron, W.M., Pritchard, D.W.: Estuaries. In: Hill, M.N. (ed.) The Sea: The Composition of Sea-water, vol. 2, Harvard University Press (1963)
2. Cai, G.: Contextualization of geospatial database semantics for mediating human-GIS dialogues. Geoinformatica 11(2), 217–237 (2007)
3. Zadeh, L.A.: Fuzzy logic and approximate reasoning. Synthese 30, 407–428 (1975)
4. Williamson, T.: Vagueness. Routledge, London (1994)

5. Fine, K.: Vagueness, truth and logic. Synthése 30, 263–300 (1975)
6. Bennett, B.: Application of supervaluation semantics to vaguely defined spatial concepts. In: Montello, D.R. (ed.) COSIT 2001. LNCS, vol. 2205, pp. 108–123. Springer, Heidelberg (2001)
7. Bennett, B.: What is a forest? on the vagueness of certain geographic concepts. Topoi 20(2), 189–201 (2001)
8. Bennett, B.: A theory of vague adjectives grounded in relevant observables. In: Doherty, P., Mylopoulos, J., Welty, C.A. (eds.) Proceedings of the Tenth International Conference on Principles of Knowledge Representation and Reasoning, pp. 36–45. AAAI Press (2006)
9. Randell, D.A., Cui, Z., Cohn, A.G.: A spatial logic based on regions and connection. In: Proc. 3rd Int. Conf. on Knowledge Representation and Reasoning, San Mateo, pp. 165–176. Morgan Kaufmann, San Francisco (1992)
10. Bennett, B.: A categorical axiomatisation of region-based geometry. Fundamenta Informaticae 46(1–2), 145–158 (2001)
11. Allen, J.F.: Towards a general theory of action and time. Artificial Intelligence 23(2), 123–154 (1984)
12. Bennett, B.: Modes of concept definition and varieties of vagueness. Applied Ontology 1(1) (2005)
13. Bennett, B.: Space, time, matter and things. In: Welty, C., Smith, B. (eds.) FOIS 2001. Proceedings of the 2nd international conference on Formal Ontology in Information Systems, Ogunquit, pp. 105–116. ACM, New York (2001)
14. Mallenby, D.: Grounding a geographic ontology on geographic data. In: Logical Formalizations of Commonsense Reasoning: papers from the AAAI Spring Symposium. AAAI Technical Report SS-07-05 (2007)
15. Santos, P., Bennett, B., Sakellariou, G.: Supervaluation semantics for an inland water feature ontology. In: Kaelbling, L.P., Saffiotti, A. (eds.) Proceedings of the 19th International Joint Conference on Artificial Intelligence (IJCAI-05), Edinburgh, pp. 564–569. Professional Book Center (2005)
16. Kokla, M., Kavouras, M.: Fusion of top-level and geographical domain ontologies based on context formation and complementarity. International Journal of Geographical Information Science 15(7), 679–687 (2001)
17. Fonseca, F., Egenhofer, M., Davis, C., Câmara, G.: Semantic granularity in ontology-driven geographic information systems. Annals of Mathematics and Artificial Intelligence 36, 121–151 (2002)
18. Smith, B., Mark, D.M.: Ontology and geographic kinds. In: Proceedings of the International Symposium on Spatial Data Handling (1998)
19. Worboys, M., Duckham, M.: Integrating spatio-thematic information. In: Egenhofer, M.J., Mark, D.M. (eds.) GIScience 2002. LNCS, vol. 2478, pp. 346–361. Springer, Heidelberg (2002)

Towards Effective Geographic Ontology Matching

Guillermo Nudelman Hess[1,2], Cirano Iochpe[1,3],
Alfio Ferrara[2], and Silvana Castano[2]

[1] Universidade Federal do Rio Grande do Sul
Instituto de Informática - Av. Bento Gonçalves, 9500, 15064 Porto Alegre - Brazil
[2] Università degli Studi di Milano
DICo - Via Comelico, 39, 20135 Milano - Italy
[3] Procempa - Empresa da Tecnologia da Informação e Comunicação de Porto Alegre
Av. Ipiranga, 1200, Porto Alegre - Brazil
{hess,ciochpe}@inf.ufrgs.br, {ferrara,castano}@dico.unimi.it

Abstract. The integration and matching of geographic ontologies is a field in which many efforts are being employed. There are many proposals, addressing a diversity of features, both at the concept as at the instance-level. In order to make clear the issues that are involved in the matching process, in this paper we present the formal definition of the heterogeneities that may occur when comparing two geographic ontologies. Some of the heterogeneities are common to the ones found in conventional ontologies integration, and some others are specific for the geographic field. Furthermore, we discuss some still open issues, neglected up to now, but very important to achieve good results in a real scenario.

1   Introduction

Since the creation of Geographic Information Systems (GIS), new fields of research are emerging due to the peculiarities of the geographic data, which is much more specific than conventional (alphanumeric) data. In fact, besides the descriptive components, geographic data is featured by at least two other characteristics, namely geometry and location [1,2]. Geographic data may also have the temporal component [3], even if this cannot be pointed as a specific feature for geographic data. In general, geographic data is stored in a Geographic Database (GDB). A GDB has all the functionalities and capabilities of a conventional database; in addition it can handle spatial relationships between two or more geographic data as well as their spatial component (geometry and location).

Actually GIS are used every day. Some examples are the Global Positioning Systems (GPS) used in cars, the Google Earth tool, maps generators on the web, and so on. Producing geographic data is time consuming and expensive. Furthermore, in many cases the data needed is already available in some other systems or organizations.

At the same time, the diffusion of the Internet allowed the interchange of information all around the world. If, on one hand, this interchange offers a lot

F. Fonseca, M.A. Rodríguez, and S. Levashkin (Eds.): GeoS 2007, LNCS 4853, pp. 51–65, 2007.

of benefits, such as the reuse of information and knowledge sharing, on the other hand it generates the need to deal with the heterogeneities among the information obtained from distinct geographic sources. This problem is difficult to solve due to poor documentation as well as implicit semantics of the data and diversity of data sets.

There are a number of works addressing the problem of the integration and matching of geographic information sources (databases, ontologies, schemas, images, and so on). However, most of them address different issues regarding the heterogeneities. Furthermore, the lack of a standard terminology, hidden assumptions, undisclosed technical details and the dearth of evaluation metrics causes difficulties in identifying the problem areas and in comprehending the solutions provided [4].

The scenario above is the starting point of this paper, in which we aim: (1) to classify the heterogeneities that may occur when matching two geographic ontologies at both concept and instance level to build a common ground for the development of geographic ontology matchers; (2) to discuss at which extent the existing conventional matchers can be used for matching geographic ontologies and what needs to be covered by specific matching functionalities; (3) to analyze what has been done in the field of geographic ontology matching and how much of the problem the existing proposals solve and; (4) to discuss some opens issues neglected so far, but extremely important in order to achieve good results when dealing with real scenarios of geographic information matching.

1.1   Motivating Example

Figures 1 and 2 presents two geographic ontologies to be compared. The rectangles with continuous lines represent concepts, the ellipses the properties representing attributes associated with a concept and the dashed rectangles the instances belonging to a concept. The arcs linking two concepts correspond to the properties which represent relationships holding between them, while the *isa* labeled arrows are the taxonomic relationships between two concepts, in which one is the specialization of the other.

The examples are rather simple, but complete for our purposes. They have both spatial concepts, such as *Park, City, Factory* and non-geographic concepts (*Administration*). There are spatial relationships in both ontologies (*crosses, inside, overlaps*) as well as conventional relationships (*hasAdministration* held between *City* and *Administration*). The geometries of the concepts are of various types. We use these two ontologies throughout this paper and in the following sections we analyze how these elements may affect the matching of geographic ontologies.

The main contribution of this paper is twofold. First, we present a formalization of the heterogeneities that may occur when comparing two geographic ontologies. Second, we discuss the open problems regarding geographic ontology matching that have been neglected so far.

The rest of the paper is organized as follows. In Section 2 we present the basic foundations and definitions regarding geographic ontologies. Section 3

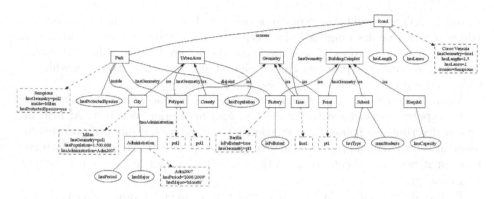

Fig. 1. Example of geographic ontology O

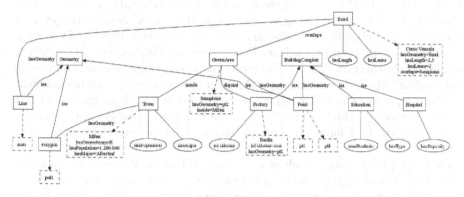

Fig. 2. Example of geographic ontology O'

formalizes the types of heterogeneities that may occur when comparing two geographic ontologies. The existing proposals addressing the integration/matching of geographic ontologies are presented in Section 4, where the need of a matcher especially tailored for geographic ontologies is also discussed. The open issues on geographic ontologies matching/integration are presented in section 5. Finally, conclusions and future directions are discussed in section 6.

2   Foundations and Definitions

Any ontology can be defined as a 4-tuple $O = <C, P, I, A>$, where C is the set of concepts, P is the set of properties, I is the set of instances, and A is the set of axioms. A concept $c \in C$ is any real world phenomenon of interest to be represented in the ontology and is defined by the term t that is used to nominate it. The name of a concept is given by the unary function $t(c)$. A property $p \in P$ is a component that is associated to a concept c with the goal of characterize it, but is defined outside the scope of a concept. It can be a data type property, which means that its range is a data type, such as string, integer, double, etc. or an object type property, meaning that the allowed range values are other concepts. A data type property can be viewed as a database attribute, while an object type property is like a database relationship.

An instance $i \in I$ is a particular occurrence of a concept c, with a value for each property p associated to the concept and an unique identifier. At last, an axiom describes an hierarchical relationship between concepts, or provides an association between a property and a concept, or associates an instance with the concept it belongs, or defines restrictions over the properties inside the context of a concept.

To handle the particularities of a geographic phenomenon, a conventional ontology is not expressive enough. Thus, we define now a geographic ontology (or geo ontology), which is an extension of a conventional ontology. It is also a 4-tuple $O =< C, P, I, A >$, where C is the set of concepts, P is the set of properties, I is the set of instances, and A is the set of axioms. However, a concept $c \in C$ is classified into domain concept, such as a *River*, a *Park* or a *Building*, geometry concept, such as *Point*, *Line* or *Polygon* or time concept, specialized in *instant* and *period*. Furthermore, a geographic domain concept gc is a specialization of a domain concept which represents a geographic phenomena. By definition, a geographic domain concept gc must have at least one associated geometric property, which is explained in the following. The geometry plays a fundamental role in defining the possible spatial relationships the concept may have.

In a geographic ontology, each property $p \in P$ can be of one of five possible types: conventional, spatial, geometric, positional or temporal. A conventional property may be even a data type property or an object type property. In the first case it represents an attribute of a domain concept. In the second case it represents an association between a domain concept (geographic or not) with a non-geographic domain concept. A spatial property (topological, directional or metric) is always an object type property, and represents an association between two geographic domain concepts. The spatial relationships have a pre-defined semantics and are already standardized in the literature [5] and by the Open GIS Consortium (OGC). The conventional relationships, on the other hand, may assume different semantics depending on the associated concepts. A geometric property (always an object type property) is an association between a geographic domain concept with a geometry concept. A positional property is a data type property that must be associated to a geometry concept, to give its location (set of coordinates). Finally, a temporal property is an association between a domain concept and a time concept.

A geographic instance $gi \in I$ is an extension of an instance i. As a geographic instance must be associated to, at least, one instance of a geometry concept, the value of the positional property (*hasLocation*) gives the spatial position (coordinates) for that geographic instance.

On the basis of this reference model, it is possible to point out at least three differences between geographic information and conventional (ontology or schema) matching:

- The spatial relationships have a pre-defined semantics and are standardized in the literature [5], while conventional relationships may assume different semantics depending the associated concepts.
- Every geographic concept has, at least, one associated geometry representing it. The geometry plays a fundamental role in defining the possible spatial relationships the concept may have.
- A geographic instance has a number of pair of coordinates (x,y) representing its spatial position over the earth surface. These coordinates are expressed in a given coordinate system.

3 A Clarification of Geographic Information Heterogeneities

In order to know what to consider when matching two geographic ontologies, in this section we define the kinds of heterogeneities that should be taken into account when comparing geographic ontologies, both at the concept-level and at the instance-level, respectively.

3.1 Concept-Level Heterogeneities

In this section the possible heterogeneities are classified regarding the comparison of a concept $c \in$ ontology O against a concept $c' \in$ ontology O'. Considering the definition of ontology, concepts, instances, properties and axioms presented in the preview section, the possible heterogeneities are defined as follows.

Name heterogeneity: The concept name heterogeneity NH occurs when given two concepts names $t(c)$ and $t(c')$ they are neither equal nor synonyms. The synonym relation $SYN(t(c), t(c'))$ is obtained by searching an external thesaurus or dictionary.

$$NH(c, c') = ((t(c) \neq t(c')) \wedge (SYN(t(c), t(c')) = false))$$

Considering the ontologies O and O', the concepts *Park* from O and *GreenArea* from O' are examples of name heterogeneity. On the other hand, *City* and *Town* do not have name heterogeneity, because even if the terms are not the same, the function $SYN(t(c), t(c'))$ returns true when searching a external dictionary.

Property heterogeneity: The concept property heterogeneity PH occurs when there is an attribute heterogeneity AH or a relationship heterogeneity RH.

The AH heterogeneity between $c \in O$ and $c' \in O'$ occurs when at least one of the attributes $a(t(p), dtp) \in P$ in ontology O, where $t(p)$ is the name of the attribute (property) and dtp is the attribute's data type, does not match with any of the attributes $a(t(p'), dtp') \in P'$ in ontology O'. The heterogeneity can be generated due to different attribute names or different attribute datatypes.

$$AH(c, c') = (\exists a(t(p), dtp) \in P | \forall a(t(p'), dtp') \in P', (t(p) \neq t(p')) \vee (dtp \neq dtp'))$$

As an example of attribute heterogeneity, lets consider the concepts *City* from O and *Town* from O'. The attribute *hasMajor* is a property of *Town*, but is not associated to *City*.

The RH heterogeneity between $c \in O$ and $c' \in O'$ is defined over the conventional relationships (not geometric nor spatial). It applies to both geographic as to non-geographic concepts. It occurs when at least one of the relationships $cr(t(p), t(c_x), minCard, maxCard) \in P$ in ontology O, where $t(p)$ is the name of the relationship (property), $t(c_x)$ is the associated concept and $minCard$ and $maxCard$ are, respectively, the minimum and maximum cardinalities of the relationship, does not have a correspondent $cr(t(p'), t(c'_x), minCard', maxCard') \in P'$ in ontology O'. The heterogeneity may occur due to a different associated concept c_x as well as due to the relationship cardinalities $minCard$ or $maxCard$. As in many times the conventional relationships names are not significant as to identify the relationship, the component $t(p)$ can be ignored.

$$RH(c, c') = (\exists cr(t(p), t(c_x), minCard, maxCard) \in P |$$
$$\forall r(t(p'), t(c'_x), minCard', maxCard') \in P', (c_x \neq c'_x) \vee (minCard \neq minCard')$$
$$\vee (maxCard \neq maxCard'))$$

The same concepts *City* from O and *Town* from O' present an example of relationship heterogeneity. The property *hasAdministration*, which relates *City* to the concept *Administration* is not in the context of the concept *Town*.

Hierarchy heterogeneity: The hierarchy heterogeneity HH between two concepts $c \in O$ and $c' \in O'$ occurs when the set of superclasses of the concept $c \in O$ is different from the set of superclasses of the compared concept $c' \in O'$. This means that at least one of the hierarchical relationships $h(t(c), t(c_x))$ from O does not have a match $h(t(c'), t(c'_x))$ in ontology O', where c_x is the superclass of c.

$$HH(c, c') = (\exists c_x \in h(c, c_x) | \forall c'_x \in h(c, c'_x), c_x \neq c'_x)$$

The concepts *City* and *Town* from O and O', respectively, are examples of hierarchy heterogeneity. The former has as superclass the concept *UrbanArea*, while the latter do not have a superclass (actually, in an ontology, all concepts are subclasses of *thing*, but for easiness of comprehension we omitted it from the ontology).

Regarding the geographic domain concepts, two additional types of heterogeneities exist, one for each type of relationship (geometry and spatial relation). The geometry itself cannot be really considered as a feature that contribute to decide if two geographic concepts have or not heterogeneity, because of the possibility of multi-representation. However, in the case of spatial relationships, specially in the case of the topological ones, the geometry counts. In [6] the equivalences between topological relationships are defined according to the geometries of the involved concepts. Following this idea, the spatial relationship heterogeneity can be divided into topological relationship heterogeneity and directional relationship heterogeneity. The metric relationships are not considered because in general they are calculated by a GIS and not defined as properties or restrictions of a concept. As the names of the spatial relationships are, in general, standardized in the literature [5], the component $t(p)$, which holds the relationship name, has to be considered.

The **directional relationship heterogeneity** DH between two geographic concepts $gc \in O$ and $gc' \in O'$ happens when there is at least one directional relationship $dr(t(p),t(gc_x),minCard,maxCard) \in P$ in ontology O without a matching $dr(t(p'),t(gc'_x),minCard',maxCard') \in P$ in ontology O', where cg_x is the associated concept, $t(p)$ is the relationship name and $minCard$ and $maxCard$ are the minimum and maximum cardinalities, respectively.

$$DH(gc,gc') = (\exists dr(t(p),t(gc_x),minCard,maxCard) \in P|$$
$$\forall dr(t(p'),t(gc'_x),minCard',minCard') \in P',(gc_x \neq gc'_x) \vee (t(p) \neq t(p')))$$

The definition of the **topological relationship heterogeneity** is a little more complicated, because of the equivalences of relationships depending on the associated geometries. Thus, given two geographic concepts $gc \in O$ and $gc' \in O'$ have topological relationship heterogeneity TH if the combination of the relationship name and the involved geometries, given by a function $top(geo,geo_x,t(p))$ and $top(geo',geo'_x,t(p'))$ are not equivalent, where geo and geo' are, respectively, the geometries of the concepts gc and gc' and $t(p)$ is the relationship name.

$$TH(gc,gc') = (\exists tr(t(p),t(gc_x),minCard,maxCard) \in P|\forall tr(t(p'),t(gc'_x),$$
$$minCard',minCard') \in P',top(geo,geo_x,t(p)) \neq top(geo',geo'_x,t(p')))$$

An example of a spatial relationship heterogeneity is the association *Road crosses Park* in ontology O and *Road overlaps GreenArea* in ontology O'. Even if we consider that *GreenArea* and *Park* could be synonyms, in ontology O the relationship name is *crosses*, while in O' $t(p')=overlaps$. As will be discussed in the paper, these relationships can be equivalent, but in a first analysis it seems that we have a spatial relationship heterogeneity.

3.2   Instance-Level Heterogeneities

As important as the matching of geographic ontology concepts is the matching of their instances. Especially in the geographic field there are many features that can influence the similarity measurement process which are not present when dealing with non-geographic ontologies. These features are, for example, the scale, spatial position, time when the instances were obtained, and so on. However, the non-spatial properties, such as the attributes (property) values, cannot be neglected. In this section we define the heterogeneities that may occur at the instance-level when comparing two geographic ontologies.

Identifier heterogeneity: When a concept in an ontology is instantiated, in general its unique identifier has a really significant value. It is not like the objectId of an instance of a class which is automatically generated. In the case of an ontology is the main way to both the user and the computer to identify the instance. When two instances $i \in O$ and $i' \in O'$ do not have the same identifier (in OWL, the ID parameter) there is an identifier heterogeneity IIH.

$$IIH(i, i') = (\exists i \in O | \forall i' \in O', id(i) \neq id(i'))$$

The concepts c and c' the instances belong are not considered because they may be not known.

Positional heterogeneity: As already stated, one of the main characteristics of the geographic data is that is has a position over the earth surface. The set of coordinates of a given instance $i \in O$ is obtained indirectly through the instance of the geometry concept that is associated to it. If two instances $i \in O$ and $i' \in O'$ do not have the same spatial position, there is a positional heterogeneity ICH. In order to simplify the formalization of the positional heterogeneity, we assume that a function $pos(i)$ gives the location of the instance. This function gets the set of coordinates from the geometry instance which is associated to the geographic instance by a geometric property.

$$ICH(i, i') = (i \in O | i' \in O' \wedge pos(i) \neq pos(i'))$$

The simple comparison of the spatial coordinates values would be a naive and simplistic definition. If the coordinate reference system and projection system of the compared instances are not the same, the harmonization of this meta information must be executed first. In the definition above we assume that the coordinate reference system and projection system are the same (originally or the translation were already performed).

Metadata heterogeneity: The metadata does not have a direct influence on the heterogeneity between two geographic instances. Instead, the influence is indirect, which means that differences on the metadata values may lead to heterogeneities regarding the other elements of the instance (coordinates and properties). For example:

1. Depending on the value for the *date* metadata, the value for some descriptive attributes may vary (for example, the population of a city). Even some spatial relationships may be different.
2. Depending on the *date* metadata the value for the spatial position of an object may change.
3. Depending on the value for the *projection* (UTM, planar) metadata, the geometry as well as coordinates of an instance change.

Attribute heterogeneity: When a property of a concept is a data type property it represents an attribute, i.e., properties which allowed values are string, float, integer, etc. When two instances $i \in O$ and $i' \in O'$ have different values for the same data type property there is an instance attribute heterogeneity IAH.

$$IAH(i, i') = (\exists at(t(c), t(p), v) \in O | \forall at(t(c), t(p'), v') \in O', (p \equiv p') \wedge (v \neq v'))$$

where i is the instance having the attribute, $t(p)$ is the name of the property p and v is the value for that property.

Relationship heterogeneity: When a property of a concept is an object type property it represents a relationship, i.e., a property which allowed values are instances of other concepts. When two instances $i \in O$ and $i' \in O'$ have associated, respectively, the instances i_x and i'_x which represent different concepts, there is a relationship heterogeneity IRH.

$$IRH(i, i') = (\exists rl(t(p), id(i_x)) \in O | \forall rl(t(p'), id(i'_x)) \in O', id(i_x \neq id(i'_x)))$$

where p is the property and $id(i_x)$ is the associated instance.

4   The State of the Art on Geographic Ontology Matching

As a geographic ontology is a special type of ontology, some of the heterogeneities may be the same as the ones found in a conventional, non-geographic ontology matching process. In these cases a conventional matcher, such as the H-MATCH[7], Prompt [8] or S-Match [9] may be used instead of developing a new matcher.

By analyzing the heterogeneities defined in section 3, at the concept-level the name heterogeneity and the taxonomy heterogeneity can be fully handled by conventional matchers. There are no geographic particularities in these features. Regarding the properties, only a partial matching can be done if using a conventional matcher. The non-geographic properties (attributes and relationships) may be matched by a conventional matcher, while especially the properties representing spatial relationships cannot, because they have a pre-defined semantics [5], which is not known by these matchers. However, these tools cannot recognize whereas a property is geographic or not, and then it is difficult to use a conventional matcher for the matching of properties.

The conventional matchers cited before do not perform the matching at the instance-level. Hence, they are not suitable for solving the instance-level heterogeneities. The proposals found in literature for conventional instance matching [10,11,12] use basically the same methods used for data integration in databases. Hence, when considering the heterogeneities defined in section 3, the identifier heterogeneity can be easily addressed. For the property heterogeneities, as the property values are the ones to be considered, if a conventional matcher is capable of dealing with attributes and relationships, it probably can deal with the geographic instance property values. The positional heterogeneity as well as the metadata heterogeneity are the specific features of a geographic instance, and clearly cannot be addressed by a conventional instance matcher.

4.1 Name Heterogeneity

The linguistic features of geographic concepts are used in many different ways. The use of natural language processing (NLP) to extract relevant information from the description (glosses) of the concepts is proposed in [13,14]. The measure of similarity based on the concepts' names is proposed by Rodriguez and Egenhofer [15], who consider two types of linguistic elements in the similarity assessment: words and meanings, and synonymy and homonymy. The resolution of the name heterogeneity aided by an external knowledge base, such as a dictionary, a thesaurus or even a mediator domain ontology is the approach presented in [16,17].

Another approach for establishing the degree of similarity regarding concepts' names is by using string-distance metrics, such as in [18]. This may be especially interesting when dealing with geographic ontologies because of the often use of acronyms or little variances when defining the concept's name.

4.2 Property Heterogeneities

Context features are explored in a number of proposals and in various different ways [15,19,16,20,21,22]. The latter considers only the properties which correspond to attributes in the context matching. The concept's attributes and properties representing part-of relationships are explored in [15,19,20]. Kavouras and Kokla [19] consider also some other properties which are explored if they can be extracted from the description of a concept, while the context in [20] is also composed by the properties representing relationships. However, in that work the authors state that for these context features there are already good matchers developed, and for this reason they do not implement a new one. Instead, the developed architecture can accomplish almost any existing matcher.

Sotnykova et al. [21] classify the context features as *user-defined properties*, *equality properties*, *spatial properties* or *temporal properties*. The user-defined properties basically correspond to attributes and relationships which are not spatial nor temporal. Equality are properties that explicitly establish the equivalence of two concepts.

4.3 Taxonomy'(Axiom) Heterogeneities

For Cruz, Sunna and Chaudhry [23] the hierarchy is the only aspect considered in the matching process, and follows a bottom-up approach. The ontologies are viewed as hierarchical structures and two concepts are considered as matching if they have matching children (same *isa* axioms). Also Worboys and Duckham [24] use the hierarchies of the concepts as the starting point for the integration process. Other works [16,20,22,15] also make use of the hierarchical structure of a ontology in the similarity assessment process.

4.4 Geographic Properties Heterogeneities

The only kind of contextual feature addressed by the group from the University of Munster are spatial relations and geometry [25]. In that proposal as well as in

Kavouras et al. [13] the spatial properties are not explicitly defined in the concept definition, but taken from its description. The spatio-temporal properties (attributes and relationships) are considered in Sotnykova et al. [21] proposal as well.

In the proposal of Quix et al. [20], for the schema's geographic features, a specific matcher was developed, which consider some geographic elements, based on the GML definition of GeometryType and supports some types of relationships. The spatial relation properties most commonly addressed are the topological ones [26].

4.5 Spatial Positioning Heterogeneities

If two instances belonging to matching concepts are in the same spatial position, i.e., have the same spatial coordinates, probably they would refer to the same real world phenomenon. This premise is applied by [18,27,24,26]. The approaches of Sehgal, Getoor and Viechnicki [18], Worboys and Duckham [24] and Beeri et al. [27], however, are limited to instances with point geometries. Besides the point geometry, Volz [26] considers also the line geometry. However, the similarity is measured in terms of length and angle instead of in terms of the instances' coordinates. In all proposals the scales and reference system must be the same for the two ontologies.

5 Open Issues

5.1 Attribute's Semantic Context

Besides the spatial properties, a geographic concept has the conventional, descriptive properties. However they might have some particularities. There may be different *geographic domain contexts* (GDC), depending on the geographic region to which a given geographic instance belongs. For example, the categorization of the values for a property "averageTemperature" of a geographic concept "city" may depend on the altitude.

In the example of Table 1 the geographic domain context is a set pairs of type GDC(averageTemperature)={(<latitude>,<altitude>)}. The proposal of a geographic context for a domain of attributes is due to the fact that the similarity rules may vary according to time and space.

Table 1. Example of context dependent attributes

Latitude x Altitude	Sea level	≤ 1000m	> 1000m
Equator	Low: < 25		
	Average: 25~38		
	High: > 38		
Between Tropics		Low: -5~12	
		Average: 12~25	
		High: > 25	

5.2   Temporality

Temporality is an intrinsic feature of the geographic phenomena. The changes may occur both regarding the spatiality and the descriptive features of the data. For example, let us consider a concept *Town*, from the ontology O', with the properties *hasPopulation, hasMajor, hasFoundationDate, hasGeometry*, defined as follows in Figure 3 (the *hasFoundationDate* was added to clarify the explanation):

$C = City$
$P = hasPopulation(Town, double)$
     $hasMajor(Town, string)$
     $hasFoundationDate(Town, date)$
     $hasGeometry(City, Polygon)$
     $hasPosition(City, string)$

If we consider the two following instances of the class *Town*

$I = instanceOf(Milan, Town)$
     $hasPopulation(Milan, 1.400.000)$
     $hasMajor(Milan,' Moratti')$
     $hasFoundationDate(Milan, 350)$
     $hasPosition(Milan,' < (45N, 10E), (45N, 12E), (48N, 15E), (45N, 10E) >')$

$I = instanceOf(Milan, Town)$
     $hasPopulation(Milan, 950.000)$
     $hasMajor(Milan,' Albertini')$
     $hasFoundationDate(Milan, 350)$
     $hasPosition(C1,' < (44N, 08E), (46N, 12E), (48N, 15E), (44N, 08E) >')$

Fig. 3. Example of the temporal influence on geographic matching process

Depending on the features considered in the matching process, the two instances may not be identified as similar, especially because the attributes *hasPopulation*, *hasMajor* and *hasPosition* are different, even if the attribute *hasFoundationDate* is equal and the instance identifier *Milan* is equal as well. These data may be captured in a different time, and the values for some of the properties (temporal properties) may vary along time. The position may change because of the creation of a new city from a former neighborhood, the population may increase or decrease and the major usually changes. However, both instances may be referring to the same real world occurrence of the city of Milan.

5.3   Geographic Metadata

For each one of the possible spatial representations of a geographic concept, the following metadata may be associated:

- capture and update date and time and, if possible, the period in which that spatial representation is valid;

- coordinate reference system, projection and scale, if exists;
- information about the data capturing system: source (satellite photo, image, aerial photo) and additional information about the capturing equipment (satellite, camera, flight, etc.);
- geometry storage format: raster or vectorial.

As the metadata for geographic concepts are almost always the same, independently of the geographic phenomenon, they have influence only in the similarity measurement among instances. If two instances to be compared are described using different metadata, probably the values of the properties which are influenced by the metadata would be different. For example, if there are two instances *Milan* of the concept *Town* of ontology O', one described using the *[latitude, longitude]* reference system, and the other using the *Universal Transverse Mercator (UTM)* reference system, the values for their *hasPosition* property would be, respectively, $< 45°20'N, 9°10'E >$ and $< 5166930.21N, 1921142.04E >$. If the metadata is ignored, a matcher would return that the two locations are not the same, while they actually are.

6 Conclusions

To perform the matching of geographic ontologies in an effective way, it is important to know exactly what to compare and the kinds of differences one may find. Because of the lack of a standard geographic ontology model, the existing approaches address different features, all of them relevant, but no one has a complete solution. This is due to the fact of the diversity of understanding of what is a geographic ontology. Thus, in this paper we informally defined a geographic ontology reference model, and based on it we formalized the possible heterogeneities that may occur when comparing two geographic ontologies, at the concept-level and at the instance-level as well.

From the heterogeneities, we discussed on what extent the existing conventional (non-geographic) matchers can be used for geographic ontology matching and justified the need for specific matchers to address the geographic features. Furthermore, by analyzing some existing approaches we can conclude that there is still a long way to go before completely solving the problem. As discussed in the open issues (section 5), to achieve good results with real data, the metadata, temporality and data context must be considered.

Finally, another important aspect to be considered, and still an open issue, regards the combination of the similarity measures obtained for the different matching features (names, properties, spatial relationships, and so on at the concept-level and identifier, property values, metadata, and so on at the instance-level). Furthermore if one chooses to use a conventional matcher for matching some features and a geographic matcher for matching the spatial features there is another challenge: how to use the results for the matching of one feature as a parameter in the matching of other feature, if one is measured by a conventional matcher and the other by the geographic matcher?

References

1. Aronoff, S.: Geographic Information Systems: A Management Perspective. WDL Publications (1991)
2. Fonseca, F.T., Davis, C.A., Camara, G.: Bridging ontologies and conceptual schemas in geographic information integration. GeoInformatica 7(4), 355–378 (2003)
3. Sotnykova, A., Cullot, N., Vangenot, C.: Spatio-temporal schema integration with validation: A practical approach. In: Meersman, R., Tari, Z., Herrero, P. (eds.) OTM 2005 Workshops. LNCS, vol. 3762, pp. 1027–1036. Springer, Heidelberg (2005)
4. Kalfoglou, Y., Schorlemmer, M.: Ontology mapping: the state of the art. Knowledge Engineering Review 18(1), 1–31 (2003)
5. Egenhofer, M.J., Franzosa, R.D.: Point set topological relations. International Journal of Geographical Information Systems 5, 161–174 (1991)
6. Belussi, A., Catania, B., Podestá, P.: Towards topological consistency and similarity of multiresolution geographical maps. In: GIS 2005. Proceedings of the 13th annual ACM international workshop on Geographic information systems, Bremen, Germany, pp. 220–229. ACM Press, New York (2005)
7. Castano, S., Ferrara, A., Montanelli, S.: Matching ontologies in open networked systems: Techniques and applications. In: Spaccapietra, S., Atzeni, P., Chu, W.W., Catarci, T., Sycara, K.P. (eds.) J. Data Semantics V. LNCS, vol. 3870, pp. 25–63. Springer, Heidelberg (2006)
8. Noy, N.F.: Tools for mapping and merging ontologies. In: Staab, S., Studer, R. (eds.) Handbook on Ontologies. International Handbooks on Information Systems, pp. 365–384. Springer, Heidelberg (2004)
9. Giunchiglia, F., Shvaiko, P., Yatskevich, M.: S-match: an algorithm and an implementation of semantic matching. In: Kalfoglou, Y., Schorlemmer, W.M., Sheth, A.P., Staab, S., Uschold, M. (eds.) Semantic Interoperability and Integration, Volume 04391 of Dagstuhl Seminar Proceedings, IBFI, Schloss Dagstuhl, Germany (2005)
10. Wang, C., Lu, J., Zhang, G.: Integration of ontology data through learning instance matching. In: Web Intelligence, pp. 536–539. IEEE Computer Society, Los Alamitos (2006)
11. Rahm, E., Do, H.H.: Data cleaning: Problems and current approaches. IEEE Data Eng. Bull. 23(4), 3–13 (2000)
12. Prabhakar, S., Richardson, J., Srivastava, J., Lim, E.: Instance-level integration in federated autonomous databases. In: Proceeding of the Twenty-Sixth Hawaii International Conference on System Sciences, vol. 3, pp. 62–69 (1993)
13. Kavouras, M., Kokla, M., Tomai, E.: Comparing categories among geographic ontologies. Computers & Geosciences 31(2), 145–154 (2005)
14. Kuhn, W.: Modeling the semantics of geographic categories through conceptual integration [29], pp. 108–118
15. Rodriguez, M.A., Egenhofer, M.J.: Determining semantic similarity among entity classes from different ontologies. IEEE Trans. Knowl. Data Eng. 15(2), 442–456 (2003)
16. Stoimenov, L., Djordjevic-Kajan, S.: An architecture for interoperable gis use in a local community environment. Computers and Geosciences 31, 211–220 (2005)
17. Visser, U., Stuckenschmidt, H., Schlieder, C.: Interoperability in gis - enabling technologies. In: Proc. of 5th AGILE Conference on Geographic Information Science, Palma, Spain, April 25-27, 2002 (2002)

18. Sehgal, V., Getoor, L., Viechnicki, P.D.: Entity resolution in geospatial data integration. In: ACM-GIS 2006, ACM Press, New York (2006)
19. Kokla, M., Kavouras, M.: Semantic information in geo-ontologies: Extraction, comparison, and reconciliation [28], pp. 125–142
20. Quix, C., Ragia, L., Cai, L., Gan, T.: Matching schemas for geographical information systems using semantic information. In: Meersman, R., Tari, Z. (eds.) OTM 2006. LNCS, vol. 4276, pp. 1566–1575. Springer, Heidelberg (2006)
21. Sotnykova, A., Vangenot, C., Cullot, N., Bennacer, N., Aufaure, M.A.: Semantic mappings in description logics for spatio-temporal database schema integration. [28], pp. 143–167
22. Dobre, A., Hakimpour, F., Dittrich, K.R.: Operators and classification for data mapping in semantic integration. In: Song, I.-Y., Liddle, S.W., Ling, T.-W., Scheuermann, P. (eds.) ER 2003. LNCS, vol. 2813, pp. 534–547. Springer, Heidelberg (2003)
23. Cruz, I.F., Sunna, W., Chaudhry, A.: Semi-automatic ontology alignment for geospatial data integration. In: Egenhofer, M.J., Freksa, C., Miller, H.J. (eds.) GIScience 2004. LNCS, vol. 3234, pp. 51–66. Springer, Heidelberg (2004)
24. Worboys, M.F., Duckham, M.: Integrating spatio-thematic information. [29], pp. 346–362
25. Schwering, A., Raubal, M.: Measuring semantic similarity between geospatial conceptual regions. In: Rodríguez, M.A., Cruz, I., Levashkin, S., Egenhofer, M.J. (eds.) GeoS 2005. LNCS, vol. 3799, pp. 90–106. Springer, Heidelberg (2005)
26. Volz, S.: Data-driven matching of geospatial schemas. In: Cohn, A.G., Mark, D.M. (eds.) COSIT 2005. LNCS, vol. 3693, pp. 115–132. Springer, Heidelberg (2005)
27. Beeri, C., Doytsher, Y., Kanza, Y., Safra, E., Sagiv, Y.: Finding corresponding objects when integrating several geo-spatial datasets. In: GIS 2005. Proceedings of the 13th annual ACM international workshop on Geographic information systems, Bremen, Germany, pp. 87–96. ACM Press, New York (2005)
28. Spaccapietra, S., Zimányi, E. (eds.): Journal on Data Semantics III. LNCS, vol. 3534. Springer, Heidelberg (2005)
29. Egenhofer, M.J., Mark, D.M. (eds.): Geographic Information Science. In: Egenhofer, M.J., Mark, D.M. (eds.) GIScience 2002. LNCS, vol. 2478, pp. 25–28. Springer, Heidelberg (2002)

An Algorithm for Merging Geographic Datasets Based on the Spatial Distributions of Their Values

Toni Navarrete and Josep Blat

Department of Information and Communication Technologies, Universitat Pompeu Fabra
Passeig de Circumval·lació, 8. 08003 Barcelona, Spain
{toni.navarrete,josep.blat}@upf.edu

Abstract. In this paper we describe an algorithm for merging ontologies from heterogeneous geographic data sources. The algorithm is based on an asymmetric similarity function that considers the spatial distribution of thematic values in the datasets. It has been used in the context of a semantic framework that provides a set of semantic services to enable external clients to find, translate and integrate thematic information from different geographic datasets in a repository. An optimised version of the algorithm is also described enabling its execution in real time, even with large datasets. The algorithm has been tested in the context of merging datasets with more than 10^8 spatial units.

1 Introduction

Geographic datasets represent the real world by assigning thematic entities to spatial elements. However, different producers structure their datasets in terms of different sets of thematic entities, which are often not precisely defined and which may be understood in different ways by different people. This semantic heterogeneity has to be addressed to achieve a meaningful integration of geographic information from diverse sources.

Kuhn defines the concept of Semantic Reference System [1] in order to deal with semantically heterogeneous geographic information. His approach is based on providing a solid framework for the thematic component of geographic information, equivalent to the existing reference systems for the spatial and temporal components. The spatial component of geographic information is represented through geographic coordinates, which are referred to a spatial reference system that specifies the reference ellipsoid, geodetic datum, map projection, reference sea level and units of measure. Methods for transforming data from one spatial reference system to another have been largely studied. Likewise, the temporal component of geographic information refers to a temporal reference system. Transformation methods between temporal reference systems also exist (for instance for transforming from one calendar to another). Semantic reference systems are conceived in a similar way as spatial and temporal reference systems, providing a framework for the thematic component. As Kuhn points out, "users of geographic information should be able to refer thematic data to semantic reference systems, just as they refer geometric data to spatial reference systems" [2]. Furthermore, methods for projecting and translating data from different semantic reference systems are necessary.

F. Fonseca, M.A. Rodríguez, and S. Levashkin (Eds.): GeoS 2007, LNCS 4853, pp. 66–81, 2007.

Semantic reference systems usually rely on the use of ontologies, since ontologies provide explicit formal definitions of thematic entities and their relations, and thus facilitate the definition of methods for projecting, translating or integrating geographic information from different sources. Ontologies are at the core of the most systems for semantic integration of geographic information, such as ODGIS [3] or BUSTER [4] among others.

We have defined in [5] a semantic framework related to the concept of semantic reference systems. The aim of our framework is to represent and to integrate geographic information from a repository of datasets. Datasets in our repository contain metadata according to the CSDGM (Content Standards for Digital Geospatial Metadata) standard [6], developed by the FGDC (Federal Geographic Data Committee). Namely, the metadata file contains the description of the application schema (entities, attributes and values) of the dataset.

Our semantic framework comprises three main elements.

1) An ontology defined to represent the thematic knowledge in the repository of geographic datasets; the thematic classes are organised in a taxonomy, and besides subsumption, other semantic relations can be set between thematic classes, which can also be defined through Description Logic axioms.
2) Different algorithms to semi-automatically add new datasets to the repository, and consequently new knowledge to the ontology.
3) A set of semantic services to enable external clients to find, translate and integrate thematic information from different datasets in the repository.

We have developed OntoGIS, an implementation of the semantic framework in Java, using the Jena framework. The ontology is expressed in OWL and the semantic services are based on Description Logic. The tool supports raster datasets in the GeoTIFF format, described through metadata according to the FGDC CSDGM standard. A screenshot of the tool can be seen in Figure 1.

In this paper we focus on the second element, and namely we describe the algorithm for merging geographic datasets based on the spatial distribution of values. When a new dataset is inserted into the repository, an expert user can connect the dataset values (extracted from the metadata file) to existing or new classes in the taxonomy of thematic values. Although this merging process can be done manually by the expert, semi-automatic algorithms have been developed to facilitate this task. The algorithm that we discuss here is based on the level of overlapping among the spatial extents of sets of values from different datasets. We define *spatial extent* of a dataset value as the union of all the spatial units in the dataset such that their main thematic variable has the indicated value. A high overlapping between the spatial extents of two different values from different datasets means that they probably refer to equivalent themes. If the spatial extent of the first value is contained in the spatial extent of the second one (in a different dataset), it probably indicates that there is a subclass relation between their thematic classes. The algorithm considers not only single values, but sets too. This is necessary to address the typical case of a thematic class specialised in different ways in two datasets. For instance, let us consider a land use dataset with values *evergreen forest* and *deciduous forest*, among others, while another dataset has the values *dense forest* and *sparse forest*, among others. Although the algorithm will probably not find any relation among individual values, it will find that the union of *evergreen forest* and *deciduous forest* is equivalent to the union of *dense forest* and *sparse forest*. The suggested action in this case will consist on defining a new class (that an expert can name as *forest*), which will have two different classifications: one comprising its subclasses *evergreen forest* and *deciduous forest*, while the other is made of its subclasses *dense forest* and *sparse forest*. It has to be noted that these classifications may comprise any number of values, not necessarily two as in the example.

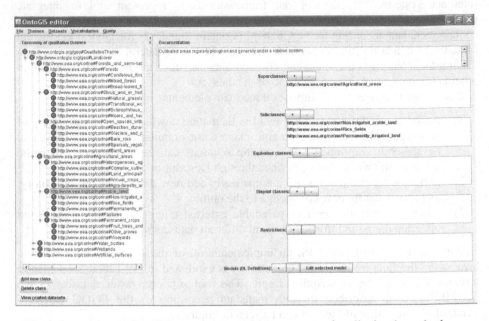

Fig. 1. Screenshot of OntoGIS: management of the taxonomy of qualitative thematic classes

In Section 2 we provide a brief overview of different approaches for merging or aligning geographic ontologies. In Section 3 we formally define our merging algorithm based on the spatial distribution of dataset values. This algorithm has an exponential execution time of $\mathcal{O}(2^{m+n})$, where m and n are the number of values in each dataset. An optimised version of the algorithm that can be run in real time is given in Section 4. An experiment where this algorithm has been used with real datasets comprising more than 10^8 spatial units is described in Section 5. Finally, Section 6 summarises the main contributions of our work and discusses some research lines for the future.

2 Related Work

Finding relations between classes or entities from different ontologies is a common problem in the areas of ontology merging and database schema integration. The main approach for finding these mappings is based on the definition of a similarity function between classes. Two examples of classifications of different alignment methods in a general context can be found in [7, 8]. Structural methods are based on similarity functions that consider the taxonomy (or graph) structure of the ontologies being aligned. Terminological methods use similarities between the names of classes or attributes, and may use external thesaurus to also consider synonymy, hypernymy and/or meronymy relations of these names. Extensional methods are based on the similarity between the individuals (instances) of the ontologies. Compound methods define a similarity function by combining several of these aspects.

However, merging or aligning geographic dataset ontologies presents some significant particularities that are not well addressed by these general methods and related tools. Dataset ontologies are usually structured in simple, almost flat, hierarchies of themes. If the dataset ontology is automatically extracted from the metadata file, it only comprises the set of values and the main topic of the map (for instance "land use"). Even in the best situations, where the dataset is organised according to a standardised vocabulary such as CORINE [9] or Anderson [10], the hierarchy of these vocabularies seldom has more than three levels. On the other hand, since the datasets being merged are usually representations of the same region, extensional methods can be defined where the instances are the spatial units, and where the spatial distribution of thematic values in the datasets provides an indication of possible relations.

Several examples of merging or aligning methods focusing on the geospatial context can be found. Some of the systems for semantic integration of geographic information are based on manual alignment methods, where an expert has to determine the mappings between classes. [11] is an example of a simple system relying on these manual methods. Other systems require a high intervention of a human expert. In the merging phase of the BUSTER system [12], the expert relates each thematic concept in a dataset with one or more terms of standard vocabularies, such as the GEMET thesaurus [13], through logical axioms in a manual process comprising several steps. A similar approach is followed by Hakimpour and Geppert [14]. The expert has to provide intensional definitions for thematic concepts in datasets, where both dataset ontologies must refer to a shared higher level ontology. Kavouras and Kokla [15] define a merging method based on Formal Concept Analysis (FCA) [16]. However, the expert has to manually determine equivalences and overlappings between the classes and to identify common attributes. Uitermark [17] defines a merging method based on considering explicit surveying rules which determine how a terrain is transformed in instances in a dataset. Nevertheless, it is an expert who, according to these surveying rules, has to relate the classes in the dataset ontologies to the classes in a higher level domain ontology. This process usually involves the definition of new classes specialising the classes in the domain ontology. The result of this merging process is what Uitermark calls the reference model.

Other approaches with a higher level of automation are based on similarity measures. A good example is the asymmetric definition of similarity of Rodríguez and Egenhofer [18]. This approach is focused on aligning large vocabularies with a hierarchical organisation in several layers. However, it cannot be applied in our context where datasets are represented through a very simple almost flat taxonomy of values. Although with a different goal but also based on similarities, it is worth mentioning the work of Universidad de Zaragoza on semantic disambiguation in geographic thesaurus in the context of spatial data infrastructures (SDIs) [19]. As in the previous case, the method cannot be applied with our very simple dataset ontologies. Schwering and Raubal [20] define an asymmetric similarity measure based on conceptual spaces [21]. A conceptual region is a representation of a concept as an *n*-dimensional convex region in a vector space, where each dimension corresponds to an attribute. Their similarity measure is obtained as the average of the minimum distance between each vector component in one conceptual region and the other conceptual region. However, this theoretical approach can be hardly put into practice, at least for the datasets in our repositories.

Very relevant and related to our approach is the work of Duckham and Worboys [22], who define an algebraic method for both merging (generating a shared thematic structure between two datasets) and integrating (obtaining a new map combining the two source datasets), based on the spatial distribution of values. The merging approach is based on the assumption that if the spatial extent of one value in one dataset is contained by that of a value in the other dataset, the first value is a subclass of the second. However, a significant limitation of this approach relates to its application to real datasets, which have a big number of spatial units. In this case, it is unlikely that one spatial extent is totally contained in another one, since different interpretations for a particular area, different generalising methods or simply small cartographic errors usually exist. Consequently, this method often fails at finding relations between two real datasets. Our approach, more flexible, uses a threshold for measuring the level of spatial containment between two spatial extents. Furthermore, another significant advantage of our approach with respect to [22] refers to the fact that our method considers different classifications of the same theme in different datasets, as in the example of the two different classifications of forests described in the previous section.

3 Formal Definition of the Merging Algorithm

We define a dataset as the tuple:

$$D = < S, V, a >$$

where S is the set of spatial units, V the set of the thematic values in the dataset, and a is the function that assigns a value to a spatial unit:

$$a: S \to V$$

Note that in the case of a raster dataset, each spatial unit in S corresponds to a cell, while in the case of a vector-based dataset, each spatial unit in S is a polygon. Also note that V can be extracted automatically from the metadata file (compliant to the FGDC CSDGM standard in our implementation). It is also important to note that this definition assumes just one thematic attribute for dataset. This is a common assumption in the majority of formal approaches in the semantic interoperability field in order to avoid discrepancies in the way the internal schema is represented (the so-called structural heterogeneity). Nevertheless, this is not really a restriction since a (typically vector) dataset with several thematic attributes can be represented as different logical datasets, one for each thematic attribute.

The spatial extent of a value or set of values is defined by means of the following function e:

$$e : \wp(V) \to \wp(S)$$
$$U \mapsto \{s \in S \mid a(s) \in U\}$$

If two datasets, $D1 = \langle S1, V1, a1 \rangle$ and $D2 = \langle S2, V2, a2 \rangle$, have to be merged, two similarity functions, $m1$ and $m2$, are defined in the following way, where $|e(U1)|$ indicates the area of the spatial extent of the set $U1$:

$$m1 : \wp(V1) \times \wp(V2) \to [0,1]$$
$$(U1,U2) \mapsto |e(U1) \cap e(U2)| / |e(U1)|$$
$$m2 : \wp(V1) \times \wp(V2) \to [0,1]$$
$$(U1,U2) \mapsto |e(U1) \cap e(U2)| / |e(U2)|$$

Given $U1 \in \wp(V1)$ and $U2 \in \wp(V2)$, $m1$ returns the quantity of the intersection of the extents of $U1$ and $U2$, $e(U1) \cap e(U2)$ that is contained in the extent of $U1$, $e(U1)$. On the other hand, $m2$ returns the quantity of the intersection of extents contained in the extent of $U2$, $e(U2)$. It has to be noted that both $m1$ and $m2$ are asymmetric similarity functions.

As it has already been stated above, a high overlapping between the spatial extents of two different values (or sets of values) means that they are probably semantically equivalent. Furthermore, if the spatial extent of one value (or set) is contained by the spatial extent of another value (or set), it is likely that a subclass semantic relation exists between them. However, in real datasets with a large number of spatial units it is unlikely that the spatial extent of a value is totally contained in another one. As it has also been mentioned above, different interpretations for particular areas, different generalizing methods or simply small cartographic errors usually appear. Our approach is based on using a threshold λ for the level of spatial containment among two spatial extents. For a similarity over this threshold, it is assumed that a subclass semantic relation exists. Consequently, if $m1(U1,U2) > \lambda$, where $U1 \in \wp(V1)$ and $U2 \in \wp(V2)$, then we can infer that $U1$ is subclass of $U2$. In the same way, if $m2(U1,U2) > \lambda$, then we can infer that $U1$ is superclass of $U2$. If both $m1(U1,U2) > \lambda$ and $m2(U1,U2) > \lambda$, then $U1$ and $U2$ are semantically equivalent. Formalising this, the merging algorithm of two datasets $D1$ and $D2$ can be written in the following way:

```
for each U1 ∈ ℘(V1) do
   for each U2 ∈ ℘(V2) do
      if m1( U1, U2 ) > λ then
          Add mapping "U1 subclass of U2" (U1 ⊑ U2)
      end if

      if m2( U1, U2 ) > λ then
          Add mapping "U2 subclass of U1" (U2 ⊑ U1)
      end if
   end for
end for
```

Note that if $m1(\,U1,\,U2\,) > \lambda$ and $m2(\,U1,\,U2\,) > \lambda$, then $U1$ is subclass of $U2$ and $U2$ is subclass of $U1$, and consequently, $U1$ is equivalent to $U2$ ($U1 \equiv U2$).

The result of this algorithm is a new ontology that contains the classes from both datasets as well as the relations provided by the algorithm. Furthermore, when either $U1$ or $U2$ is not a singleton set, a new class is added to the merged ontology corresponding to the union of all its elements.

Let us consider the following example that involves two simple raster datasets, each comprising four spatial units, and where s_{1i} corresponds to the same region than s_{2i}, for $i \in \{1,\ldots,4\}$.

s_{11}	s_{12}
A	B
s_{13}	s_{14}
B	C

Dataset D1

s_{21}	s_{22}
X	Y
s_{23}	s_{24}
Z	Z

Dataset D2

$D1 = \,< S1, V1, a1 >$

$S1 = \{\, s_{11}, s_{12}, s_{13}, s_{14} \,\}$

$V1 = \{\, A, B, C\}$

$a1: S1 \rightarrow V1$

$\quad a1(s_{11}) = A$

$\quad a1(s_{12}) = B$

$\quad a1(s_{13}) = B$

$\quad a1(s_{14}) = C$

$e1: \wp(V1) \rightarrow \wp(S1)$

$\quad e1(\,\{A\}\,) = \{s_{11}\}$

$\quad e1(\,\{B\}\,) = \{s_{12}, s_{13}\}$

$\quad e1(\,\{C\}\,) = \{s_{14}\}$

$\quad e1(\,\{A,B\}\,) = \{s_{11}, s_{12}, s_{13}\}$

$\quad e1(\,\{A,C\}\,) = \{s_{11}, s_{14}\}$

$\quad e1(\,\{B,C\}\,) = \{s_{12}, s_{13}, s_{14}\}$

$\quad e1(\,\{A,B,C\}\,) = S1$

$m1: \wp(V1) \times \wp(V2) \rightarrow [0,1]$

$\quad m1(\,\{A\},\{X\}\,) = 1$

$\quad m1(\,\{A\},\{Y\}\,) = 0$

$\quad m1(\,\{B\},\{Y\}\,) = 0.5$

$\quad m1(\,\{B\},\{Z\}\,) = 0.5$

$\quad m1(\,\{A\},\{X,Y\}\,) = 1$

$\quad m1(\,\{A,B\},\{X\}\,) = 0.33$

$\quad m1(\,\{B,C\},\{Y,Z\}\,) = 1$

$\quad \ldots$

$D2 = \,< S2, V2, a2 >$

$S2 = \{\, s_{21}, s_{22}, s_{23}, s_{24} \,\}$

$V2 = \{\, X, Y, Z\}$

$a2: S2 \rightarrow V2$

$\quad a2(s_{21}) = X$

$\quad a2(s_{22}) = Y$

$\quad a2(s_{23}) = Z$

$\quad a2(s_{24}) = Z$

$e2: \wp(V2) \rightarrow \wp(S2)$

$\quad e2(\,\{X\}\,) = \{s_{21}\}$

$\quad e2(\,\{Y\}\,) = \{s_{22}\}$

$\quad e2(\,\{Z\}\,) = \{s_{23}, s_{24}\}$

$\quad e2(\,\{X,Y\}\,) = \{s_{21}, s_{22}\}$

$\quad e2(\,\{X,Z\}\,) = \{s_{21}, s_{23}, s_{24}\}$

$\quad e2(\,\{Y,Z\}\,) = \{s_{22}, s_{23}, s_{24}\}$

$\quad e2(\,\{X,Y,Z\}\,) = S2$

$m2: \wp(V1) \times \wp(V2) \rightarrow [0,1]$

$\quad m2(\,\{A\},\{X\}\,) = 1$

$\quad m2(\,\{A\},\{Y\}\,) = 0$

$\quad m2(\,\{B\},\{Y\}\,) = 1$

$\quad m2(\,\{B\},\{Z\}\,) = 0.5$

$\quad m2(\,\{A\},\{X,Y\}\,) = 0.5$

$\quad m2(\,\{A,B\},\{X\}\,) = 1$

$\quad m2(\,\{B,C\},\{Y,Z\}\,) = 1$

$\quad \ldots$

If we consider 0.9 as the λ threshold, we can get from *m1* and *m2* the following relations:

- A is equivalent to X (A is subclass of X, and X of A). Obviously, A is also subclass of $\{X,Y\}$, and X of $\{A,B\}$
- Y is subclass of B
- $\{B,C\}$ is equivalent to $\{Y,Z\}$: they conform two different classifications of the same class
- ...

In our implementation, the λ threshold has a default value, but can be changed by the expert user. If the datasets being merged present homogenous thematic and spatial structures, a high threshold close to 1 can be chosen. For instance, in the evaluation experiments described in Section 5, we have chosen a threshold of 0.95. On the other hand, if datasets present very different thematic categorizations, a lower threshold may be needed to find semantic relations. Likewise, if datasets present very different spatial structures, for instance having different scales (and resolutions), a lower threshold should also be considered.

The quality of the results of this algorithm relies on the statistical value of the datasets being merged. They have to contain enough spatial units for each value. Otherwise, the mappings generated by the algorithm may not be semantically valid. It is worth noting that this is a semi-automatic approach, where an expert user should confirm whether the relations provided by the algorithm are really meaningful or not.

4   An Optimised Version of the Algorithm

The algorithm described in the previous section has two main problems. The first problem refers to the fact that comparing the spatial extents of two sets of values from different datasets may be a slow process. In the case of raster datasets, it requires a cell-by-cell comparison of the whole dataset. In the case of vectors, it may require to execute spatial operators between a great number of polygons. Furthermore, the process has to be repeated each time that a value belongs to a set being compared, and thus, it becomes extremely inefficient. 2^{m+n} comparisons of dataset values will have to be made, where m and n are the number of values in each dataset. Since the number of spatial units in a dataset may be huge (more than 100 million spatial units in our experiment described in Section 5), it is necessary to reduce the number of this type of comparisons.

The second problem of the algorithm is that it produces redundant mappings. For instance if there is a mapping "A subclass of X", the algorithm will also generate subclass mappings between A and any set of values containing X, that are clearly redundant. To reduce this list of possible mappings with non-redundant mappings we only allow a class to be involved in one equivalence mapping. Furthermore, although we consider that any type of these mapping relations (equivalence, subclass or superclass) between two individual values is relevant, we assume that only equivalence provides meaningful information for non-atomic sets of values. Note that in this case, an equivalence indicates that a common concept has been specialised in different ways in both ontologies. This way, the algorithm generates any type of relation (subclass, superclass or equivalence) for 1-to-1 mappings, while only equivalences for 1-to-many and many-to-many mappings.

We propose here a variation of the previous algorithm that uses a $m \times n$ matrix M, where m and n are respectively the number of values in two datasets $D1$ and $D2$. $M(i,j)$ contains the area of the overlapping space (measured for instance in m^2) between the i-th value of dataset $D1$, and the j-th value of dataset $D2$. Note that only atomic values (not sets) have to be compared to fill M, and consequently, only $m \times n$ comparisons of spatial extents are required, instead of 2^{m+n} in the previous algorithm. In the case of raster datasets with a common tessellation, they have to be traversed only once to fill M, comparing cell by cell. If the datasets being inserted have X cells, this solution only requires X cell comparisons, while the previous one required $X \cdot 2^{m+n-1}$.

In our tool we have only implemented support for raster datasets, but the algorithm can also be used with vector-based datasets. In fact, once the M matrix has been filled, the algorithm works in the same way for vectors and rasters. In fact, it is also possible to integrate a raster dataset with a vector one if the corresponding topological functions are developed.

The following pseudo-code describes the process of filling the M matrix in the case of two raster datasets with a common tessellation, $D1 = < S, V1, a1 >$ and $D2 = < S, V2, a2 >$, where $V1 = \{v_{11}, \ldots, v_{1m}\}$ and $V2 = \{v_{21}, \ldots, v_{2n}\}$, and where M is initialised with 0's, and X and Y are respectively the number of rows and columns in $D1$ and in $D2$:

```
for x=1 to X do
    for y=1 to Y do
        value1 = a1( cell(x,y) in D1 )
        v1 = transformIndex(D1, value1)
        value2 = a2( cell(x,y) in D2 )
        v2 = transformIndex(D2, value2)
        M(v1,v2) = M(v1,v2) + sizeOfCell
    end for
end for
```

where *transformIndex* transforms a value in its position in the list of values of its dataset, that is, an index between 1 and the number of values in the dataset.

It has to be noted that, since datasets may be big and filling M is the most expensive part of the algorithm, the tool permits the user to save and load M matrices. Furthermore, it is also important to note that once the M matrix has been filled, it is possible to compare whatever sets of values without accessing the datasets. This way, the similarity functions $m1$ and $m2$ are now obtained in the following way, for $U1 \in \wp(V1)$ and $U2 \in \wp(V2)$:

$$|U1| = \sum_{i,v1i \in U1} \sum_{j=1}^{n} M(i,j) \qquad\qquad |U2| = \sum_{j,v2j \in U2} \sum_{i=1}^{m} M(i,j)$$

$$|U1 \cap U2| = \sum_{i,v1i \in U1} \sum_{j,v2j \in U2} M(i,j)$$

$$m1(U1,U2) = \frac{|U1 \cap U2|}{|U1|} \qquad\qquad m2(U1,U2) = \frac{|U1 \cap U2|}{|U2|}$$

As in the previous algorithm, if $m1(U1,U2)$ is greater than the λ threshold, it suggests that $U1$ is subclass of $U2$, while if $m2(U1,U2)$ is greater than the λ threshold, it suggests that $U1$ is superclass of $U2$. Consequently, if both are greater than λ, it suggests that $U1$ and $U2$ are equivalent.

Two new matrices $M1$ and $M2$ are generated from M. They represent the ratio of the spatial extent of one value contained in that of another. This way, $M1(i,j)$ contains the similarity $m1(\{v_{1i}\},\{v_{2j}\})$, while $M2(i,j)$ contains $m2(\{v_{1i}\},\{v_{2j}\})$. Note that the sum of the values of a row of $M1$ is always 1, while the sum of the values of a column of $M2$ is also always 1.

Besides the use of matrices, the original algorithm is also modified in order to avoid the 2^{m+n} comparisons between every set of values in $\wp(V1)$ with every set in $\wp(V2)$ and to avoid the generation of redundant mappings. A greedy approach is proposed, which firstly processes those values having the highest similarities $m1$ or $m2$. Once a value is involved in an equivalence mapping, it is not considered in other sets to be mapped. However, it could be involved in other mappings of atomic values, where the relation is not an equivalence.

The new algorithm first selects the highest value in $M1$ and $M2$, and its position (i,j). The i-th value from $D1$, v_{1i}, and the j-th value from $D2$, v_{2j}, are considered as the best candidates to be mapped. If $M1(i,j)$ is greater than the λ threshold, v_{1i} is suggested to be a subclass of v_{2j}. Likewise, if $M2(i,j)$ is greater than the λ threshold, v_{1i} is suggested to be a superclass of v_{2j}. Consequently, if both values are greater than the λ threshold, an equivalence is suggested between them.

In the case of not obtaining an equivalence, the algorithm adds v_{1i} to $U1$ and v_{2j} to $U2$, and it starts the process of searching an equivalence between sets. The maximum among the values in the i-th row of $M1$ and j-th column of $M2$ is selected as the best candidate. If the maximum is obtained from $M1$ at position (i,k), then the k-th value from $D2$ (v_{2k}) is added to $U2$. Otherwise, if the maximum is obtained from $M2$ at position (k,j), the k-th value from $D1$ (v_{1k}) is added to $U1$. The similarities $m1(U1,U2)$ and $m2(U1,U2)$ are obtained again, and an equivalence is suggested if they both are greater than the λ threshold. Otherwise, the process continues adding values to either $U1$ or $U2$ until either an equivalence is obtained or no more values can be added. The function $compareSets$ recursively compares two sets of values and adds more values to the sets until an equivalence is found or no more values can be added. The function uses the sets $Eq1$ and $Eq2$ which are respectively the sets containing the values in $D1$ and $D2$ that have been previously assigned to other equivalence mappings.

```
function compareSets ( U1 ∈ ℘(V1), U2 ∈ ℘(V2) )
    let max1 be the maximum of M1 and i,j its position
       such that v₁ᵢ ∈ U₁, v₂ⱼ ∉ U2 and v₂ⱼ ∉ Eq2
    let max2 be the maximum of M2 and p,q its position
       such that v₂ᵩ ∈ U2, v₁ₚ ∉ U1 and v₁ₚ ∉ Eq1
    if max1 > 0 or max2 > 0 then
       if max1 > max2 then
          U2 = U2 ∪ { v₂ⱼ }
       else
          U1 = U1 ∪ { v₁ₚ }
       end if
```

```
if m1(U1,U2) > λ and m2(U1,U2) > λ then
    add mapping "U1 equivalent to U2"
    Eq1 = Eq1 ∪ U1
    Eq2 = Eq2 ∪ U2
else
    compareSets(U1,U2)
    end if
  end if
end function
```

Finally, a particular case has to be further analyzed. When $U1$ and $U2$ are suggested as equivalent in this way, the algorithm may miss mappings for values with small spatial extents. Let us consider a value v_{1i} such that $v_{1i} \notin U1$ and that has a small spatial extent. Even if there exists a value in $U2$, v_{2j}, such that $m1(\{v_{1i}\},\{v_{2j}\})=1$, note that $m2(\{v_{1i}\},\{v_{2j}\})$ is probably very small. But, perhaps the similarity between $U1$ and $U2$, which is already greater that the λ threshold, would grow if v_{1i} was added. Formalising this idea, once an equivalence mapping is found between two sets $U1$ and $U2$, the remaining values have to be analyzed. In particular, those values v_{1i} from $D1$ such that $m1(\{v_{1i}\},U2) > \lambda$ and those values v_{2j} from $D2$ such that $m2(U1,\{v_{2j}\}) > \lambda$ will be considered. This way, if value v_{1i} satisfies that $m1(U1\cup\{v_{1i}\},U2) > m1(U1,U2)$ and $m2(U1\cup\{v_{1i}\},U2) > m2(U1,U2)$, then v_{1i} is added to $U1$. Likewise, if value v_{2j} satisfies that $m1(U1,U2\cup\{v_{2j}\}) > m1(U1,U2)$ and $m2(U1, U2\cup\{v_{2j}\}) > m2(U1,U2)$, then v_{2j} is added to $U2$.

5 Experiment

In this section we describe an experiment where the optimised algorithm has been executed to merge real datasets. We have used a set of land cover/land use datasets from the Global Land Cover project of the USGS (http://edcsns17.cr.usgs.gov/glcc/glcc.html). More specifically, we have used the Eurasia Land Cover Characteristics Data Base, which consists of different land cover/land use maps of Eurasia, each with a different thematic classification. All these datasets have a common tessellation ($169 \cdot 10^6$ cells), with the same resolution (1 pixel = 1 km^2) and projection (Lambert azimuthal equal area, optimised for Europe).

This experiment consists in finding the relations between pairs of these datasets that are organised according to different thematic classifications or vocabularies. The results are compared to those obtained using the algorithm defined by Duckham and Worboys [22]. It has to be noted that all the tests have been conducted with the OntoGIS tool, and that in all the cases the list of mappings has been obtained in real time on a laptop PC with an Intel® Core™ 2 Duo processor at 2.0 GHz with 2GB of RAM. We have used a λ threshold of 0.95 in all the tests.

In the first test we have merged the dataset classified according to the USGS vocabulary (a variation of the Anderson vocabulary) and the dataset classified according to the International Geosphere Biosphere Programme (IGBP) classification. The list of the obtained mappings is displayed in Table 1, while Figure 2 presents a screenshot showing how the mappings for these datasets are generated in the OntoGIS tool.

Table 1. Results of merging IGBP and USGS datasets

	Classes in IGBP	Relation	Classes in USGS
1	Evergreen Needleleaf Forest	equivalent	Evergreen Needleleaf Forest
2	Evergreen Broadleaf Forest	equivalent	Evergreen Broadleaf Forest
3	Deciduous Needleleaf Forest	equivalent	Deciduous Needleleaf Forest
4	Deciduous Broadleaf Forest	equivalent	Deciduous Broadleaf Forest
5	Mixed Forest	equivalent	Mixed Forest
6	Union of - Closed Shrublands - Open Shrublands - Woody Savannas - Nonwoody Savannas	equivalent (*)	Union of - Shrubland - Mixed Shrubland/Grassland - Savanna - Wooded Tundra
7	Grasslands	equivalent	Grassland
8	Permanent Wetlands	equivalent	Union of - Herbaceous Wetland - Wooded Wetland
9	Croplands	equivalent	Union of: - Dryland Cropland and Pasture - Irrigated Cropland and Pasture
10	Urban and Built-up	equivalent	Urban and Built-Up Land
11	Cropland/Natural Vegetation Mosaic	equivalent	Union of - Cropland/Grassland Mosaic - Cropland/Woodland Mosaic
12	Snow and Ice	equivalent	Snow or Ice
13	Barren or Sparsely Vegetated	equivalent	Union of - Barren or Sparsely Vegetated - Mixed Tundra - Bare Ground Tundra
14	Water Bodies	equivalent	Water Bodies

(*) 6.a: *Nonwoody Savannas* is subclass of *Savanna*.

The algorithm of Duckham and Worboys generates fewer semantic relations. It obtains the equivalence mappings 1, 2, 3, 4, 5, 7, 10, 12 and 14. Furthermore, it also obtains the subclass relation 6.a, the superclass relation between *Croplands* and *Dryland Cropland and Pasture* (part of mapping 9), the superclass relation between *Cropland/Natural Vegetation Mosaic* and *Cropland/Woodland Mosaic* (part of mapping 11), the two superclass relations for *Permanent Wetlands* (part of mapping 8), and the three superclass relations for *Barren or Sparsely Vegetated* (part of

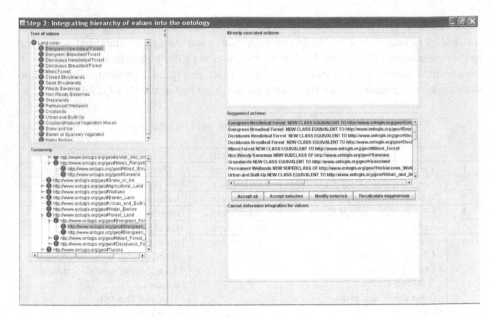

Fig. 2. Screenshot of the OntoGIS tool: merging IGBP and USGS datasets

mapping 13). It does not find some relations because they are not satisfied in 100% of cells (although they are very close to this percentage). And it does not obtain any equivalence mapping involving sets of values (6, 8, 9, 11 and 13).

In the second test we have merged the dataset classified according to the USGS vocabulary and the dataset classified according to the Simple Biosphere Model 2 (SBM2) classification. The list of the obtained mappings is displayed in Table 2.

Since the difference in the organisations of these datasets is greater than in the previous test, there are fewer semantic relations that are satisfied in 100% of cells, and consequently the Duckham and Worboys algorithm retrieves far fewer mappings than our algorithm. It only obtains two equivalence mapping, 3 and 10. It also obtains the superclass mappings 5 and 6 obtained by our algorithm, as well as the mappings 2.a, 2.b and 2.c. Finally, it also obtains the superclass mapping between *Broadleaf Evergreen Trees* and *Evergreen Broadleaf Forest* (part of mapping 1); the superclass mappings between *Agriculture or Grassland* and *Urban and Built-Up Land, Dryland Cropland and Pasture, Irrigated Cropland and Pasture* and *Cropland/Grassland Mosaic* (which are part of mapping 8); the superclass mappings between *Shrubs with Bare Soil* and *Barren or Sparsely Vegetated* and *Bare Ground Tundra* (which are part of mapping 4); and the two superclass mappings for *Ice Cap and Glacier* related to equivalence 9.

More tests can be found in [5] involving these and other datasets organised according to other classifications.

Table 2. Results of merging SBM2 and USGS datasets

	Classes in SBM2	Relation	Classes in USGS
1	*Broadleaf Evergreen Trees*	equivalent	*Evergreen Broadleaf Forest*
2	Union of - *Broadleaf Deciduous Trees* - *Broadleaf and Needleleaf Trees* - *Needleleaf Evergreen Trees* - *Needleleaf Deciduous Trees*	equivalent (*)	Union of - *Deciduous Broadleaf Forest* - *Deciduous Needleleaf Forest* - *Evergreen Needleleaf Forest* - *Mixed Forest*
3	*Short Vegetation*	equivalent	*Savanna*
4	*Shrubs with Bare Soil*	equivalent	Union of - *Shrubland* - *Mixed Shrubland/Grassland* - *Barren or Sparsely Vegetated* - *Bare Ground Tundra*
5	*Dwarf Trees and Shrubs*	superclass	*Wooded Tundra*
6	*Dwarf Trees and Shrubs*	superclass	*Mixed Tundra*
7	*Dwarf Trees and Shrubs*	superclass	*Wooded Wetland*
8	*Agriculture or Grassland*	equivalent	Union of - *Urban and Built-Up Land* - *Dryland Cropland and Pasture* - *Irrigated Cropland and Pasture* - *Cropland/Grassland Mosaic* - *Cropland/Woodland Mosaic* - *Grassland*
9	*Water, Wetlands*	equivalent	Union of - *Water Bodies* - *Herbaceous Wetland*
10	*Ice Cap and Glacier*	equivalent	*Snow or Ice*

(*) 2.a: *Broadleaf Deciduous Trees* is subclass of *Deciduous Broadleaf Forest*; 2.b: *Needleleaf Deciduous Trees* is superclass of *Deciduous Needleleaf Forest*; 2.c: *Needleleaf Evergreen Trees* is superclass of *Evergreen Needleleaf Forest*.

6 Conclusions and Further Work

In this paper we have formally defined two asymmetric similarity functions that are the basis of our algorithm for merging geographic datasets. The algorithm generates 1-to-1 equivalence, subclass and superclass mappings, as well as equivalence mappings between sets of classes (1-to-many and many-to-many relations). Although 1-to-many and especially many-to-many mappings are not considered in other approaches in the literature, they are needed to represent the typical situation of a thematic class specialised in different sets of classes in two datasets.

The use of a threshold to determine mappings makes our approach more flexible than others, enabling us to use it in the context of merging geographic datasets with a large number of spatial units.

Furthermore, the definition of an optimised version of the algorithm makes it possible to merge large datasets with more than 10^8 spatial units in real time, as it has been described in the experiment in Section 5. This experiment has shown that our algorithm obtains more meaningful relations than others in the literature.

As a future line of work, we would like to explore how the neighbourhood can be considered in the definition of the similarity function. The current similarity function is based on comparing the global spatial extents of two values. However, this approach does not consider whether the percentage of non overlapping extents between two values corresponds to partially overlapping spatial units or to completely separated ones.

An alternative that has to be further analysed too is the introduction of fuzzy logic in the definition of the merged ontology. While in the current approach semantic relations between thematic classes are crisp (they exist or not), the similarity function between the involved classes could be used for the definition of fuzzy relations.

References

1. Kuhn, W.: Semantic Reference Systems. International Journal of Geographic Information Science. 17(5), 405–409 (2003)
2. Kuhn, W.: Semantic Reference Systems. Seminar at Universitat Pompeu Fabra, Barcelona, Spain (2003)
3. Fonseca, F., Egenhofer, M., Agouris, P., Câmara, G.: Using Ontologies for Integrated Geographic Information Systems. Transactions in GIS 6(3) (2002)
4. Visser, U., Stuckenschmidt, H., Wache, H., Vögele, T.: Using Environmental Information Efficiently: Sharing Data and Knowledge from Heterogeneous Sources. In: Rautenstrauch, C., Patig, S. (eds.) Environmental Information Systems in Industry and Public Administration, pp. 41–73. IDEA Group, Hershey, USA & London, UK (2001)
5. Navarrete, T.: Semantic Integration of Thematic Geographic Information in a Multimedia Context. PhD Thesis. Technology Department. Universitat Pompeu Fabra, Barcelona, Spain (2006)
6. FGDC: Content Standard for Digital Geospatial Metadata, v2.0. Federal Geographic Data Committee, Metadata Ad Hoc Working Group (1998)
7. Rahm, E., Bernstein, P.A.: A Survey of Approaches to Automatic Schema Matching. VLDB Journal 10(4), 334–350 (2001)
8. Knowledge Web Consortium: State of the Art on Ontology Alignment (2004)
9. Bossard, M., Feranec, J., Otahel, J.: CORINE Land Cover Technical Guide - Addendum 2000. European Environment Agency (EEA), Copenhagen, Denmark (2000)
10. Anderson, J.R., Harby, E., Roach, J., Witmer, R.: A Land Use And Land Cover Classification System For Use With Remote Sensor Data. Washington, DC, USA (1976)
11. Cruz, I.F., Rajendran, A.: Semantic Data Integration in Hierarchical Domains. IEEE Intelligent Systems 18(2), 66–73 (2003)
12. Schuster, G., Stuckenschmidt, H.: Building Shared Terminologies for Ontology Integration. In: Künstliche Intelligenz (KI), Vienna, Austria (2001)
13. EEA: GEMET version 2001 (General Multilingual Environmental Thesaurus). European Environment Agency, European Topic Centre on Catalogue of Data Sources (ETC/CDS) (2001)
14. Hakimpour, F., Geppert, A.: Resolving Semantic Heterogeneity in Schema Integration: an Ontology Based Approach. In: Formal Ontology in Information Systems, Ogunquit, Maine, USA, ACM, New York (2001)

15. Kavouras, M., Kokla, M.: A Method for the Formalization and Integration of Geographical Categorizations. International Journal of Geographic Information Science 16(5), 439–453 (2002)
16. Ganter, B., Wille, R.: Formal Concept Analysis. Springer, Heidelberg (1999)
17. Uitermark, H.: Ontology-Based Geographic Data Set Integration. PhD Thesis. Universiteit Twente, Deventer, The Netherlands (2001)
18. Rodríguez, M.A., Egenhofer, M.J.: Determining Semantic Similarity Among Entity Classes from Different Ontologies. IEEE Transactions on Knowledge and Data Engineering 15(2), 442–456 (2003)
19. Nogueras-Iso, J., Lacasta, J., Bañares, J.A., Muro-Medrano, P.R., Zarazaga-Soria, F.J.: Exploiting Disambiguated Thesauri for Information Retrieval in Metadata Catalogs. In: Conejo, R., Urretavizcaya, M., Pérez-de-la-Cruz, J.-L. (eds.) Current Topics in Artificial Intelligence. LNCS (LNAI), vol. 3040, pp. 322–333. Springer, Heidelberg (2004)
20. Schwering, A., Raubalm, M.M.: Measuring Semantic Similarity between Geospatial Conceptual Regions. In: Rodríguez, M.A., Cruz, I.F., Egenhofer, M.J., Levashkin, S. (eds.) GeoS 2005. LNCS, vol. 3799, pp. 90–106. Springer, Heidelberg (2005)
21. Gärdenfors, P.: Conceptual Spaces: the Geometry of Thought. MIT Press, Cambridge, Massachusetts, USA (2000)
22. Duckham, M., Worboys, M.F.: An Algebraic Approach to Automated Information Fusion. International Journal of Geographical Information Science 19(5), 537–557 (2005)

Structure-Based Methods to Enhance Geospatial Ontology Alignment[*]

William Sunna and Isabel F. Cruz

Department of Computer Science
University of Illinois at Chicago
851 S. Morgan St. (M/C 152), Chicago, IL 60607, USA
{wsunna,ifc}@cs.uic.edu

Abstract. In geospatial applications with heterogeneous classification schemes that describe related domains, an ontology-driven approach to data sharing and interoperability relies on the alignment of concepts across different ontologies. To enable scalability both in the size and the number of the ontologies involved, the alignment method should be automatic. In this paper, we propose two fully automatic alignment methods that use the structure of the ontology graphs for contextual information, thus providing the matching process with more semantics. We have tested our methods on a set of geospatial ontologies pertaining to the domain of wetlands and on four sets that belong to an ontology repository that is becoming the standard for testing ontology alignment techniques. We have compared the effectiveness and efficiency of the proposed methods against two previous approaches. The effectiveness results that we have obtained with at least one of the new methods are as good or better than the results obtained with the previously proposed methods.

1 Introduction

Geospatial data and metadata are highly dependent on the regions for which they have been defined. Such heterogeneity can, for example, be caused by the autonomic and often uncoordinated development of classification schemes by diverse local government organizations or even by different countries. Other causes include the adaptation of those schemes to particular characteristics of the regions that they describe. Therefore, geospatial data sharing and interoperability will require the matching of metadata concepts across a variety of classification schemes.

In our work, classification schemes are represented by ontologies and the matching of concepts is achieved by aligning those ontologies. Ontology alignment encompasses a wide variety of techniques, which include the matching of single concepts [1,2,3], the matching of several concepts at a time taking into

---

[*] This research was supported in part by the National Science Foundation under Awards ITR IIS-0326284 and IIS-0513553.

F. Fonseca, M.A. Rodríguez, and S. Levashkin (Eds.): GeoS 2007, LNCS 4853, pp. 82–97, 2007.

account the structure of the ontologies [4,5,6], or even the data associated with the ontological concepts [7,8,9]. In this paper, we concentrate on the structure of the ontologies that we want to align.

Two types of architecture can be considered: a *centralized architecture* and a *peer-to-peer architecture*. In the former case, each of the ontologies associated with the heterogeneous data sources is mapped to the global ontology. In the latter case, mappings are established between pairs of ontologies, as needed. In both cases, the ontology from which the mapping is defined is called the *source* and the other ontology is called the *target*. Once a pair of ontologies is mapped, queries posed in terms of one of the ontologies can be automatically translated to the other ontology. A full discussion of these architectures and associated query mechanisms has been presented elsewhere [10,11].

As ontologies grow in size or the number of ontologies grows, their alignment should ideally be automatic or require minimum user intervention. Much attention has been recently placed on the automatic alignment of ontologies. For example, the Ontology Alignment Evaluation Initiative (OAEI) [12] promotes the comparison of automatic alignment methods by publishing every year sets of ontologies so as to compare the effectiveness (in terms of recall and precision) of the methods proposed by the contestants. Each set contains a source ontology, a target ontology, and the expected alignment results between them.

In our previous work, we have explored ontology alignment for geospatial applications leading to a multi-layered approach [13,10,14], which consists currently of four layers [15]. Two of these layers use automatic methods, one uses a semi-automatic method, and the other one uses only a manual method. The overall process is supervised by a domain expert.

In the first layer, an *automatic mapping by definition* process is undertaken that compares each concept in the first ontology to each concept in the second ontology according to their definition, as provided by a dictionary. A similarity measure from 0% (no match) to 100% (exact match) between the concepts being compared is returned. If a dictionary is not consulted, the procedure will be performed by comparing only the concept names and any associated descriptions or properties of the concepts.

In this paper, we propose an enhancement to our first layer of mapping by introducing two (fully) automatic structure-based methods: the *Descendants' Similarity Inheritance (DSI)* method, which uses the relationships between ancestor concepts, and the *Sibling's Similarity Contribution (SSC)* method, which uses the relationships between sibling concepts.

Our chosen application domain of wetlands demonstrates the importance of ontology alignment in the geospatial domain. Organizations monitoring the wetlands data inventory have an interest in sharing data. The lack of standard classification has long been identified as an obstacle to the development, implementation, and monitoring of wetland conservation strategies both at the national and regional levels [16]. In defining wetlands, the United States adopts the "Cowardin" Wetland Classification System [17]. In contrast, European nations use the International Ramsar Convention Definition (www.ramsar.org) and

South Africa uses the National Wetland Classification Inventory [16]. Most classifications recognize the need for regionalization because of the variations in climate, geology, soils, and vegetation. Regionalization is designed to facilitate three activities: (i) planning, where it is necessary to study management problems and potential solutions on a regional basis, (ii) organization and retrieval of data gathered in a resource inventory, and (iii) interpretation of inventory data, including differences in indicator plants and animals among the regions. It can thus be concluded that it is extremely difficult to have a standardized classification system between nations and also between regions of a country with a large geographic area [17].

We implemented our proposed methods and tested them against our previous technique [15], which provides us with a "base case". In addition, we tested our methods against the implementation of a structure-based algorithm, the Similarity Flooding algorithm by Melnik *et al.* [4]. Our experiments involve aligning five pairs of ontologies. In particular, we have covered in detail the alignment of ontologies describing the classification schemes of wetlands, so as to illustrate the main principles that underlie our structure-based methods. Our experiments show that at least one of our structure-based methods is as effective or better than both our base case method and the Similarity Flooding algorithm.

The rest of this paper is organized as follows. In Section 2, we give an overview of related work in the area. We present a brief description of our multi-layered approach to ontology alignment and an overview of our alignment tool in Section 3. In Section 4, we present our automatic structure-based methods that support the first layer of mapping in our multi-layered approach along with the experimental results of applying these methods on five ontology sets. Finally, in Section 5, we draw conclusions and outline future work.

2   Related Work

In their survey paper, Shvaiko and Euzenat [12] provide a comparative review of recent schema and ontology matching techniques in the context of a new classification system they propose, where the techniques are classified as element level or structure level. In the element level category, the techniques can be based on strings, language, linguistic considerations, constraints, or alignment reuse. In the structure level category, the techniques are further classified as graph-based, taxonomy-based, or model-based. In order to derive mappings between concepts during the alignment process, the element level techniques consider the labels of concepts, their definitions, the language they are expressed in, and any possibility to reuse previous mappings to derive new ones. The structure level techniques consider the location of the concept in the ontology structure (e.g., tree, graph) and how the mappings of concepts can contribute to the mappings of adjacent concepts. According to their classification system, our alignment techniques fall into their element level category because of our definition mapping layer (base technique), and structure level category because of our mapping by context layer [15] and of the new methods proposed in this paper.

OLA is an alignment tool, whose main purpose is to align ontologies expressed in OWL [18]. OLA offers parsing and visualization of OWL-Lite and OWL-DL ontologies. In addition it offers similarity computations between concepts of the ontologies being aligned. OLA employs linguistic element level and structure level techniques and supports both manual mappings and automated mappings. The available knowledge about the concepts in the aligned ontologies is taken into consideration prior to the alignment process by allowing appropriate alignment methods to be chosen. OLA tries to achieve the highest level of automation, by letting users provide a minimal set of parameters at the initial steps of the alignment process and then leaving it to the tool to end the alignment. Unlike in our approach, similarities between concepts do not contribute to the similarities of their neighbors.

RiMOM (Risk Minimization based Ontology Mapping) is a system that intends to combine different strategies to achieve optimal alignment from a source ontology to a target ontology [19]. There are two types of defined strategies in the system: linguistic-based techniques (includes edit-distance and statistical-learning), and structure-based techniques (includes similarity-propagation, property-to-property propagation, and concept-to-property propagation). RiMOM first examines the structural similarity of the ontologies and the label similarity of the concepts in the ontologies to determine which strategies to use in the alignment process. For example, if there is high similarity in the labels, RiMOM will rely more on linguistic-based strategies to find the matchings between concepts. RiMOM then applies the selected alignment strategies; each strategy outputs its own independent results and the results are then combined using a linear-interpolation method. Finally, RiMOM applies a refinement procedure to prune alignments that are not considered good. Compared to our approach, we are also using multiple matching techniques and allowing for the determination of which techniques will play a more important role for each matching. However, we offer structure level matching, whereas RiMOM does not.

Silva *et al.* discuss the situation when different mapping agents establish different semantic bridges between the concepts in the source and target ontologies [20]. Due to the inherent and subjective nature of ontologies, different agents establish different semantic bridges for the same set of ontologies. This may cause conflicts. To address this issue, they propose an approach to ontology mapping negotiation where various agents are able to achieve consensus among them. In our approach, multiple alignment layers are supported, such that each layer proposes a set of mappings between the source ontology and the target ontology. In our case, the consolidation mapping layer (the fourth layer in our architecture) is used where it is up to the mapping expert to specify the priority scheme across the different layers.

Melnik *et al.* propose a simple structural model-based level technique, the Similarity Flooding algorithm, that can be used in matching a variety of data structures (referred to as models) [4]. Models can be data schemas, data instances, or a mixture of both. In their approach, models are converted to directed labeled graphs. For their algorithm to work, they rely on the fact that concepts from the two graphs are similar when their adjacent concepts on the graphs are similar. The algorithm starts by obtaining initial mappings between concepts in the two input graphs using a string matching function that returns initial similarities between matched concepts. Having established the initial mappings, the algorithm proceeds iteratively to establish more mappings between other concepts based on the assumption that whenever any two concepts in the input models match with some similarity measure, the similarity of their adjacent concepts increases. The iterations continue "flooding" the similarities across the concepts in the graphs until a fixed point is reached where similarity measures for all concepts have been stabilized. Of the matching techniques that we surveyed, this one is the closest to our vision of what a structure level approach should be, hence we have implemented their algorithm so as to compare its results with those of the methods that we propose in this paper.

3   The AgreementMaker Framework

We have been working on a framework that supports the alignment of two ontologies. In our framework, we introduce an alignment approach that uses different matching techniques between the concepts of the aligned ontologies. Each matching technique is embedded in a mapping layer [15]. As mentioned in Section 1, we have currently four layers in our framework with the possibility of adding more mapping layers in the future. The motivation behind our framework is to allow for the addition of as many mapping layers as possible in order to capture a wide range of relationships between concepts.

Our mapping layers use element-based alignment techniques (first layer) and structure-based alignment techniques (first and third layers). In addition, domain experts can use their knowledge and contribute to the alignment process (second and third layers).

We have developed a tool, the AgreementMaker, which implements our approach. The user interface of our tool displays the two ontologies side by side as shown in Figure 1. After loading the ontologies, the domain expert can start the alignment process by mapping corresponding concepts manually or invoking procedures that map them automatically (or semi-automatically). The mapping information is displayed in the form of annotated lines connecting the matched nodes. Many choices were considered in the process of displaying the ontologies and their relationships [15].

4   Automatic Similarity Methods

In order to achieve a high level of confidence in performing the automatic alignment of two ontologies, a thorough understanding of the concepts in the ontologies is highly desired. To this end, we propose methods that investigate the ontology concepts prior to making a decision on how they should be mapped. We consider both the labels and the definitions of the ontology concepts and the relative positions of the concepts in the ontology tree. Our alignment method

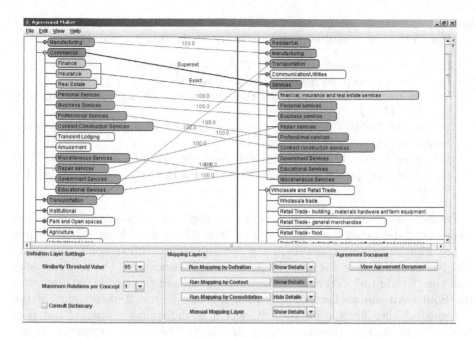

Fig. 1. Results of running three of the mapping layers

onablos the uscr to sclcct onc of thc following thrcc matching mcthods: (1) applying the base similarity calculations only, (2) applying the base similarity calculations followed by the Descendant's Similarity Inheritance *(DSI)* method, or (3) applying the base similarity calculations followed by the Sibling's Similarity Contribution *(SSC)* method. Both the *DSI* and the *SSC* methods have been introduced to enhance the alignment results that were obtained from using the base similarity method previously proposed [15]. We apply one of these methods in our first mapping layer.

4.1 Base Similarity Calculations

The very first step in our approach is to establish initial mappings between the concepts of the source ontology and the concepts of the target ontology. These initial mappings will be a starting point for both the *DSI* and *SSC* methods. We try to find matching concepts in the target ontology for each concept in the source ontology. This is achieved by defining a similarity function that takes a concept in the source ontology and a concept in the target ontology and returns a similarity measure between them. If the similarity measure is equal or above a certain threshold decided by the domain expert, then the two concepts match each other. In order to find the base similarity measure between two concepts, we utilize the concepts' labels and definitions as provided by a dictionary [15].

In what follows, we present the details of finding the base similarity between a concept in the source ontology and a concept in the target ontology:

- Let S be the source ontology and T be the target ontology.
- Let C be a concept in S and C' be a concept in T.
- We use function $base\_sim(C, C')$ that yields a similarity measure M, such that $0 \leq M \leq 1$.
- Parameter TH is a threshold value such that C' is matched with C when $base\_sim(C, C') \geq TH$.
- For every concept C in S, we define the mapping set of C, denoted $MS(C)$, as the set of concepts C' in T that are matched with C (i.e., $base\_sim(C, C') \geq TH$).

Establishing base similarities between concepts of the source ontology and concepts of the target ontology may not be sufficient to achieve a high degree of precision in relating concepts in the two ontologies. To exemplify this point, we give an example in the domain of wetland classification. The first ontology uses the "Cowardin" wetland classification system and the second ontology uses the South African wetland classification system. Figure 2 shows part of the "Cowardin" classification on the left, which is the source ontology, and part of the South African classification on the right, which is the target ontology. When calculating the base similarities between concepts of the two ontologies, the concept *Reef* that belongs to the *Intertidal* wetland subsystem in the source ontology, will yield a base similarity measure of 100% with the concept *Reef* that belongs to the *Intertidal* wetland subsystem in the target ontology. Furthermore, it will also yield a base similarity measure of 100% with the concept *Reef* that belongs to the *Subtidal* wetland subsystem in the target ontology. This example shows that the base similarity measure is misleading because it does not correctly express the true meaning of the relationship between the two concepts, which should not be related because they belong to different wetland subsystems.

In order to eliminate such situations, we propose the Descendant's Similarity Inheritance *(DSI)* method, which reconfigures the base similarity between the concepts based on the similarity of their parent concepts.

4.2   Descendant's Similarity Inheritance (DSI) Method

We define the *DSI* reconfigured similarity between a concept C in S and a concept C' in T as $DSI\_sim(C, C')$. In what follows, we present the details on how to determine $DSI\_sim(C, C')$:

- Let $path\_len\_root(C)$ be the number of edges between the concept C in S and the root of the ontology tree S. For example, in Figure 3, $path\_len\_root(C) = 2$. Similarly, we define $path\_len\_root(C')$ with respect to T. For example, in Figure 3, $path\_len\_root(C') = 2$.
- Let $parent_i(C)$ be the *ith* concept from the concept C to the root of the source ontology S, where $0 \leq i \leq path\_len\_root(C)$. Similarly define $parent_i(C')$ with respect to T. For example, in Figure 3, $parent_1(C) = B$ and $parent_1(C') = B'$.

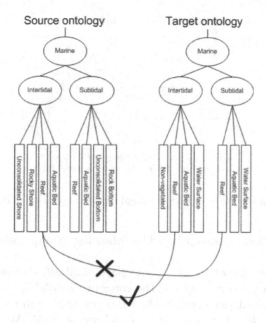

Fig. 2. An example of a case where misleading mappings may occur when two concepts have the same label

- Define *MCP* as the *main contribution percentage*, which is the fraction of the similarity measure between C and C' that will be used in determining the overall *DSI_sim(C,C')*.
- We compute *DSI_sim(C, C')* as follows:

$$MCP \cdot base\_sim(C, C') + \frac{2(1 - MCP)}{n(n+1)} \sum_{i=1}^{n} (n+1-i) \, base\_sim(parent_i(C), parent_i(C')))$$

where $n = \min(path\_len\_root(C), path\_len\_root(C'))$

The main characteristic of the *DSI* method is that it allows for the parent and in general for any ancestor of a concept to play a role in the identification of the concept. Intuitively, the parent of a concept should contribute more to the identity of the concept than its grandparent. This is achieved by assigning a relatively high value to *MCP*. The grandparent concept contributes more than the great grandparent, and so on, until the root is reached. This can be demonstrated by considering the example in Figure 3. In the figure, we show how the *DSI* similarity is determined between the concept C in the source ontology S (shown left) and the concept C' in the target ontology T (shown right) when applying the *DSI* method using an *MCP* value of 75%. The *DSI* similarity is determined by adding 75% of the base similarity between C and C' to 17% of the base similarity of their immediate parents (B and B') and finally to 8% of the base similarity of their grandparents (A and A'). Experiments have shown that 75% for the value of the *MCP* factor works well (in fact, any values in that

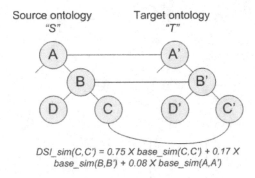

$$DSI\_sim(C,C') = 0.75\ X\ base\_sim(C,C') + 0.17\ X$$
$$base\_sim(B,B') + 0.08\ X\ base\_sim(A,A')$$

Fig. 3. Applying the *DSI* method to calculate the similarity between C and C'

neighborhood performed similarly). The following example illustrates just one such case.

Considering the case of Figure 2, the base similarity between the concepts *Intertidal* in the source ontology and the concept *Subtidal* in the target ontology is 37%. The base similarity between the concepts *Marine* in the source ontology and the concept *Marine* in the target ontology is 100%. When applying the *DSI* method with an *MCP* value of 75%, the *DSI* similarity between the concept *Reef* that belongs to the *Intertidal* wetland subsystem in the source ontology and the concept *Reef* that belongs to the *Subtidal* wetland subsystem in the target ontology will be 88%. Applying the *DSI* method again between the concept *Reef* that belongs to the *Intertidal* wetland subsystem in the source ontology and the concept *Reef* that belongs to the *Intertidal* wetland subsystem in the target ontology will yield a similarity of 100%. Therefore, we conclude that the last match is the best one (in fact the optimal one). This is just one example that shows how the *DSI* method can be useful in determining more accurate similarity measures between concepts.

4.3   Sibling's Similarity Contribution (SSC) Method

In this method, siblings of a concept contribute to the identification of the concept. This may further enhance the quality of the automatic alignment process. Similarly to the *DSI* method, the *SSC* method reconfigures the base similarities between concepts. We define the *SSC* similarity between a concept C in S and a concept C' in T as $SSC\_sim(C, C')$. In what follows, we present the details on how to determine this similarity.

- Let *sibling_count*(C) be the number of sibling concepts of concept C in S. For example, in Figure 4, *sibling_count*(C) = 2.
- Let *sibling_count*(C') be the number of sibling concepts of concept C' in T. For example, in Figure 4, *sibling_count*(C') = 3.
- Let $SS(C)$ be the set of all the concepts that are siblings of C in S and $SS(C')$ be the set of all the concepts that are siblings of C' in T.

- Let S_i be the *ith* sibling of concept C where $S_i \in SS(C)$, and $1 \leq i \leq$ *sibling_count(C)*.
- Let S'_j be the *jth* sibling of concept C' where $S_j \in SS(C')$, and $1 \leq j \leq$ *sibling_count(C')*.
- Define MCP as the *main contribution percentage*, which is the fraction of the similarity measure between C and C' that will be used in determining the overall $SSC\_sim(C, C')$.
- If both $SS(C)$ and $SS(C')$ are not empty, we define $SSC\_sim(C, C')$ as follows:

$$MCP \cdot base\_sim(C, C') + \frac{1 - MCP}{n} \sum_{i=1}^{n} \max(base\_sim(S_i, S'_1), \ldots, base\_sim(S_i, S'_m))$$

where $n = sibling\_count(C)$ and $m = sibling\_count(C')$.

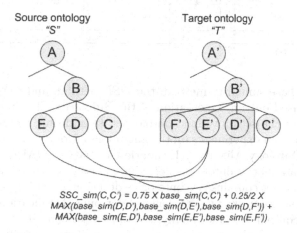

SSC_sim(C,C') = 0.75 X base_sim(C,C') + 0.25/2 X
MAX(base_sim(D,D'),base_sim(D,E'),base_sim(D,F')) +
MAX(base_sim(E,D'),base_sim(E,E'),base_sim(E,F'))

Fig. 4. Applying the SSC method to calculate the similarity between C and C'

The main characteristic of the SSC method is that it allows for the siblings of a given concept to play a role in the identification process of the concept. In Figure 4 we show how the SSC similarity is determined between the concept C in the source ontology S (shown on the left) and the concept C' in the target ontology T (shown on the right) when applying the SSC method with an MCP value of 75% . The SSC similarity is determined by adding 75% of the base similarity between C and C' to (1) 12.5% of the maximum base similarity between D and D', D and E', and D and F' and to (2) 12.5% of the maximum base similarity between E and D', E and E', and E and F'. As for the DSI method, the value of 75% for the MCP factor was found to work well in practice.

4.4 Evaluation

To validate our approach from the point of view of efficiency and of effectiveness, we have aligned the two geospatial wetland ontologies mentioned in Section 4.2

Table 1. Depth and number of concepts in the ontology sets

Ontology set	Depth	Number of concepts in the source ontology	Number of concepts in the target ontology
Wetlands	5	29	29
Weapons	6	153	213
People and pets	4	65	93
Computer networks	5	90	89
Russia	5	86	87

Table 2. Performance results for the base similarity, *DSI*, *SCC*, and Similarity Flooding algorithms in milliseconds

Algorithm	Wetlands	Weapons	People and pets	Computer networks	Russia
Base Similarity	125	1516	236	391	484
DSI	156	3656	562	579	844
SSC	172	4344	938	719	1891
Similarity Flooding	187	2266	703	906	1796

using our own base similarity method, the *DSI* method, and the *SSC* method. We have also used our implementation of the Similarity Flooding algorithm in the alignment of the set of wetland ontologies. In addition, to further evaluate our methods, we run experiments on the alignment of four sets of ontologies provided by the Ontology Alignment Evaluation Initiative (OAEI) [12]. Of these, the first set contains two ontologies describing classifications of various weapon types, the second set contains two ontologies describing attributes of people and pets, the third set contains two ontologies describing classifications of computer networks and equipments, and, finally, the fourth set contains general information about Russia. Each set contains a source ontology, a target ontology, and the expected alignment results between them. Table 1 displays the depth and number of concepts in the five ontology sets we consider.

Similarly to the Similarity Flooding algorithm [4], both our *DSI* and *SSC* methods depend on establishing initial similarities between concepts before they can be executed. However, unlike the Similarity Flooding algorithm, our *DSI* and *SSC* methods do not run in multiple iterations that keep reconfiguring the similarities between concepts until the similarities become stable.

We conducted experiments to determine the running time of all the four methods (base similarity, *DSI, SSC*, and Similarity Flooding) for the previously mentioned five ontology sets. We have implemented all the methods using Java and have run them on an 1.6 GHz Intel Centrino Duo with 1GB of RAM, running Windows XP. The results are shown in Table 2.

Looking at the performance results of Table 2, the running time for the *DSI, SSC*, and Similarity Flooding algorithms include the running time for the base similarity method because they rely on it to run. Therefore, the base similarity algorithm takes the least amount of time. Examining the results with the exclusion of the base similarity method, the *DSI* method has the best run time performance for four of the test cases, while the Similarity Flooding algorithm

Table 3. Applying the base similarity, *DSI*, *SSC*, and Similarity Flooding algorithms to align the geospatial wetland ontologies

Algorithm	Total correct relations	Discovered relations	Correct relations	Precision	Recall
Base similarity	54	39	24	61.54%	44.44%
DSI	54	39	37	94.87%	68.52%
SSC	54	39	29	74.36%	53.70%
Similarity flooding	54	39	36	92.31%	66.67%

has the best running time performance for one test case only. The *SSC* method has the worst performance in three test cases while it performs better than the Similarity Flooding algorithm in two test cases. As compared to the Similarity Flooding algorithm, the *DSI* method only runs once to complete, whereas the Similarity Flooding algorithm will need several iterations to complete. The *SSC* method depends on the number of siblings for a given concept, therefore the larger the number of siblings the worse it performs. In other words, if the ontology trees are wide, then the performance of *SSC* will suffer. Similarly, the running time of the *DSI* method degrades for deep ontology trees. In our future work we are planning to examine ways to improve the running time of the *DSI* and *SSC* methods.

To compare the effectiveness of the four methods, we started by aligning the set of ontologies for the wetlands as described in Section 4.2 and did the same for the other four sets of ontologies. In the wetlands example, we have captured the number of discovered relations between the concepts of the source ontology ("Cowardin") and the concepts of the target ontology ("South African") for each method. Each relationship represents a mapping from a concept C in the source ontology S to a matching target ontology concept $C' \in MS(C)$ with the highest similarity measure. We note that there may be concepts in S that are not mapped to any concepts in the target ontology (corresponding to an empty mapping set). After capturing the discovered relations, we count how many of these relations are valid when compared with the expected alignment results. Having figured the number of correct relations, we calculate both the precision and the recall values. The precision is calculated by dividing the number of discovered valid relations to the total number of discovered relations, the recall is calculated by dividing the number of discovered valid relations to the total number of valid relations as provided by the expected alignment results.

In the alignment of the wetland ontologies, the *DSI* method yielded slightly higher precision and recall values than the Similarity Flooding algorithm, which in turn yielded higher values than the *SSC* method. Overall, these three methods significantly enhanced the precision and recall values obtained by applying the base similarity method only. Table 3 shows the complete results for this test case. The following tests pertain to the four sets of ontologies of the OAEI initiative. In the alignment of the ontologies in the first OAEI set (Weapons), the *DSI* method yielded slightly higher precision and recall values than both the *SSC* and the Similarity Flooding methods as shown in Table 4.

Table 4. Applying the base similarity, *DSI, SSC*, and Similarity Flooding algorithms on the ontology set describing weapons

Algorithm	Total correct relations	Discovered relations	Correct relations	Precision	Recall
Base similarity	73	78	64	82.05%	87.67%
DSI	73	78	66	84.62%	90.41%
SSC	73	78	65	83.33%	89.04%
Similarity flooding	73	78	65	83.33%	89.04%

Table 5. Applying the base similarity, *DSI, SSC*, and Similarity Flooding algorithms on the ontology set describing people and pets

Algorithm	Total correct relations	Discovered relations	Correct relations	Precision	Recall
Base similarity	74	81	49	60.49%	66.22%
DSI	74	81	49	60.49%	66.22%
SSC	74	81	49	60.49%	66.22%
Similarity flooding	74	81	49	60.49%	66.22%

Table 6. Applying the base similarity, *DSI, SSC*, and Similarity Flooding algorithms on the ontology set describing computer networks

Algorithm	Total correct relations	Discovered relations	Correct relations	Precision	Recall
Base similarity	29	23	17	73.91%	58.62%
DSI	29	23	14	60.87%	48.28%
SSC	29	23	19	82.61%	65.52%
Similarity flooding	29	23	18	78.26%	62.07%

Table 7. Applying the base similarity, *DSI, SSC*, and Similarity Flooding algorithms on the ontology set about Russia

Algorithm	Total correct relations	Discovered relations	Correct relations	Precision	Recall
Base similarity	117	51	45	88.24%	38.46%
DSI	117	51	48	94.12%	41.03%
SSC	117	51	46	90.20%	39.32%
Similarity flooding	117	51	46	90.20%	39.32%

All four methods yielded the same results for recall and precision in the alignment of the second OAEI set (People and pets) as shown in Table 5. This is an indication that the locality of all the concepts in the ontologies of the second set are irrelevant in distinguishing their identity.

The *SSC* method yielded better recall and precision results than the Similarity Flooding algorithm, which in turn yielded better results the the *DSI* method when aligning the third OAEI set (Computer networks) as shown in Table 6. Finally, as shown in Table 7, in the alignment of the fourth OAEI set (Russia), the *DSI* method yielded the highest results for precision and recall than either the *SSC* method or the Similarity Flooding algorithm.

The differences found in the recall and precision values for a given method when applied across different test cases are mainly due to the characteristics of the ontologies. For example, in the first OAEI set (Weapons) and the second OAEI set (People and pets), the relations between the concepts, their parents, and their siblings do not contribute to refining the base similarity results. However, the relationships between the concepts and their siblings added value in refining the base similarity results when aligning the third OAEI set (Computer networks). The relationships between the concepts and their parents added value in refining the results when aligning the fourth OAEI set (Russia). Therefore, the selection of an appropriate matching method should be done after a preliminary examination of the concepts in the ontologies and how they relate to each other. Mochol *et al.* [21] present a methodology on how to select an appropriate matching method for a specific alignment case by having a domain expert fill a questionnaire about the nature of the ontologies to be aligned.

5 Conclusions

The subject of automatic ontology alignment has been receiving a lot of attention recently. In this paper, we have proposed two methods that will enhance our multi-layer approach to ontology alignment, which is supported by a visual interface. Our methods use the structure of the ontology graph for contextual information thus providing the matching process with more semantics.

The two methods that we propose, the *Descendants' Similarity Inheritance (DSI)* method and the *Sibling's Similarity Contribution (SSC)* method use respectively the information associated with the descendants and with the siblings of each concept. Our main test case is provided by a geospatial domain application for wetlands. Other ontologies were also tested in the spirit of the Ontology Alignment Evaluation Initiative (OAEI) [12], which currently does not include geospatial ontologies in their repository of ontologies, but is widely regarded as the repository with which to study the effectiveness of ontology alignment methods. The pairs of ontologies in the OAEI repository have associated with them the correct mappings that should be derived by any automatic alignment method, thus enabling an objective effectiveness comparison.

In addition to implementing our own methods, we have also implemented the Similarity Flooding algorithm [4] and tested our new methods against: (1) our base technique that uses a similarity comparison among individual concepts and (2) the Similarity Flooding algorithm. The experimental results show that from an effectiveness viewpoint at least one of our new methods is as good or better than the results obtained with the previously proposed methods.

Much work remains to be accomplished in the general area of ontology alignment and in the particular area of geospatial ontology alignment. A research subject involves the determination of which methods to use depending on the ontologies involved and on their particular topologies. For example, the fact that the most effective method is not always the same and that sometimes all the four methods have similar results shows that: (1) the best method depends on the topology of the ontology graph and (2) for certain topologies, structure-based methods do not play an important role. Both of these conclusions have been arrived at by others [21] and they further justify our multi-layered approach where several techniques can be used and combined [15].

The knowledge of the best method to apply will directly impact our consolidation layer in which priority weights are given to the different matching layers. If such priority weights can be automatically determined, then our overall approach will further attain automation. Another subject of research would be the "fusion" in the same method of different techniques (e.g., *DSI, SCC,* and Similarity Flooding), where such fusion could be guided again by the characteristics of the topologies at hand. A comparison of these two alternatives can then be undertaken.

Many more test cases and studies are needed: the introduction of geospatial ontologies in the OAEI repository will allow for a wide variety of researchers to explore their methods in the geospatial domain; also, there is the need for many more geospatial ontologies to become available. In particular, initiatives such as the Open Geospatial Consortium (http://www.opengeospatial.org/) will likely bring about a plethora of standardized and much larger ontologies that must be semantically aligned to promote data sharing and interoperability.

Acknowledgments

We would like to thank Sarang Kapadia for his help with the implementation of the Similarity Flooding algorithm.

References

1. Bergamaschi, S., Castano, S., Vincini, M.: Semantic Integration of Semistructured and Structured Data Sources. SIGMOD Record 28(1), 54–59 (1999)
2. Castano, S., Antonellis, V.D., di Vimercati, S.D.C.: Global Viewing of Heterogeneous Data Sources. IEEE Transactions on Knowledge and Data Engineering 13(2), 277–297 (2001)
3. Palopoli, L., Saccà, D., Ursino, D.: An Automatic Techniques for Detecting Type Conflicts in Database Schemes. In: 7th International Conference on Information and Knowledge Management (CIKM), pp. 306–313 (1998)
4. Melnik, S., Garcia-Molina, H., Rahm, E.: Similarity Flooding: A Versatile Graph Matching Algorithm and its Application to Schema Matching. In: 18th International Conference on Data Engineering (ICDE) (2002)
5. Noy, N.F., Musen, M.A.: Anchor-PROMPT: Using Non-local Context for Semantic Matching. In: Workshop on Ontologies and Information Sharing at the 17th International Joint Conference on Artificial Intelligence (IJCAI) (2001)
6. Rodríguez, M.A., Egenhofer, M.J.: Determining Semantic Similarity among Entity Classes from Different Ontologies. IEEE Transactions on Knowledge and Data Engineering 15(2), 442–456 (2003)
7. Doan, A., Madhavan, J., Domingos, P., Halevy, A.Y.: Learning to Map between Ontologies on the Semantic Web. In: 11th International World Wide Web Conference (WWW), pp. 662–673 (2002)

8. Fossati, D., Ghidoni, G., Eugenio, B.D., Cruz, I.F., Xiao, H., Subba, R.: The Problem of Ontology Alignment on the Web: a First Report. In: 2nd Web as Corpus Workshop (associated with the 11th Conference of the European Chapter of the ACL) (2006)

9. Ichise, R., Takeda, H., Honiden, S.: Rule Induction for Concept Hierarchy Alignment. In: Workshop on Ontologies and Information Sharing at the 17th International Joint Conference on Artificial Intelligence (IJCAI) (2001)

10. Cruz, I.F., Sunna, W., Ayloo, K.: Concept Level Matching of Geospatial Ontologies. In: GISPlanet Second Conference and Exhibition on Geographic Information, Estoril, Portugal (2005)

11. Cruz, I.F., Xiao, H.: The Role of Ontologies in Data Integration. Journal of Engineering Intelligent Systems 13(4), 245–252 (2005)

12. Shvaiko, P., Euzenat, J.: A Survey of Schema-Based Matching Approaches. In: J. Data Semantics IV. LNCS, vol. 3730, pp. 146–171. Springer, Heidelberg (2005)

13. Cruz, I.F., Rajendran, A., Sunna, W., Wiegand, N.: Handling Semantic Heterogeneities using Declarative Agreements. In: ACM GIS 10th International Symposium on Advances in Geographic Information Systems, pp. 168–174 (2002)

14. Cruz, I.F., Sunna, W., Chaudhry, A.: Semi-Automatic Ontology Alignment for Geospatial Data Integration. In: Egenhofer, M.J., Freksa, C., Miller, H.J. (eds.) GIScience 2004. LNCS, vol. 3234, pp. 51–66. Springer, Heidelberg (2004)

15. Cruz, I.F., Sunna, W., Makar, N., Bathala, S.: A Visual Tool for Ontology Alignment to Enable Geospatial Interoperability. Journal of Visual Languages and Computing 18(3), 230–254 (2007)

16. Dini, J., Gowan, G., Goodman, P.: South African National Wetland Inventory, Proposed Wetland Classification System for South Africa (1998), http://www.ngo.grida.no/soesa/nsoer/resource/wetland/inventory_classif.htm

17. Cowardin, L.M., Carter, V., Golet, F.C., LaRoe, E.T.: Classification of Wetlands and Deepwater Habitats of the United States. U.S. Department of the Interior, Fish and Wildlife Service, Washington, D.C. Jamestown, ND: Northern Prairie Wildlife Research Center Online (Version 04DEC1998) (1979), http://www.npwrc.usgs.gov/resource/wetlands/classwet/index.htm

18. Euzenat, J., Guégan, P., Valtchev, P.: OLA in the OAEI 2005 Alignment Contest. In: Integrating Ontologies 2005, K-CAP Workshop on Integrating Ontologies. vol. 156 of CEUR Workshop Proceedings, Banff, Canada (2005)

19. Tang, J., Li, J., Liang, B., Huang, X., Li, Y., Wang, K.: Using Bayesian Decision for Ontology Mapping. Journal of Web Semantics 4(4), 243–262 (2006)

20. Silva, N., Maio, P., Rocha, J.: An Approach to Ontology Mapping Negotiation. In: Integrating Ontologies 2005, K-CAP Workshop on Integrating Ontologies. Volume 156 of CEUR Workshop Proceedings, Banff, Canada (2005)

21. Mochol, M., Jentzsch, A., Euzenat, J.: Applying an Analytic Method for Matching Approach Selection. In: International Workshop on Ontology Matching (OM-2006) collocated with the 5th International Semantic Web Conference (ISWC), Athens, Georgia, USA (2006)

Geographic Information Retrieval by Topological, Geographical, and Conceptual Matching

Felix Mata

PIIG Lab – Centre for Computing Research
National Polytechnic Institute
Av. Juan de Dios Bátiz s/n, 07738, México, D.F., Mexico
migfel@sagitario.cic.ipn.mx

Abstract. Geographic Information Science community is recognized that modern Geographic Information Retrieval systems should support the processing of imprecise data distributed over heterogeneous repositories. This means the search for relevant geographic results for a geographic query (Q_G) even if the data sources do not contain a result that matches exactly the user's request and then approximated results would be useful. Therefore, GIR systems should be centred at the nature and essence of spatial data (their relations and properties) taken into consideration the user's profile. Usually, semantic features are implicitly presented in *different* data sources. In this work, we use three heterogeneous data sources: vector data, geographic ontology, and geographic dictionaries. These repositories usually store *topological relations*, *concepts*, and *descriptions* of geographical objects under certain scenarios. In contrast to previous work, where these layers have been treated in an isolated way, their integration expects to be a better solution to capture the semantics of spatial objects. Thus, the use of spatial semantics and the integration of different information layers improve GIR, because adequate retrieval parameters according to the nature of spatial data, which emulate the user's requirements, can be established. In particular, we use topological relations {*inside, in*}, semantic relations {*hyperonimy, meronimy*}, and descriptions {*constraints, representation*}. An information extraction mechanism is designed for each data source, while the integration process is performed using the algorithm of ontology exploration. The ranking process is based on similarity measures, using the previously developed confusion theory. Finally, we present a case study to show some results of integrated GIR (iGIR) and compare them with Google's ones in a tabular form.

1 Introduction

Geographic Information Retrieval (GIR) is becoming increasingly popular task of using Geographic Information Systems (GIS). Due to the nature of geographic data, these are usually distributed over numerous heterogeneous repositories that makes the task challenging. Several proposals have been cited as methods to perform this task [25], but existing methodologies do not handle the variety of data sources in order to solve the problem adequately.

Usually, an approach is centered at just one of them; see e.g. [26]. In contrast, we believe that only an approach, integrating different information sources can essentially

F. Fonseca, M.A. Rodríguez, and S. Levashkin (Eds.): GeoS 2007, LNCS 4853, pp. 98–113, 2007.

improve GIR. The present paper is based on this belief, presenting a systemic approach to GIR.

Our approach consists of retrieving geographic information by processing queries, which can be split into a triplet <what, relation, where>, where *"what"* denotes a geographic object, *"where"* can be a spatial reference or a geographic object, and *"relation"* denotes a spatial relation linking *"where"* and *"what"*. These queries have been used in other works such as [23]. The approach is based on a retrieval strategy that uses three types of matching: the first one is a topological matching, i.e. topological relations extracted from overlaying data layers such as {*in, contain*}; the second one is a geographical matching, i.e. constraints obtained from dictionaries such as {*Airports represented by points or polygons*}; and the last one is a conceptual matching, given by a geographic ontology such as {*type of Airport*}. Thus, we use three heterogeneous data sources: vector files, dictionaries, and geographic ontology. A motivation to use these data sources is that they store different relations and properties depicting the nature of spatial data. These data sources have been also used in previous works [2][3][4] but separately. In contrast, we design herein an integrated system, which use all three sources, seeking for more powerful GIR. Hence, iGIR integrates a few processes (to be described in the following): querying, retrieving, and the integration and ranking. Figure 1 shows the framework of approach and the overall retrieval strategy.

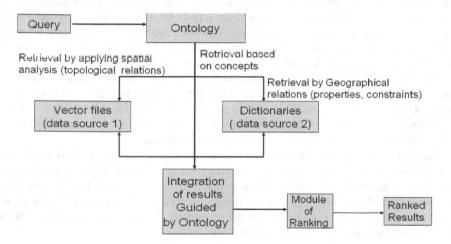

Fig. 1. Framework to retrieve geographic information

In the query processing, each data source allows associating geographic objects or spatial relations to each item of the query. This process starts by submitting a query into the system; the query is processed and all elements of the previously described triplet are identified. Then, a priori constrained ontology is explored to find the concepts, which correspond to the triplet's elements. The goal is to determine what other relations and objects should be required to be searched for and which data sources should be used; this is achieved by extracting context, where a context consists of neighbour relations of a particular concept. Next, retrieval process is performed in the corresponding data source, and the answers are integrated into a set of results.

The final processes are the integration and ranking: a set of results is ranked according to their similarity using the confusion theory [1]. In essence, iGIR is based on integration of retrieval results: if a data source does not contain a relevant result, the other sources either provide an answer or, in the best case, each one adds a relevant answer to final set of results.

The first one of the data sources used in present research is an a priori and manually developed *ontology*. It contains the knowledge about geographic objects of a particular domain (e.g. *hotels in tourism domain, rivers in hydrology domain*). More generally, ontologies contain concepts and semantic relations between them {e.g. *meronimy, hyperonimy*}. In addition, a context is defined as the set of relations, which link a concept with other concepts. For example, in our ontology the concept *"Agave"* is linked to the concepts *"Jalisco"* and *"Plant"* by relations *"is-a"*, and *"grows"* respectively, thus the context = { *"is-a", "grows", "Plant", "Soil", "Weather", "Country"*}. Ontology is implemented in XML. The XML structure allows using or integrating other ontologies and, thus, our ontology can be enhanced. The systemic use of ontology in the retrieval process is described in section 3.1.

The second one of data sources is *vector files*. They are used to obtain the topological relations between data layers by means of overlaying operation (e.g. a layer of roads overlaid with airports, generated the new layer *"roads connect airports"*). In addition, other spatial relations can be discovered. To achieve this, a project (set of layers) is used: according to the parameters of the request, the appropriate layers or attributes are retrieved. A similar approach using spatial Bayesian learning is described in [3]. The processing of vector data is pointed out in section 3.3.

The last one of data sources is dictionaries. We use the dictionaries of INEGI-Mexico (National Institute of Statistics, Geography and Informatics). They contain descriptions, representations, scale, and constraints of particular objects (section 3.2).

Summing up, iGIR uses three data sources, described in above paragraphs, to retrieve geographic information by means of three matching.

The rest of the paper is structured as follows: Section 2 outlines related work. Section 3 describes the retrieval strategy: Sections 3.1 to 3.3 explain the mechanisms of conceptual, geographical, and topological matching, respectively. In addition, these sections describe the ontology design, the characteristics of dictionaries and vector data. In Section 4 some retrieval results are presented. Finally, in section 5 the conclusions as well as a future work are sketched out.

2   Related Work

To date, GIR presents several challenges; some of them have been treated using different approaches. For example, Rule-based methods and Data-driven methods are described in [5]; this article presents several heuristics to access data resources. Other proposed approach is a geographic search using a query-expansion [6]; the authors used a Google API. However, one of the serious disadvantages of this approach is that the query expansion (number of query terms) is constrained by the search engine.

Even if, it is possible to use an added term in order to disambiguate the words of query, this can also add more ambiguity. Thus, the retrieval process is not produced good results, if many terms are needed, because a number of terms required for disambiguation are a priori unknown.

Other proposals are focused on solving the problem of words ambiguity (words which describe geographic objects). The proposed solutions are based on a knowledge representation such as hierarchies of terms, taxonomies, and ontologies. Most of them use textual or syntactic properties; while others describe query processing, missing, however, spatial relations; see e.g. [7]. Inside this group of works, several semantic approaches have been also proposed; one of the main contributions consists of including ontologies and semantic annotation into the retrieval process; an example of such approaches is described in [8].

Although, the GIS community suggested and made emphasis on the use and treatment of spatial relations, only a few studies have been addressed these issues; see [9] [10]. A recent work focused on qualitative spatial reasoning; an example can be found in the often-cited model of topological relations between point sets [11]. Taking into account the above analysis of the state-of-the art, we use vector files, because they are very rich in spatial relations.

Ontologies [12] [13] have been widely used in several semantic approaches. They are now applied in many domains and in particular in GIS [14] [15]. Nevertheless, the proposed approaches do not consider processes and algorithms to explore ontologies. This would be, however, useful, because ontology describes domain theories and the explicit representations of the data semantics [16]. Thus, ontology can be used to discover the semantics of geographic objects involved in a query. Moreover, the algorithms to explore these ontologies and their semantics are required. Thus, we use herein ontologies and propose an algorithm to extract the semantics and domain knowledge stored into them.

On the other hand, many approaches in Information Retrieval (IR) are used the term-based Vector Space Model (VSM) [17]. They are based on lexicographic term matching. While, in iGIR the matching is performed by conceptual matching, topological relations, and descriptions of geographic objects according to the semantics of spatial data.

IR systems use models, techniques, and mechanisms to extract information that has already been processed and stored (e.g. plain text files, databases, XML files). In these systems, the fast processing of queries is possible, because the index structure has been previously built. The same idea is applied in GIR; see e.g. [18].

Besides this, the index structure is also used in domain dictionaries. Thus, we use dictionaries to extract properties and constraints of geographic object. These dictionaries are trusted and consensual sources, because they are designed by specialized and large institutions such as INEGI-Mexico or NASA-USA.

Our method is based on information retrieval guided by ontology, using *geographic queries*. For example, ontology describes where a plant grows, its type, and so on. Thus, we search for this plant either in dictionaries or in vector data. Next, the integration is guided by the relations between geographic objects. Finally, the ranking process is based on the confusion theory, measuring and controlling the dissimilarity between retrieved results. The overall system is described in the subsequent sections.

3 Strategy of Retrieval

For each data source, the goal consists of retrieving results according to semantic relations defined by ontologies, dictionaries, and vector files. Once the results are found, the integration is performed using spatial relations, and the final set of results is generated. Next, the ranking process is applied and the retrieval ends. We describe the overall retrieval process and the integration of three matching layers (conceptual, geographical, and topological) in the following subsections.

3.1 Conceptual Matching

This is the first step of the retrieval strategy, in which we use ontology. It plays the role of an expert in a specific domain, simulating the user's knowledge about this domain. Ontology allows guiding the retrieval, indicates which data should be searched for and where. In other words, ontology describes the way to retrieve relevant results according to semantic relations between geographic objects. For example, considering the following query Q_{G1} = { *"Hotels near Airport Benito Juarez"*} submitted by two types of users: a GIS user and a GIS neophyte. In both cases, the expected results are different: the GIS user wants to find digital data (vector files), while the neophyte wants to find the locations where the hotels stand near the Airport, and other information such as lodging prices, services, and so on. Varying the number of data sources used in our system, we can satisfy these two requirements. Moreover, we require knowing what type of data should be searched for. Thus, ontology defines the properties and relations of each geographic object (Geo_{obj}) i.e. it describes what is a Hotel, its type of representation, its properties and relations to other objects. Other geographic objects and relations involved in query are processed in the same way.

Ontology has manually built using articles from *Wikipedia*. The categories and links contained in each article have been considered as parameters to define relations and concepts of ontology. The semantic relations are classified according to their *meronimy* and *hyperonimy*. Wikipedia is a free online encyclopedia http://wikipedia.org/. In other works it has been used: 1) as data resource [27]; 2) for ontologies design [19]; 3) for words disambiguation [20].

Figure 2 shows a fragment (in Spanish) of the Wikipedia article and a fragment of the generated ontology, according to query Q_{G2} = {*"Agave grows Country"*}.

According to semantic relations between geographic objects, figure 2 depicts a fragment of ontology generated from the Wikipedia article (*Agave*). Note that *Jalisco* and *Plant* are linked through the Agave concept. Corresponding properties are extracted from words in bold and the relations are obtained from verbs, which link the concepts. In addition, the classes are defined and one of the properties is a list of synonyms. These synonyms are also extracted from the Wikipedia articles. In general, ontology is used to explore a data structure of the ontology tree. The goal is to find a matching for a particular concept. For example, assume that a user wants to know where the Agave grows, and submits the following query Q_{G2}= {*"Agave grows Country"*}. In the process of term's identification (triplet) the result is: **what = {Agave}; rel = {grows}; where= {Country}.** To find a matching for the overall query, we will classify it into four types: *atomic, partial, complete,* and *null.* A

Agave (planta)

De Wikipedia, la enciclopedia libre

Los **agaves** son plantas suculentas pertenecientes a una extensa familia botánica del mismo nombre: _Agavaceae_. Se le conoce además con los nombres de **pita, maguey o cabuya**. Proceden principalmente de México (la región de Tequila, en el estado de Jalisco es la máxima productora de tequila, la bebida nacional mexicana) y también se localizan en la zona meridional y occidental de Estados Unidos y en zonas centrales y tropicales de Sudamérica. Los agaves requieren un clima semiseco con **temperatura promedio de 20 °C**, generalmente a una **altitud entre 1.500 y 2.000** msnm. Las condiciones del suelo: **arcilloso, permeable y abundante en elementos derivados del basalto y riqueza en hierro**, preferentemente volcánico. Es muy importante la exposición al sol, y no debe haber más de 100 días nublados al año y preferentemente sólo 65. Otros nombres: **Maguey, Fique**.

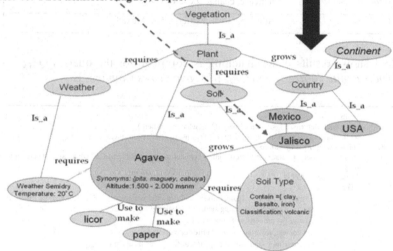

Fig. 2. Fragment of ontology generated from the Wikipedia article

matching is _atomic_, if one of the elements of query's triplet is identified in ontology. For example, if the object "_Agave_" is found, but neither class "_Country_" nor relation "_grows_" are found then we have an _atomic_ matching. A matching is _partial_, if the geographical objects are not found, but relation or relations are presented in ontology. For example, if the class or object "_Country_" is not found, but the relation "_grows_" is found, then we have a partial matching. A matching is _complete_, if all three elements of query's triplet are identified in ontology. Finally, a matching is _null_, if none of the elements of query's triplet is identified. In this case, the retrieval system returns a Geo_{obj} that is processed by using the algorithm of confusion [1]. This algorithm measures the dissimilarity (confusion) for each element of triplet and returns the concept or relation somewhat close to the expressed term. For example, if a user searches for "_rivers_", then the algorithm of confusion will return "_bodies of water_" as a farther matching, and "_lakes_" as a closer matching. Note that this process can be automatically controlled up to given error (confusion).

Table 1. Results of ontology exploration

Query (Q$_{G1}$): "Hotels near Airport Benito Juarez"	
Concept	*Geo$_{obj}$ returned if ontology matching is complete*
Near	periphery , time, distance
Hotel	Tourism, lodging
Airport	hangar, transport
Query (Q$_{G2}$): "Agave grows Country"	
Concept	*Geo$_{obj}$ returned if ontology matching is complete*
Agave	Plant, Desert Plant, Vascular Plant
Grow	Increase, change, develop
Country	Administrative district, USA, Mexico

Table 1 shows results if the matching is complete for the queries Q$_{G1=}$ { *"Hotels near Airport Benito Juarez"*} and Q$_{G2=}$ { *"Agave grows Country"*}.

---

1. **Begin**
2. Select each element of geographic query (Q$_N$) {what, rel, where}
3. Set the topological relation according to rel (Tr) (using association rules)
4. **Retrieving guided by ontology:**
5. Search (by concept name) the corresponding concept to geographic objects "what" and "where" into ontology
6. If (there_are_matches)
7. Then
   a. Extract the context {parent nodes, child nodes, neighbourhood nodes, and the instances for the main concept} by using semantic relations
   b. Extract the properties of concept (P$_{concept}$)
   c. Set context, relations and properties into array Ont$_{results}$
   d. Generate new queries (New$_{Query}$)according to elements in Ont$_{results}$
8. Else
9. Request Q$_N$ to Dictionary and vector Data
10. **Retrieving Dictionary Data:**
11. Request data to dictionary using values of neighbourhood nodes
12. Return response (Resp$_{Ont}$)
13. **Retrieving Vector Data:**
14. Using the response, request de vector data (layers)
   i. If (there_are_ matches (vector data))
   ii. Then
   iii. Request the corresponding relation (Tr)
   iv. Return Resp$_{Topological}$
15. **Integration and Ranking:**
   i. Set the Resp$_{Topological}$ and Resp$_{Ont}$ into final set (FS $_{geobj}$)
   ii. Rank the FS $_{geobj}$ and show to the user
16. Else
17. Select each element of geographic query (Q$_N$) {what, rel, where}
18. Search into dictionary the objects what and where
19. If (there_are_matches)
20. Then
21. Extract properties and relations
22. Set into array PR$_{dict}$
23. Return PR$_{dict}$
24. Else
25. Search into vector data, geographic objects *what* and *where*
26. Return the matches

Fig. 3. Algorithm of ontology exploration used in the retrieval strategy

Table 1 also shows several relations and geographic objects for concepts found by the algorithm of ontology exploration. We use this algorithm to search for the relevant ontology concepts and relations, and then apply the matching according to the submitted geographic query. A fragment of algorithm is described in Figure 3.

Figure 3 describes the steps to process the query, where we explain the functionality and processing using the query Q_{G2} = {"*Agave grows Country*"}. According to the algorithm, each triplet's element is identified, and for the relation *grows*, the overlay spatial operation is applied. This operation is defined according to a set of rules. These are described in section 3.2. Then results for the lines *5* to *7d* are:

Context= {
Parents (Plant);
Neighbourhood (soil, weather, country)
}

Using the context, new queries are generated by combining the elements of context. Thus, the generated queries are the following:

Q_{G3}= *{Agave grows Mexico}*
Q_{G4}= *{Agave grows USA}*
Q_{G5}= *{Plants grows Mexico}*
Q_{G6}= *{Plants grows USA}*

The queries Q_{G3} to Q_{G6} are searched for into the dictionary and vector data. If a term of query is found, then its properties, constraints, representation, and relations are integrated into a set of results. In the worst case, there is no matching for an element of query. In this case, the queries are submitted and processed by the confusion module, where the new queries are generated and resubmitted to the initial process. Therefore, in a successful scenario, each object is found according to the previously established criteria (see lines 4-9 of Figure 3), in which each one of them are requested in the source of vector data which fulfil these criteria. Sections 3.2 and 3.3 are described in detail the rest of the process (geographical and topological matching).

3.2 Geographical Matching

The next step consists of making a geographical matching. This process uses as a data source the information dictionaries. They represent a consensual agreement between the GIS specialists and contain the scale, properties, constraints, and relations, etc. of geographic objects. The dictionaries are initially in *PDF* afterwards to be semiautomatically transformed into XML files, using the API, *PJX* (see http://java-source.net/open-source/pdf-libraries/pjx). Figure 4 shows a fragment of dictionary in PDF format and the corresponding XML file. In particular, the fragment describes the object *Airport*. Due to didactical reasons and available data, we explain the process of geographical matching using a query which contains airports.

Figure 4 depicts the sections of dictionary; each section is to be transformed into nodes of XML file. Now, we explain the process to extract the information required to improve GIR. Consider the generated queries of section 3.1 (Q_{G3} to Q_{G6}) or original

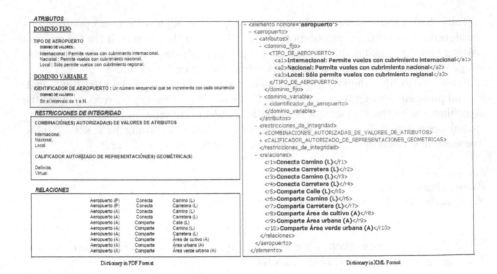

Fig. 4. INEGI dictionaries used in the geographical matching

query Q_{G2}. In this case, due to availability of data, we use the query Q_{G1}= { *"Hotels near Airport Benito Juarez"*}. This query is processed to obtain the triplet: what = { *"Hotels"*}; relation = { *"near"*}; where = { *"Airport Benito Juarez"*}. These properties and relations are extracted from the dictionary for the object *"Airport"*, where the results are: Relations = {*"connect"*, *"sharing"*} Properties= {Type = {*"Local"*, *"National"*, *"International"*}} Constraints= {primitive of representation = { *"point"*, *"polygon"*}}. The object *Hotel* has no occurrences in the dictionary, therefore, the process continues by using only the object *Airport*. The constraints are needed because of possible semantic changes. For instance, if an airport is depicted by a point feature, then it represents a building of operations, while if it is depicted by a polygon, then it represents the area, infrastructure and services of air navigation. Additionally, we define a set of rules for processing each relation. These rules are established in our previous work [24]. The main idea is to associate a topological relation to the relation expressed in query (e.g. *"near"* is associated to relation *"connect"*). In this case one of the rules of association to relation *"near"* is the following:

R_1 (NEAR) = {**X near Y, if X and Y are connected by Z**} where X, Y, are geographic objects (Geo$_{Obj}$), represented by points or polygons, while Z is a geographic object represented by an arc. Moreover, its length is less than 1 kilometer (e.g. Z is a road). Finally, *"connected"* is a relation between X and Y.

Therefore, the retrieval in this step consists of searching for documents which fulfill the above rule. Therefore, the parameters of searching (P$_{search}$) are:

$P_{search1}$ = {**Geo$_{ObjX}$ connects Geo$_{ObjY}$**} where Geo$_{ObjY}$ can be a point or a polygon.
$P_{search2}$ = {**Geo$_{ObjZ}$ sharing Geo$_{Objw}$**} where Geo$_{ObjY}$ can be a point or a polygon.

These parameters represent the search performed by using vector files. This is the point where the next step (topological matching) starts. This process is described in section 3.3.

3.3 Topological Matching

The process of topological matching is based on topological relations between geographic objects. This process uses vector files as a data source. The data of this representation model is provided by INEGI and SCT (Secretary of Communications and Transportation). The data are processed to obtain a proprietary format file called herein *Topologyfile*. In particular, these files store the topological relations between two geographic objects. Table 2 shows the structure of a *Topologyfile* (File.geo).

Table 2. Structure of *Topologyfile*

ID	Id_GeoObj_1	LAYER_BELONG	Id_GeoObj_2	LAYER_BELONG	RELATION
1	5	Airport.geo	2	Roads.geo	C
2	10	Hotels.geo	2	UrbanArea.geo	S

C= Connects, S= sharing

Table 2 shows an identifier for each record which fulfils $P_{search1}$ and $P_{search2}$ (section 3.2). The following columns allow identifying each geographical object and its corresponding layer. The attributes indicate the type of relations between two geographical objects (e.g. *"connect"*, *"inside"*). These relations are obtained by applying the overlay spatial operation (e.g. overlaying *roads* and *airports* = road A connects airport B). The details on how to get spatial relations into tables are described in [21]. Therefore, for the query Q_{G1}, the sources which include layers: *"Airports"*, *"Hotels"* and *"roads"* are explored to find that *"Airport Benito Juarez"* is connected by several streets and avenues, and some Hotels are also connected by the same streets and avenues. Then, the objects linked by the topological relations (*"connect"*, *"sharing"*) are retrieved. Figure 5 shows a table in which an example of topological matching is presented.

ID	ID_OBJ_1	LAYER_BELONG	ID OBJ_2	LAYER_BELONG	RELATION
0	5	Airport.geo	2	Roads.geo	C

ID	ID_OBJ_1	LAYER_BELONG	ID OBJ_2	LAYER_BELONG	RELATION
12	7	Hotels.geo	2	Roads.geo	C

Fig. 5. Topological matching: Object 5 and object 2 are connected. Object 2 is connected to object 7.

Figure 5 shows how a result, where object 5 (*Airport Benito Juarez*) represented by a point, containing a relation *"connect"* to object 2 (*Avenue "Circuito Interior"*), is retrieved. Then, the next step is to find Hotels, where object 2 appears (in the best case, it will connect by the same relation). This way, we find that object 7 (*Hotel Holiday Inn*) has a relation to object 2 (*Avenue "Circuito Interior"*). This means that

Hotel Holiday Inn is connected by Avenue Circuito Interior and the Avenue is connected to the Airport Benito Juarez. These results are subsequently submitted to the integration and ranking module.

Finally, each result is integrated to a final set of results and submitted to the ranking module. Here the retrieval process ends and the ranking process starts. The ranking process is based on the similarity measures between concepts and relations called in [1] *confusion*.

4 Experiments

In this section, we present some screenshots of the query $Q_{G1} = \{$ *"Hotels near Airport Benito Juarez"* $\}$ processing, applying the algorithm of ontology exploration that returns the list of classes and objects to be searched for into dictionaries and vector files. In addition, the values of properties for each geographic object are visualized. These results are addressed to GIS neophytes. Figure 6 shows the result for the class Airport.

Figure 6 shows a retrieved document with the properties and attributes, which define the Airport class. The definition is based on Wikipedia documents (in Spanish)

```
Aeropuerto
Es un: área
Función: llegada, salida y movimiento en superficie de aeronaves
Tipo: nacional e internacional.
Sinónimos: aeródromo.
Tiene:

                Pistas de aterrizaje
                Calles de rodaje
                Terminales de pasajeros y carga
                        Tienen: vestíbulos de salidas y llegadas, control de pasaportes, salas de embarque, zonas c
                Plataformas de  estacionamiento
                        Área destinada a dar cabida a las aeronaves durante operaciones de embarque y desembarc
                        mantenimiento y limpieza de aeronave
                Hangares de mantenimiento

Operaciones
          Por aire
                Aplican a
                        Plataformas de Terminal
                                Operación:
                                        Rodaje de las aeronaves hasta/desde las pistas y el
                                        Despegue y aterrizaje de las aeronaves.

                        Plataformas remotas
                        Aeronaves
                        Requerimientos de aeronaves
          Por tierra
                Aplican a
                        Pasajeros
                        Necesidades de pasajeros
                        Edificios terminales
                                Operaciones
                                Conexión con vehículos, autobuses, tren, metro)
                Centro de control de área
                                Operación:
                                        Dirigir y controlar todo el movimiento de aeronaves
                                        bajo su jurisdicción.
```

Fig. 6. The result of the algorithm of ontology exploration searching for the Airport class; this result is addressed to GIS neophytes

which contain the NL commonalities and popular use of geographic objects such as Airport. Therefore, these documents can be useful for travelers, businessmen, etc.

According to the algorithm of ontology exploration the contextual results are sent to the module of geographic matching. The goal is to find other objects related to the original query according to spatial relations. These results are addressed to the GIS specialists. Figure 7 shows the results of conceptual and geographical matching.

Figure 7 also shows the relations retrieved for the object *"Airport"* from dictionaries. In this case, some relations are: {*connect* and *sharing*} and some associated classes are: {*"highway"*, *"street"*, *"urban area"*, etc}. The classes are represented as a

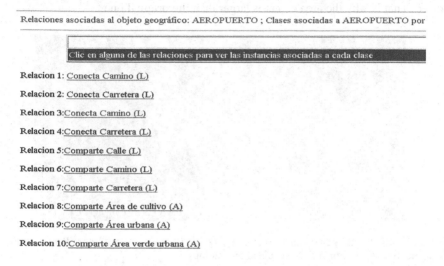

Relaciones asociadas al objeto geográfico: AEROPUERTO ; Clases asociadas a AEROPUERTO por

Clic en alguna de las relaciones para ver las instancias asociadas a cada clase

Relacion 1: Conecta Camino (L)

Relacion 2: Conecta Carretera (L)

Relacion 3:Conecta Camino (L)

Relacion 4:Conecta Carretera (L)

Relacion 5:Comparte Calle (L)

Relacion 6:Comparte Camino (L)

Relacion 7:Comparte Carretera (L)

Relacion 8:Comparte Área de cultivo (A)

Relacion 9:Comparte Área urbana (A)

Relacion 10:Comparte Área verde urbana (A)

Fig. 7. Results of the query Q_{G1} = {*"Hotels near Airport Benito Juarez"*} processing by using dictionaries and guided by ontology

Laboratorio de Procesamiento Inteligente de Información Geoespacial

Usted busco: Aeropuerto Conecta Camino

Aeropuerto				
OBJECTID	TIPO	ENTIDAD	FC	CAPA DATOS (LAYER)
1	Internacional	AEROPUERTO	643	airport1 shp
2	Internacional	AEROPUERTO	643	airport2.geo
3	Internacional	AEROPUERTO	643	airport1.shp
4	Internacional	AEROPUERTO	643	airport1.shp
5	Internacional	AEROPUERTO	643	airport2.geo
6	Internacional	AEROPUERTO	643	airport2.geo
7	Internacional	AEROPUERTO	643	airport2.geo
8	Internacional	AEROPUERTO	643	airport2.geo
9	Internacional	AEROPUERTO	643	airport3.geo
10	Internacional	AEROPUERTO	643	airport3.geo
11	Internacional	AEROPUERTO	643	airport3.geo
12	Internacional	AEROPUERTO	643	airport3.geo
13	Internacional	AEROPUERTO	643	airport3.geo

Fig. 8. Results of the query Q_{G1}= {*"Hotels near Airport Benito Juarez"*} processing by using dictionaries, vector files, and guided by ontology

link, because they are related to instances of these classes. These instances are re-trieved from vector data (e.g. searching for *highways, streets* with a relation such as *connect, sharing*).

Figure 8 shows the final results by selecting one of the links (shown in figure 7) for Airports. In this case, the relation is *"connect"*.

The last column depicts the vector file associated to each object. In addition, this process generates a KML file if available vector data contain the latitude and longi-tude of spatial objects. Thus, figure 9 shows the KML file generated for the query Q_{GI}= {*"Hotels near Airport Benito Juarez"*}. Figure 9 also shows the area where the airport Benito Juarez is located. In addition, the subway stations near to it are dis-played. That is why the area extent is larger than the original one.

Fig. 9. The KML file generated for the query Q_{GI} = {*"Hotels near Airport Benito Juarez"*}, using dictionaries, vector files, and guided by ontology

Finally, we test our approach comparing it with the results provided by *Google*. The results have been validated by the group of postgraduate students of the PIIG Lab. In general, the results have matched their expectations. In our test, the results of retrieval process have been classified into three types: *somewhat relevant, relevant, and irrele-vant*. Relevant and irrelevant results are eloquent, while a result is somewhat relevant if it either contains a property of the geographic objects or an object related to them. For

Table 3. Statistics of final results

Geographic query	System Used	Somewhat relevant	Relevant	Irrelevant
1	*Google*	2	5	1
	iGIR	3	7	3
2	*Google*	0	2	5
	iGIR	6	7	2
3	*Google*	2	3	4
	iGIR	6	5	2
4	*Google*	1	2	4
	iGIR	4	5	2

example, retrieving *"Agave"* is somewhat relevant result, if *"tequila"* concept is searched for. This classification is based on the confusion measures [1].

Table 3 shows the experimental results, in which the number in columns *somewhat relevant*, *relevant*, and *irrelevant*, represents the number of documents, which have been found. The results are satisfactory and generally match the user's expectations. Nevertheless, additional tests are required, using other data and methods to measure the relevance of results. These are another part of work in progress. Finally, the ranking process is applied, using a set of previously defined geographical objects.

5 Conclusions

This work describes an approach to perform geographic information retrieval based on integration of three sources of geographic information (iGIR system). The main idea is to extract and process the properties and relations of the geographic objects which appear in the data sources (the former store descriptions, constraints, topological and geographic relations). The approach is based on the algorithm of ontology exploration. A method to match the concepts of geographic objects by their relations and properties, not only syntactically but also semantically, is developed as well. The retrieval is guided by ontology. It is manually designed and based on Wikipedia articles. Wikipedia is a free online encyclopedia – a trusted and consensual information resource also used in other GIR works. Ontology helps to decide where and what should be searched for into other two data sources – geographic dictionaries and vector files, thus, simulating the user's judgement. INEGI-Mexico dictionaries have been used in this work. These contain descriptions, properties, and relations of particular geographic objects at certain scale. Vector files are used in form of a proprietary file format called herein *Topologyfile*. These files represent topological relations such as *adjacent*, *in*, etc. This work is primarily different from others (e.g. query expansion), because the geographic information retrieval is made by matching concepts using the algorithm of ontology exploration, and their integration with geographic dictionaries and vector data. The purpose of integration is to provide adequate search parameters and in consequence improve the overall retrieval process.

The paper exposes some results of processing the geographical queries over heterogeneous repositories. The retrieved results are addressed to two types of users: GIS specialists and GIS neophytes according to either their requirements or their profiles. Nevertheless, additional testing is needed to validate the overall approach as well as its components; especially on large document collections (e.g. the test of the ranking module requires such large collections, while present test used only 20 documents).

Acknowledgments

The author of this paper wishes to thank his scientific advisor Dr. Serguei Levachkine, the Centre for Computing Research (CIC), SIP-IPN, National Polytechnic Institute (IPN), and the Mexican National Council for Science and Technology (CONACYT) for their support.

References

1. Levachkine, S., Guzman-Arenas, A.: Hierarchy as a New Data Type for Qualitative Variables. Expert Systems with Applications: An International Journal 32(3), 899–910 (2007)
2. Gazetteer Development at the Alexandria Digital Library Project. available at: http://www.alexandria.ucsb.edu/gazetteer
3. Walker, A.R., Pham, B., Moody, M.: Spatial Bayesian Learning Algorithms for Geographic Information Retrieval. In: Proceedings of the 13th annual ACM international workshop on Geographic information systems, GIS 2005, 42, Bremen, Germany (2005)
4. Clough, P.: Extracting Metadata for Spatially-Aware Information Retrieval on the Internet. In: GIR 2005. Proceedings of the Workshop on Geographic Information Retrieval, Bremen Germany, pp. 25–30. ACM Press, New York (2005)
5. Egenhofer, M.: Interaction with Geographic Information Systems via Spatial Queries. Journal of Visual Languages and Computing 1(4), 389–413 (1990)
6. Delboni, T.M., Borges, K.A., Laender, A.H., Davis, C.A.: Semantic Expansion of Geographic Web Queries Based on Natural Language Positioning Expressions. Transactions in GIS 11(3), 377–397 (2007)
7. Maedche, A., Staab, S., Stojanovic, N., Studer, R., Sure, Y.: SEAL A Framework for Developing Semantic Web Portals. In: Read, B. (ed.) Advances in Databases. LNCS, vol. 2097, Springer, Berlin (2001)
8. Arpinar, I.B., Sheth, A., Ramakrishnan, C., Usery, L., Azami, M., Kwan, M.: Geospatial Ontology Development and Semantic Analytics. Transactions in GIS 10(4), 551–576 (2006)
9. Heinzle, F., Kcopczynsky, M., Sester, M.: Spatial Data Interpretations for the Intelligent Access to Spatial Information in the Internet. In: Proceedings of the 21st International Cartographic Conference, Durban, South Africa (2002)
10. Shilder, F., Versley, Y., Habel, C.: Extracting Spatial Information: rounding, Classifying and Linking Spatial Expressions. In: Proceedings of the ACM SIGIR workshop on Geographic Information Retrieval, Sheffield, UK (2004)
11. Egenhofer, M., Franzosa, R.: Point-Set Topological Spatial Relations. International Journal of Geographical Information Systems 5(2), 161–174 (1991)
12. Guarino, N.: Formal ontology and information systems. In: Proceeding of FOIS 1998, Trento, Italy, pp. 3–15. IOS press, Amsterdam (1998)
13. Gruber, R.: A translation approach to portable ontology specifications. Knowledge Acquisition 5(2), 199–220 (1993)
14. Harding, J.: Geo-ontology Concepts and Issues, Report of a workshop on Geo-ontology, Ilkley UK (2003)
15. Hammond, B., Sheth, A., Kochut, K.: Semantic Enhancement Engine: A Modular Document Enhancement Platform for Semantic Applications over Heterogeneous Content. In: Kashyap, V., Shklar, L. (eds.) Real World Semantic Web Applications, pp. 29–49. IOS Press, Amsterdam (2002)
16. Koo, S., Lim, S., Lee, S.: Building ontology based on hub words for information retrieval. In: Proceedings of the IEEE/WIC International Conference on Web Intelligence, IEEE Computer Society, Los Alamitos, 466 (2003)
17. Baeza-Yates, R., Ribeiro-Neto, B.: Modern Information Retrieval. ACM Press, Addison-Wesley, New York (1999)
18. Jones, C.-B., Abdelmoty, A.-I., Finch, D., Fu, G., Vaid, S.: The Spirit Spatial Search Engine: Architecture, Ontologies and Spatial Indexing. In: Proceedings of Third International Conference, Maryland, USA. LNCS, vol. 3234, pp. 125–139. Springer, Berlin (2004)

19. Buscaldi, D., Rosso, P., García, P.: Inferring geographical ontologies from multiple resources for geographical information retrieval. In: Proceedings of 3erd Int. SIGIR Workshop on Geographic Information Retrieval, SIGIR, Seattle, pp. 52–55. ACM Press, New York (2006)
20. Martins, B., Silva, M.J., Silveira, M.: Challenges and Resources for Evaluating Geographical IR. In: Proceedings of the 2005 workshop on Geographic Information Retrieval, pp. 65–69. ACM Press, New York (2005)
21. Martinez, M.: Topologic Descriptor for Topographic Maps, Master of Science Thesis, Mexico, PIIG Lab, Centre for Computing Research, in Spanish (2006)
22. Wordnet: A Lexical Database for the English Language. available at http://www.wordnet.com
23. Martins, B., Silva, M., Freitas, S., Afonso, A.: Handling, Locations in Search Engine Queries. In: Proceedings of GIR 2006, the 3erd Workshop on Geographic Information Retrieval (2006)
24. Mata, F., Levachkine, S.: Semantics of Proximity in Locative Expressions. In: Rodríguez, M.A., Cruz, I., Levashkin, S., Egenhofer, M.J. (eds.) GeoS 2005. LNCS, vol. 3799, Springer, Heidelberg (2005)
25. Yee, W.G., Beigbeder, M., Buntine, W.: SIGIR06 Workshop Report: Open Source Information Retrieval systems (OSIR06). SIGIR Forum 40(2), 61–65 (2006)
26. Petras, V., Fredric, C.G., Larson, R.: Domain-Specific CLIR of English, German and Russian Using Fusion and Subject Metadata for Query Expansion. In: Peters, C., Gey, F.C., Gonzalo, J., Müller, H., Jones, G.J.F., Kluck, M., Magnini, B., de Rijke, M., Giampiccolo, D. (eds.) CLEF 2005. LNCS, vol. 4022, pp. 226–237. Springer, Heidelberg (2006)
27. Cardoso, N., Martins, B., Chaves M., Andrade, L., Silva, M.: The XLDB group at Geo-CLEF 2005. In: GeoCLEF 2005 Workshop, Poster Session (2005)

A Rule-Based Description Framework for the Composition of Geographic Information Services

Michael Lutz, Roberto Lucchi, Anders Friis-Christensen, and Nicole Ostländer

European Commission – Joint Research Centre (JRC)
Institute for Environment and Sustainability, Spatial Data Infrastructures Unit
Via E. Fermi 1, 21020 Ispra, Italy
{michael.lutz,roberto.lucchi,anders.friis,
nicole.ostlaender}@jrc.it

Abstract. SDIs offer access to a wealth of distributed data sources through standardised service interfaces. Recently, also geoprocessing capabilities are offered as services in SDIs. Combining data sources and processing services in service chains enable the generation of information that is tailored to the users' needs. In this paper, we present a rule-based description framework and an associated discovery and composition method that helps service developers to create such service chains from existing services. The goal of the description framework is to describe services at a conceptual level rather than closely mirroring specific implementation details. It consists of a simple top-level ontology as well as a domain ontology, which provide the basic vocabulary for creating descriptions of both services and the information the service chain is to produce as a result. The composition method uses these descriptions to discover appropriate services and compose them into a service chain that can produce the required information. The method is illustrated using an example from the domain of risk management.

1 Introduction

The main goal of spatial data infrastructures (SDIs) is to offer access to distributed data sources based on the service-oriented architecture (SOA) principle. SDIs provide a framework for optimizing the creation, maintenance and distribution of Geographic Information (GI) services at different organization levels [1] over a distributed computing platform, typically the Web. In such a scenario, where resources are distributed and controlled by different organizations, catalogue services provide a means for describing the services' locations and capabilities. They store meta-information and support users in discovering and using these resources. Consequently, SDI-based applications enable an efficient sharing and reuse of geographic data among heterogeneous user groups [1, 2].

Recently, SDIs are also providing capabilities that have traditionally been offered by monolithic GIS [3], including the capture, modelling, manipulation and analysis of geospatial data. In this paper, we are particularly concerned with two kinds of GI services: (1) services that provide geographic data (data access services) and (2) services that analyse (and manipulate) geospatial data (geoprocessing services). While

F. Fonseca, M.A. Rodríguez, and S. Levashkin (Eds.): GeoS 2007, LNCS 4853, pp. 114–127, 2007.
© Springer-Verlag Berlin Heidelberg 2007

standardised interfaces for data access services [4, 5] already exist for several years and have been widely adopted, geoprocessing capabilities have only recently been made available as services in SDIs. A service interface for such services has recently been adopted as a standard by the Open Geospatial Consortium (OGC): The Web Processing Service (WPS) Specification [6].

Since in SDIs data is created by a variety of domain-specific applications, in many cases this data cannot be directly re-used within other domains without further processing. This can be achieved by combining different existing data and geoprocessing services into a value-added *service chain* [7, 8]. Service chains can be described using some orchestration language (e.g. WS-BPEL [9]), which can then be deployed on a corresponding orchestration engine. Creating such service chain descriptions, a task which we term *service composition* in this paper, involves discovering appropriate services for data access and geoprocessing and combining them in a way such that they are capable of generating the required results. Both of these (sub)tasks are currently executed manually by application developers. In this paper we present a method that supports application developers in creating such service chains.

The method is based on a rule-based framework for describing services and the information required by the application developer that stays at the conceptual level and thus abstracts away from specific implementation details. The composition method uses the conceptual service descriptions to discover and (semi)automatically compose a service chain that can provide the information required by the application developer.

In order to illustrate and exemplify the presented methodology throughout the paper, we use the scenario of an application developer, who is requested to deliver a service application that provides information on forest fire density.

The remainder of the paper is structured as follows. In Section 2, we describe previous work in the area of (automatic) service composition and rule-based GI discovery. In Section 3, we present the methodology proposed for supporting the composition of service chains for the given scenario. In Section 4, we conclude the paper and point out topics for future research.

2 Related Work

In this section, we discuss existing approaches for automating the process of service composition and present an approach for discovery of geographic data based on rules, which we extend in this paper to allow also the discovery and composition of services.

Automatic Service Composition. Discovery of services and service functionality is a new task within SDIs. Well-tested specifications of what metadata is required in order to perform this task are lacking. Thus, it can be difficult to understand the functionality of services from their metadata and hence to understand how to combine several services to obtain a certain result [10]. A service composition task can be oriented towards solving different kinds of problems: i) fulfilling preconditions, ii) generating multiple effects and iii) overcoming a lack of knowledge [11]. We are addressing the latter problem in this paper, i.e. we are concerned with cases where a service providing the required information exists but some of the necessary input parameters are not directly available and, therefore, they must be obtained by using additional services. One possible technique to address this kind of problem is backward chaining [11]. The basic idea of this method is to start by selecting a service which meets the user requirements and place it at the end of the chain (so that it is the last one to be executed). Then, for each input this service requires, services providing such information are added in the chain before the service. The method is iterated until all the necessary input information is available in the chain. In such an approach, what is characterizing the solution is the way user requirements are expressed as well as how the corresponding service selection is done.

In the automatic service composition research line there are a number of related approaches; in the following we mention the ones closely related to the proposal we present. A method for automatic service composition based on backward chaining is presented in [12] where services functionalities are described using ontologies, based on OWL [13] and DAML-S (now OWL-S [14]). Such functionalities are used to express user requirements and, by means of inference engines, to discover and compose services. The semantic descriptions mainly focus on the functionality supplied by the service without expressing in detail the input and output information types as well as the interdependencies between the two. This aspect penalises the approach in scenarios where these details are necessary, for instance in the case where a services will provide different outputs depending on the input provided. E.g., a simple *division* service can return a *density* when its inputs represent a *mass* and a *volume*, or a *velocity* when provided with a *distance* and a *time period*. Furthermore, the types used to describe inputs and outputs are very close to the implementation. Often they are simply modelled as *strings* or simple enumeration types (e.g. *language*). More complex concepts, such as are required to describe spatial data (e.g. feature collections or coverages of a certain type) are not modelled in the examples provided in the application described.

Rule-based GI Discovery. In recent years, ontologies, i.e. formal explicit representations of conceptualizations [15], have been used extensively to model domain-specific knowledge. There are also a number of proposed approaches to use ontologies for the discovery of geographic data [16-19] and GI services in SDIs [10]. Many of these approaches are based on Description Logics (DL) [20] and subsumption reasoning between DL concepts.

In contrast to these approaches, in [21] a *rule-based* approach to discovering data sources within an SDI is presented. It distinguishes two types of rules: *Schema mapping rules* describe a mapping between a local schema and the (global) domain ontology, i.e. they represent the local data using a shared domain vocabulary. *Domain rules* describe domain knowledge, thus complementing the DL concept definitions in the domain ontology. They can then be used to derive implicit knowledge based on existing facts in a knowledge base [22]. Based on these rules, the presented methodology enables the discovery of those data sources within an SDI that contain facts relevant for deriving an answer to the user's question. The rules are traversed backwards, from the goal specified by the user, through domain and schema mapping rules to the data sources. Two alternative approaches for this backward chaining are presented. The first one only considers class atoms (i.e. atoms representing feature types), while the second one also considers relations.

We base the methodology for service discovery and composition, which is presented in detail in the next section, on the approach presented in [21]. Similarly to this approach, rules are used to describe the resources to be discovered (in this paper, these resources are services rather than data sources). A further similarity is the inclusion of domain rules representing background domain knowledge in the composition approach. Furthermore, both approaches use similar backward chaining approaches, in which the starting point is a goal specified by the user and rules matching (parts of) this goal are used to consecutively discover (and in our case compose) appropriate resources. The approach presented in this paper additionally takes into account how the discovered services have to be combined (while in [21], the order in which data sources are discovered is of no concern). Finally, our approach focuses on (service) descriptions at the conceptual level (while in [21], data sources were described by rules connecting the logical and conceptual levels), and rules are used only for generating a service chain matching the goal specified by the user (rather than also inferring new knowledge as in [21]).

3 An Approach to Semi-automatic Service Composition

In this chapter, we present a framework for describing services and the information required by the user, together with an associated method for supporting GI service composition.

While existing approaches were strongly focussed on implementation details like the input/output types [12] or very detailed functionality descriptions [10], the goal for the presented description framework and composition method is to stay at the conceptual level. Thus, we want to abstract away from application details at the logical level, which can be very diverse and thus lead to incompatible descriptions if the description follows the logical structure too closely.

In this Section, after giving a general overview (Section 3.1), we present the conceptual description framework (Section 3.2). This consists of a top level ontology of the basic concepts and relations as well as a domain ontology. These are the basic building blocks for creating the description of services and the composition goal. Finally, we describe the composition method based on the presented service and goal descriptions (Section 3.3).

3.1 Overview

In this section, we present an approach that supports service developers in service composition. The goal of the presented approach is to support *generating information* of a particular kind. Hence, we do not consider services that have (real-world) side effects, e.g. reserving a hotel room or charging a credit card, but only services that consume and produce information. In the geospatial domain, these services are either data access services (e.g. WFS [4], WCS [5]) or geoprocessing services[1] (WPS [6]), which therefore are the focus of our research. For these information generating services, we make the assumption that the output (in relation to the provided input) of such a service can be considered to be the same as its functionality. For example, a service which provides a (Euclidian) distance between two points has the functionality

---

[1] While the WPS specification allows arbitrary (i.e. also non-spatial) processing to be provided by a WPS, we restrict our focus to those WPS that provide spatial data as output.

"compute Euclidian distance". In contrast, a hotel booking service could have the functionality of booking a hotel room (and charging your credit card) while returning a booking confirmation document as an output. We therefore propose a method for supporting service composition that focuses on *service outputs*.

We adopt a backward chaining approach (cf. Section 2) for supporting the user in composing a service chain. In the following, we give a more detailed overview of this approach (Fig. 1).

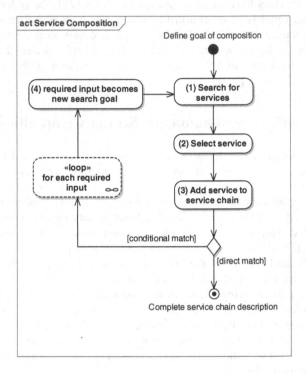

Fig. 1. UML activity diagram illustrating the steps of the proposed approach

The basis for the method is a conceptual description framework. The framework will be used to describe both the services available in the infrastructure and the information required by the user, i.e. the output of the application to be created. The latter is goal and starting point for the service composition method. The description framework is presented in Section 3.2.

The first step in the composition process is to search for services that can generate information as specified by the user. In some cases, data access services providing such data might be directly available (*direct match* in Fig. 1). In most cases, however, only geoprocessing services will be available that can *potentially* generate the required data, *provided that they are given appropriate input data* (*conditional match* in Fig. 1). In this case, further searches have to be performed to discover services that can provide appropriate input data.

In each discovery step, the user has to select one of the discovered services (step 2 in Fig. 1). In order to help the user in selecting the most appropriate service, the discovered services could be ranked. The ranking could be based on specific service properties that are specified by the user during the request formulation, e.g. service security or trust in a certain service or service provider. In addition, some heuristics for the quality of the match can be used, e.g. some measure of how similar the output of a discovered is to the goal. Developing such methods for ranking the discovered services will be part of our future research.

The selected service is then added to the service chain description (step 3). The service is connected to the service already present in the chain for whose input the search was performed. Thus, the service chain is built back-to-front.

If the discovered service requires additional input data, these become the search goal for these further searches (step 4). The discovery (steps 1-3) has to be repeated, until data access services have been discovered for all required input data.

3.2 A Conceptual Description Framework for Services and Composition Goal

Discovery in the presented composition approach is based on a conceptual description framework for both the composition goal and the services available in the infrastructure. For this description framework, we extend the rule-based approach for GI discovery presented in Section 2. In the following sections, we define a simple top level ontology for service descriptions and illustrate how a domain ontology can be derived for a certain domain of interest. Finally, we show how concepts and relations from these ontologies can be used to create service descriptions.

In this chapter and the remainder of the paper, we use the following basic terminology. A *term* is a constant or variable symbol. In this paper, constants are written starting in upper case (e.g. *Event*), while variables start in lowercase (e.g. *ec*). If R is a predicate symbol of arity n and $t_1; ...; t_n$ are terms, $R(t_1; ...; t_n)$ and $\neg R(t_1; ...; t_n)$ are called *literals*. A *rule* is a disjunction of literals with *exactly* one positive literal and at least one negative literal, i.e. it has the form $h \vee \neg b_1 \vee ... \vee \neg b_m$ which is equivalent to $b_1 \wedge ... \wedge b_m \rightarrow h$. $b_1 \wedge ... \wedge b_m$ is called the *body* of the rule, and h is called the *head*. Note that in this paper, we also use rules with more than one head literal. As these head literals do not include existential quantification, a rule of the form $b_1 \wedge ... \wedge b_m \rightarrow h_1 \wedge ... \wedge h_n$ can be trivially reduced, without loss of meaning, to n rules of the form: $b_1 \wedge ... \wedge b_m \rightarrow h_1; b_1 \wedge ... \wedge b_m \rightarrow h_2; ...; ; b_1 \wedge ... \wedge b_m \rightarrow h_n$. A rule with one positive literal and no negative literals, i.e. one that only contains a head and no body, is called a *fact*. [23]

Top level ontology. For the limited domain considered for our method (only data access and geoprocessing services), we have developed a simple top level ontology, which provides a basis for the domain ontology (described in the next section) as well as the service description rules and the specification of a composition goal by the user. The main concepts in this ontology mirror the duality between the field and object views of geographic information [3]. Thus, we distinguish the following main concepts (unary predicates) and relationships (n-ary predicates):

- **entity.** Unary predicate denoting an entity (or object)
- **entityCollection.** Unary predicate denoting a collection of entities.

- **field.** Binary predicate denoting a field (or spatio-temporal function). The first argument is the (value) range of the field; the second argument is its (spatio-temporal) domain.
- **hasMember.** Binary predicate stating that an entity collection contains an entity.

The top level ontology is deliberately kept simple. It will be extended and aligned with geographic top level ontologies such as presented in [24] and [25] as part of our future work.

Domain ontology. Further rules describing relationships in a domain of interest can be considered in the methodology (called domain rules in [21]). These describe types of entities and fields used in the domain and the relationships between them. They can also include implicit domain knowledge that can be used for inferences during discovery and composition. Domain rules can thus be used as a shared vocabulary [26] by users and service providers. Note that while currently domain rules are not labelled, labels could be used to mark different sets of domain rules (e.g. from different user communities). These labels could then be used to identify the sets of rules used for deriving the service chain.

In Fig. 2, some examples for domain rules are given. Lines 1 and 2 state that a *density d* (respective a *frequency f*) of an entity collection *ec* with respect to a spatio-temporal domain *std* is a *field* with *f* as its range and *std* as its domain. Line 3 asserts that a *forestFire* is an *entity*. Lines 4 and 5 state that entity collections whose members are *adminUnit*s (respective *postCode*s) can be considered as *tessellation*s.

```
1   density(d, std, ec) → field(d, std)
2   frequency(f, std, ec) → field(f, std)
3   forestFire(e) → entity(e)
4   entityCollection(ec), (∀e : hasMember(ec,e) → adminUnit(e)) → tessellation(ec)
5   entityCollection(ec), (∀e : hasMember(ec,e) → postCode(e)) → tessellation(ec)
```

Fig. 2. Examples for domain rules: *density* and *frequency* are *field*s (1-2); *forestFires* are *entities* (3); all feature collections that contain *adminUnit* or *postCode* features are tessellations (4-5)

Service descriptions. In the approach presented in [21], the focus was on describing feature types provided by data access services and general domain knowledge. In contrast, in this paper, the goal is to describe operations provided by data access or geoprocessing services. An operation is described by a rule specifying the relationship between the operation's inputs and its outputs. Such a rule can be interpreted as follows: "If input of a certain type is available, then (using the operation associated with this rule) output of a certain type can be provided". To enable the association of a rule with an operation, the rules are labelled. These labels will later also be used to link the operation description to (other) service metadata stored in a catalogue service (cf. Section 4).

One basic assumption for our approach is that at the logical level we only consider services that provide features, feature collections or coverages as outputs (cf. Section 3.1). The services naturally also need to be able to consume these types of parameters, as well as simple datatypes like strings or integers. In this paper, we consider operation signatures at the logical level only for illustration purposes. We have therefore chosen a simplified representation that is not based on a formal model like the top level ontology used for the descriptions at the conceptual level. Such a formal model allowing to refer to the actual XML datatypes used in the implementations (e.g. GML types like `gml:AbstractFeatureCollectionType` or simple XML schema datatypes like `xs:string`) will be developed as part of our future work. This model will then be used for expressing mappings from the logical to the conceptual level.

Some examples for operation signatures that are useful in the context of our example scenario are given in Fig. 3. The *getFeature* operation (of a WFS) returns a feature collection of a certain feature type for a given query. Similarly, the *getCoverage* operation (of a WCS) returns the requested coverage based on the given query. The other three operations shown in Fig. 3 are geoprocessing operations. The *count* operation counts for each feature f in $fc2$ the number of features of $fc1$ that are spatially contained within f. The count value is added as an attribute to f and the new feature collection is returned as a result. The *normalizeByArea* operation normalizes (for each feature f in the feature collection fc^2) the value of the attribute identified by the parameter *attToNorm* by the area of the geometry of f. The normalized value is added as an additional attribute to the feature collection, which is returned as a result.

```
1   getFeature(featureType:string, q:query):FeatureCollection
2   getCoverage(coverage:string, q:query):Coverage
3   count(fc1:FeatureCollection, fc2:FeatureCollection):FeatureCollection
4   normalizeByArea(fc:FeatureCollection, attToNorm:string):FeatureCollection
```

Fig. 3. Examples for operations at the logical level

Example descriptions for each of these operations at the conceptual level are shown in Fig. 4 and Fig. 5. Note that, based on the assumption described above, only relationships between those parameters of an operation that can be considered as entities, entity collections or fields are described at the conceptual level. The inputs and outputs are described using concepts from a domain ontology, e.g. *density* or *frequency*.

Data access services and geoprocessing services are described slightly differently. While every instance of a processing service will exhibit the same functionality, i.e. provide the same output given the same input, different instances of a data access service will provide different outputs, depending on the feature types or coverages stored in its underlying data store. Therefore, data access services have to be described at the instance level, while geoprocessing services can be described at the type level.

For data access services, different rules describe the different feature types and coverages that can be provided by a specific instance of a WFS or WCS. As these services do not have any inputs (that are included in the description at the conceptual level) they can be described as facts, i.e. as rules without a body. For example, the rules in Fig. 4, lines 1-2 state that the described WFS *getFeature* operation can

---

[2] fc is assumed to consist of polygon features forming a tessellation.

provide *entityCollection* all of whose members are *forestFires* and *adminUnits*. The rule in line 3 describes a WCS whose *getCoverage* operation can provide an *elevation* field (with domain *d* and range *e*). Feature types and field types should already be defined in the domain ontology as rules such as those shown in Fig. 2, lines 1-3.

1 **getFeature**: → entityCollection(ec), (∀e : hasMember(ec,e) → forestFire(e))
2 **getFeature**: → entityCollection(ec), (∀e : hasMember(ec,e) → adminUnit(e))
3 **getCoverage:** → elevation(e,d)

Fig. 4. Examples for service description rules for data access services

Also, for geoprocessing services, there can be several rules describing the same operation. Each of these rules expresses a different conceptualization. Thus, two rules could e.g. express that the *normalizeByArea* operation can be used to derive a *density* from a *frequency* (line 2 in Fig. 5) as well as a *fraction* from an *area* (line 3).

1 **count**: entityCollection(ec1), tessellation(ec2) → frequency(f, ec2, ec1)
2 **normalizeByArea**: frequency(f, std, ec) → density(d, std, ec)
3 **normalizeByArea**: area(a, std) → fraction(f, std)

Fig. 5. Examples for service description rules for geoprocessing services

The most intuitive mapping of types at the logical to types at the conceptual level would be from feature collections to entity collections and from coverages to fields. However, there might also be cases where a parameter, while represented as a feature collection at the logical level, can also be conceptualized as a field, or, more rarely, where a coverage can be represented conceptually as an entity collection. We will illustrate this using the *normalizeByArea* operation listed in Fig. 3, line 4. Conceptually, the input of the operation could also be seen as a coverage, whose domain consists of the set of geometries of all features in *fc*, and whose range is the set of the corresponding attribute values of *attToNorm*.

We believe that feature collections are best mapped to *fields* in this way if only one attribute of its features is of interest for the processing (in the case of the collection being an input parameter) or for the result of the processing (if the collection is the output). Conversely, a feature collection should be mapped to a (subconcepts of) *entityCollection* if its features are used as a whole in the processing, e.g. the features in *fc1* in the *count* operation (Fig. 3, line 3). If new features (entities with a different identity) are created in the processing, e.g. a buffer zone around a feature, this output feature collection should also be mapped to an *entityCollection*.

The representation of an operation's parameters as a field or entity collection will of course affect its discovery. To alleviate this decision, rules could be included that map between both views (e.g. "if *ec* is an entityCollection, and every entity *e* that is a member of *ec* has a *temp* property *t* and a *geometry* property *g*, then the collection of all pairs *(t,g)* is a *temp* field"). Such statements will also need to be part of the mapping rules between the logical and conceptual level that we will develop as part of our future work.

3.3 Rule-Based Service Discovery and Composition

Based on the description framework described in the previous section, services can be discovered and composed into a service chain that is able to produce the information defined as the composition goal by the user. To illustrate the presented method, we show how a service chain can be composed for the scenario introduced in Section 1. For this walk-through, we assume the service description rules listed as examples in Fig. 4 and Fig. 5 and the domain rules given in Fig. 2.

Goal description. It is the aim of the proposed composition method to create a service chain that can provide the information required by the user without requiring any further input. This means that the chain behaves like a data access service, and therefore the composition goal is specified in the same way, i.e. as a fact. The application developer's goal ("forest fire density") can be represented as a density d of an entity collection ec (all of whose members are *forestFires*) with respect to a spatio-temporal domain std:

$$density(d, std, ec), (\forall e : hasMember(ec,e) \rightarrow forestFire(e))$$

Discovery and composition. During each discovery step, the current goal is compared with the rules describing the services. A rule is considered to be a match for a goal if it contains (part of) the goal in its head. Rules that have no body (i.e. describe data access services) represent direct matches and are endpoints in the composition process (cf. Section 3.1). Rules that have a body represent conditional matches and require additional searches. The head literal is replaced by the body literals, and the thus generated rule becomes the new goal. When there is a match (and the rule is not a domain rule), the corresponding operation (which can be derived from the label) is added to the service chain.

Fig. 6 shows each of the discovery steps, in which both (labeled) service description rules and a domain rule are used.

The first query for this goal to the does not return any direct matches. However, a conditional match would be discovered: The *normalizeByArea* operation – under the condition that forest fire *frequency* data can be provided as input. This operation is added to the service chain, and the new goal for the next query is defined as a frequency f of an entity collection ec (all of whose members are *forestFires*) with respect to a spatio-temporal domain std:

$$frequency(f, std, ec), (\forall e : hasMember(ec,e) \rightarrow forestFire(e))$$

This search only returns one (conditional) match, the *count* operation – under the condition that a collection of *forest fires* and a feature collection that represents a *tessellation* of space can be provided as inputs:

$$tessellation(std), entityCollection(ec), (\forall e : hasMember(ec,e) \rightarrow forestFire(e))$$

When searching for these two data sets, two direct matches are discovered: The getFeature operations providing *forestFire* and *adminUnit* entity collections. Note that for discovering the *adminUnit* entityCollection, a domain rule (stating that *adminUnit* entity collections represent *tessellations*) has to be used as an intermediate step.

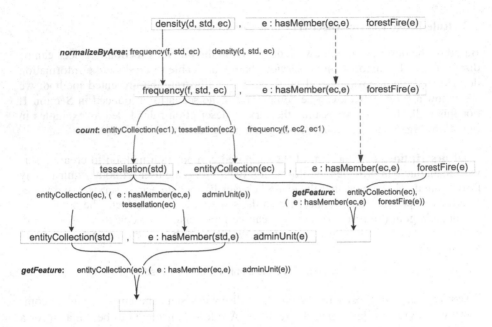

Fig. 6. Example composition of a service chain providing forest fire density information

The data flow in the resulting service chain is depicted in Fig. 7. Note that as our composition methodology only uses the service descriptions at the conceptual level, it does not create a service chain description that is directly executable using some workflow engine. Deriving such an executable description, which might include further (e.g. coordinate or schema) transformation steps between each of the component services, will be part of future work.

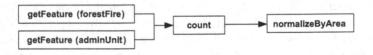

Fig. 7. Data flow in the composed service chain

4   Conclusions and Future Work

In this paper, we have presented a method for supporting users in creating service chain descriptions on a conceptual level. The method builds on a rule-based description framework for data access and geoprocessing services. We have illustrated how this framework can be used to describe service functionalities on a conceptual level by linking their inputs and outputs. The presented method uses these descriptions to semi-automatically derive a service chain description. The presented approach will be extended in several directions in our future research:

- **Creating executable service chain descriptions.** In this paper, we only consider composition at the conceptual level, at which services are described. This service chain description cannot directly be executed by a workflow engine. To achieve this, a formal model for mapping the operations at the logical level to the descriptions at the conceptual level has to be developed. An important requirement for the mapping language to be developed is to (intuitively) support mapping between the field and entity views on GI. The presented composition method will be extended to take these mappings into account in order to create an executable service chain description.

- **SDI architecture.** Based on this extended composition method, an SDI architecture and prototype will be developed. Most importantly, these will combine the service discovery steps at the conceptual level with service discovery using catalogue services. To realize this, a suitable reference between conventional metadata and the rule-based descriptions. E.g., when using ISO 19119 metadata [8], this reference could be included in the *operationDescription* attribute of the *SV_OperationMetadata* item [27].

- **Enhancing ontologies.** In this paper, we have presented a simple top level ontology as a basis for the description framework. This ontology should be compared and, if possible, aligned with other geographic top level ontologies (e.g. [24, 25]). Also, we have not yet developed a full-fledged domain ontology, which is the main basis for service providers to describe their services' functionalities (and therefore a crucial factor during discovery). The presented concepts will be extended with further examples from the domain of disaster management (e.g. flood damage assessment).

- **Enhancing the discovery and composition method.** Finally, the presented composition method can be enhanced in several ways. In order to avoid introducing loops in the service chain, services that are already part of the service chain (in the same branch) should not be presented as results during discovery[3]. Also, heuristics could be introduced for ranking the discovered services and thus helping the user in selecting the most appropriate service. For example, several further discovery steps could be automatically executed (in the background) to find out how many more services would have to be added to the discovered service before a direct match is found. Other ranking criteria could be some measure of how similar the output of a discovered is to the goal. For such a ranking, a similarity metric would have to be developed.

Acknowledgements

The work presented in this paper has been supported by the European Commission through the ORCHESTRA project (grant number IST-2002-511678).

References

1. Nebert, D.D.: Developing Spatial Data Infrastructures: The SDI Cookbook. GSDI (2004)
2. McKee, L.: Who wants a GDI? In: Groot, R., McLaughlin, J. (eds.) Geospatial Data Infrastructure - Concepts, cases, and good practice, pp. 13–24. Oxford University Press, Oxford (2000)

---

[3] The same service occurring in parallel branches is not a problem.

3. Worboys, M., Duckham, M.: GIS – A Computing Perspective. CRC Press (2004)
4. OGC: Web Feature Service (WFS) Implementation Specification, Version 1.1. Open Geo-spatial Consortium, OGC 04-094 (2004)
5. OGC: Web Coverage Service (WCS) Implementation Specification, Version 1.1. Open Geospatial Consortium, OGC 06-083r8 (2006)
6. OGC: Web Processing Service (WPS) Implementation Specification, Version 1.0.0. Open Geospatial Consortium, OGC 05-007r6 (2007)
7. ISO: Information Technology - Open Distributed Processing - Reference Model: Overview. International Organization for Standardization, ISO/IEC 10746-1 (1998)
8. ISO: Geographic Information - Services. International Organization for Standardization, ISO 19119:2005 (2005)
9. OASIS: Web Services Business Process Execution Language (WS-BPEL), Version 2.0. Organization for the Advancement of Structured Information Standards (2007)
10. Lutz, M.: Ontology-based Descriptions for Semantic Discovery and Composition of Geoprocessing Services. GeoInformatica 11, 1–36 (2007)
11. Küster, U., Stern, M., König-Ries, B.: A Classification of Issues and Approaches in Automatic Service Composition. In: Proc. First International Workshop on Engineering Service Compositions (WESC05) at Third International Conference on Service Oriented Computing (ICSOC05) (2005)
12. Sirin, E., Hendler, J., Parsia, B.: Semi-automatic Composition of Web Services using Semantic Descriptions. In: Proc. 1st Workshop on Web Services: Modeling, Architecture and Infrastructure (2002)
13. Antoniou, G., Van Harmelen, F.: Web Ontology Language: OWL. In: Staab, S., Studer, R. (eds.) Handbook on Ontologies, pp. 67–92. Springer, Heidelberg (2003)
14. Martin, D., Paolucci, M., McIlraith, S., Burstein, M., McDermott, D., McGuinness, D., Parsia, B., Payne, T., Sabou, M., Solanki, M., Srinivasan, N., Sycara, K.: Bringing Semantics to Web Services: The OWL-S Approach. In: Cardoso, J., Sheth, A.P. (eds.) SWSWPC 2004. LNCS, vol. 3387, pp. 111–123. Springer, Heidelberg (2004)
15. Gruber, T.: A Translation Approach to Portable Ontology Specifications. Knowledge Acquisition 5, 199–220 (1993)
16. Klien, E., Lutz, M., Kuhn, W.: Ontology-Based Discovery of Geographic Information Services - An Application in Disaster Management. Computers, Environment and Urban Systems 30, 102–123 (2006)
17. Lutz, M., Klien, E.: Ontology-Based Retrieval of Geographic Information. International Journal of Geographical Information Science 20, 233–260 (2006)
18. Bowers, S., Ludäscher, B.: An Ontology-Driven Framework for Data Transformation in Scientific Workflows. In: Rahm, E. (ed.) DILS 2004. LNCS (LNBI), vol. 2994, Springer, Heidelberg (2004)
19. Yue, P., Di, L., Yang, W., Yu, G., Zhao, P.: Semantics-based Automatic Composition of Geospatial Web Service Chains. Computers & Geosciences 33, 649–665 (2007)
20. Baader, F., Nutt, W.: Basic Description Logics. In: Baader, F., Calvanese, D., McGuinness, D., Nardi, D., Patel-Schneider, P. (eds.) The Description Logic Handbook. Theory, Implementation and Applications, pp. 43–95. Cambridge University Press, Cambridge (2003)
21. Lutz, M., Kolas, D.: Rule-based Discovery in Spatial Data Infrastructures. Transactions in GIS, Special Issue on the Geospatial Semantic Web 11, 317–333 (2007)
22. Kolas, D., Hebeler, J., Dean, M.: Geospatial Semantic Web: Architecture of Ontologies. In: Rodríguez, M.A., Cruz, I., Levashkin, S., Egenhofer, M.J. (eds.) GeoS 2005. LNCS, vol. 3799, pp. 183–194. Springer, Heidelberg (2005)

23. Russell, S.J., Norvig, P.: Artificial Intelligence. A Modern Approach. Prentice-Hall, Upper Saddle River (2003)
24. Bittner, T.: From Top-level to Domain Ontologies: Ecosystem Classifications as a Case Study. In: Proc. Conference on Spatial Information Theory (COSIT 2007). LNCS, vol. 4736, pp. 61–77. Springer, Heidelberg (2007)
25. Probst, F., Espeter, M.: Spatial Dimensionality as Classification Criterion for Qualities. In: Proc. International Conference on Formal Ontology in Information Systems (FOIS 2006) (2006)
26. Wache, H., Vögele, T., Visser, U., Stuckenschmidt, H., Schuster, G., Neumann, H., Hübner, S.: Ontology-Based Integration of Information — A Survey of Existing Approaches. In: Proc. IJCAI-01 Workshop: Ontologies and Information Sharing (2001)
27. Lutz, M.: Ontology-Based Service Discovery in Spatial Data Infrastructures. In: GIR 2005. Proc. ACM Workshop on Geographic Information Retrieval, pp. 45–54. ACM, New York (2005)

Algorithm, Implementation and Application of the SIM-DL Similarity Server

Krzysztof Janowicz, Carsten Keßler, Mirco Schwarz, Marc Wilkes, Ilija Panov,
Martin Espeter, and Boris Bäumer

Institute for Geoinformatics
University of Muenster, Germany
{janowicz,carsten.kessler,mirco.schwarz,marc.wilkes,
i.panov,m.espeter,boris.baeumer}@uni-muenster.de

Abstract. Semantic similarity measurement gained attention as a
methodology for ontology-based information retrieval within GIScience
over the last years. Several theories explain how to determine the similarity between entities, concepts or spatial scenes, while concrete implementations and applications are still missing. In addition, most existing
similarity theories use their own representation language while the majority of geo-ontologies is annotated using the Web Ontology Language
(OWL). This paper presents a context and blocking aware semantic similarity theory for the description logic $\mathcal{ALCHQ}$ as well as its prototypical
implementation within the open source SIM-DL similarity server. An
application scenario is introduced showing how the Alexandria Digital
Library Gazetteer can benefit from similarity in terms of improved search
and annotation capabilities. Directions for further work are discussed.

1   Introduction and Motivation

Semantic similarity measurement has become a major research topic within geographic information science during the last years, aiming at improved methods
for information retrieval and integration of heterogeneous spatial data sources.
The utilization of findings on similarity measurement from psychology [1] promises
user interfaces and search results with an improved cognitive plausibility. However, existing similarity theories aiming at the geospatial domain [2,3,4] mostly
lack compatibility with current widespread knowledge representation languages
such as the Web Ontology Language (OWL). The similarity theories require the
knowledge to be present in specific formats, ignoring the applicability to existing
(geo-)ontologies. To overcome this gap between semantic similarity theories on the
one hand, and existing ontologies on the other hand, we present the description
logic (DL) based SIM-DL theory [5].

The relevance of a similarity framework, however, is not only depending on its
applicability to existing knowledge representations, but also on its adaptation to
technical prerequisites. The DIG[1] interface has been established as a standard

---

[1] Description Logic Implementation Group, http://dig.sourceforge.net/

F. Fonseca, M.A. Rodríguez, and S. Levashkin (Eds.): GeoS 2007, LNCS 4853, pp. 128–145, 2007.
© Springer-Verlag Berlin Heidelberg 2007

interface for communication between applications such as ontology editors and reasoners. To ensure compatibility with this de-facto standard, we extend the DIG interface by a group of similarity functions. The open source SIM-DL server is introduced as a reference implementation of the SIM-DIG interface.

Existing similarity theories for the geospatial domain have been evaluated in specially designed application scenarios, without implementation in real-world applications. Beyond the SIM-DL theory and server, we also present a gazetteer application to demonstrate the benefits of similarity based applications. Current gazetteers are mostly based on semi-formal feature[2] type thesauri, defining feature types in terms of a hierarchy with a restricted number of relations. We present a novel Web interface for the Alexandria Digital Library gazetteer that makes use of the SIM-DL server and retrieves its information from a feature type ontology to provide an intuitive work flow and enhanced support for novice users.

The remainder of this paper is organized as follows: we first present related work on similarity measurement and description logics, and then introduce an extended version of the SIM-DL theory [5] and framework. The server prototype is discussed, followed by a description of the application scenario and an outlook on future work.

2   Related Work

This section gives a brief overview of related work concerning semantic similarity and introduces the description logic $\mathcal{ALCHQ}$ and its normalization. Only such aspects which are necessary for the understanding of the SIM-DL similarity theory and implementation are described; for further details see [6].

2.1   Semantic Similarity Measurement

The notion of similarity originated in psychology and was established to determine why and how entities are grouped into categories, and why some categories are comparable to each other while others are not [1,7]. The main challenge with respect to *semantic* similarity measurement is the comparison of meanings as opposed to purely structural comparison. A language has to be specified to express the nature of entities and metrics are needed to determine how (conceptually) close the compared entities are. While entities can be expressed in terms of attributes, the representation of entity types is more complex. Depending on the expressivity of the representation language, types are specified as sets of features, dimensions in a multidimensional space, or formal restrictions specified on sets using various kinds of description logics. While some representation languages have an underlying formal semantics (e.g. model theory), the grounding of several representation languages remains on the level of an informal description.

---

[2] It is important to distinguish between *geographic* features as organized in gazetteers, and the features—i.e. properties, parts and functions—used for concept comparison in certain similarity theories (see section 2.1).

Because similarity is measured between entity types which are representations of concepts in human minds, similarity depends on what is said (in terms of computational representation) about these types. This again is connected to the chosen representation language, leading to the fact that most similarity measures cannot be compared. Beside the question of representation, context is another major challenge for similarity assessments. In many cases meaningful notions of similarity cannot be determined without defining in respect to what similarity is measured [8,7,9].

Similarity has been widely applied within GIScience over the past few years. Based on Tversky's feature model [10], Rodríguez and Egenhofer [2] developed an extended model called Matching Distance Similarity Measure (MDSM) that supports a basic context theory, automatically determined weights, and asymmetry. Raubal and Schwering [3,4] used conceptual spaces [11] to implement models based on distance measures within geometric space, while Janowicz and Raubal [12] combined model theoretic and geometric aspects to determine similarity based on affordances. Several measures [13,14,5] were developed to close the gap between (geo-)ontologies described by various kinds of description logics, and similarity theories that had not been able to handle the expressivity of such languages. Other similarity theories [15,16] have been developed to determine the similarity between spatial scenes. The ConceptVISTA[3] ontology management and visualization toolkit uses similarity for concept comparison.

2.2   Description Logics and DIG Interface

Description Logics are a family of knowledge representation languages used to model concepts and entities in a knowledge base. Such a knowledge base consists of a TBox containing the terminology, i.e. the vocabulary describing a given domain, and an ABox storing assertions (about named entities). Description logics distinguish two kinds of symbols, logical and non-logical symbols. The former have a pre-defined meaning grounded in set theory, while the latter are domain specific. Logical symbols are either[4] constructors ($\sqcap, \sqcup, \exists, \forall, \leq, \geq$) used to compose complex concepts out of primitive ones or connectives such as equality ($\equiv$) or inclusion ($\sqsubseteq$). Same as for first order logic, the formal semantics of description logics is given by its interpretation. An interpretation $\Im$ is defined as a tuple $\langle \Delta^{\mathcal{I}}, \mathcal{I} \rangle$. $\Delta^{\mathcal{I}}$ denotes a non-empty set called the domain of interpretation, whereas $\mathcal{I}$ describes the interpretation function mapping from non-logical symbols to elements and (binary) relations over $\Delta^{\mathcal{I}}$. The subset $A^{\mathcal{I}}$ of $\Delta^{\mathcal{I}}$ associated with a concept A is also called its extension. Within this paper the term description or specification of a concept denotes the statements (phrased using the DL language; see Table 1) used to represent a concept in our mind, not its extension.

$\mathcal{ALCHQ}$ used as representation language for the SIM-DL similarity measure is an expressive description logic that supports intersection, union, full existential quantification, value restriction, full negation and qualified number restrictions

---

[3] http://www.geovista.psu.edu/ConceptVISTA
[4] Leaving punctuation and numbers aside.

Table 1. Syntax and semantics of $\mathcal{ALCHQ}$

Syntax	Semantics	Name
$\top$	$\Delta^{\mathcal{I}}$	Top
$\bot$	$\emptyset$	Bottom
A	$A^{\mathcal{I}} \subseteq \Delta^{\mathcal{I}}$	Atomic concept
R	$R^{\mathcal{I}} \subseteq \Delta^{\mathcal{I}} \times \Delta^{\mathcal{I}}$	Atomic role
$\neg C$	$\Delta^{\mathcal{I}} \setminus C^{\mathcal{I}}$	(Full) negation
$C \equiv D$	$C^{\mathcal{I}} = D^{\mathcal{I}}$	Concept equality
$C \sqsubseteq D$	$C^{\mathcal{I}} \subseteq D^{\mathcal{I}}$	Concept inclusion
$R \equiv S$	$R^{\mathcal{I}} = S^{\mathcal{I}}$	Role equality
$R \sqsubseteq S$	$R^{\mathcal{I}} \subseteq S^{\mathcal{I}}$	Role inclusion
$C \sqcap D$	$C^{\mathcal{I}} \sqcap D^{\mathcal{I}}$	Concept intersection
$C \sqcup D$	$C^{\mathcal{I}} \sqcup D^{\mathcal{I}}$	Concept union
$\forall R.C$	$\{a \in \Delta^{\mathcal{I}} \vert \forall b.(a,b) \in R^{\mathcal{I}} \rightarrow y \in C^{\mathcal{I}}\}$	Value restriction
$\exists R.C$	$\{a \in \Delta^{\mathcal{I}} \vert \exists b.(a,b) \in R^{\mathcal{I}} \wedge y \in C^{\mathcal{I}}\}$	Existential quantification
$\leq nR.C$	$\{a \in \Delta^{\mathcal{I}} \vert \vert\{b \in \Delta^{\mathcal{I}} \vert (a,b) \in R^{\mathcal{I}} \wedge b \in C^{\mathcal{I}}\}\vert \leq n\}$	Qualified max. number restriction
$\geq nR.C$	$\{a \in \Delta^{\mathcal{I}} \vert \vert\{b \in \Delta^{\mathcal{I}} \vert (a,b) \in R^{\mathcal{I}} \wedge b \in C^{\mathcal{I}}\}\vert \geq n\}$	Qualified min. number restriction

to inductively construct complex concepts out of primitive ones and roles (binary predicates). In the following sections the letters A and B are used to represent atomic concepts, R and S for roles and C and D for complex (composed) concepts. X and Y are used for general statements about similarity and alignment that hold for both, concepts and roles. Additional background information about $\mathcal{ALCHQ}$ and related description logics is discussed in [6].

The Web Ontology Language (OWL) comes in different flavors: OWL-Lite is based on the description logic $\mathcal{SHIF}$, while OWL-DL corresponds to $\mathcal{SHOIN}(\mathcal{D})$. The extended new version OWL 1.1 matches the expressivity of $\mathcal{SROIQ}(\mathcal{D})$. For this paper we have chosen the description logic $\mathcal{ALCHQ}$ because it is close enough to OWL-DL, leaving aspects that are not relevant for similarity aside. $\mathcal{ALCHQ}$ even supports qualified number restrictions which are part of OWL 1.1. The main difference between $\mathcal{ALCHQ}$ and the OWL logics is the missing support for several role axioms such as role inclusion in $\mathcal{ALCHQ}$ (a similarity measure for role intersection was discussed in [5]), role transitivity and inverse roles on the one hand as well as nominals and datatype properties on the other hand. While it is hard to find a meaningful notion of similarity for role axioms such as transitivity, the similarity between nominals (and simple datatypes) boils down to instance similarity.

The DIG interface is an API specification for reasoning in DL systems [17]. The DIG 1.1 specification provides an interface for reasoning services based on the $\mathcal{SHOIN}(\mathcal{D})$ language. The specification provides an XML-encoded HTTP interface. Clients communicate with a server via HTTP POST, with requests and responses encoded based on the underlying DIG XML Schema[5]. DIG distinguishes between different types of messages and operations. The reasoner's *identification* message is comparable to OGC's getCapabilities requests: the server responds which language and services it supports. This is especially important because of the variety of DL languages, i.e. not every DIG server will support all constructs that are part of the specification (the basic constructs are

---

[5] The DIG XML Schema can be found at: http://dl-web.man.ac.uk/dig/2003/02/

compulsory, however). The *management* operation creates or releases a knowledge base (KB) that is further identified with an unique URI. *Tells* operations insert assertions into the reasoner's KB, while *Asks* operations allow the client to perform reasoning tasks on the KB (see [17] for details).

3   Similarity Framework and Theory

By studying several similarity theories (including feature, geometric and model driven approaches) we found generic patterns which jointly form a framework for measuring similarity between concepts (see also [5,18]). This section describes the framework and applies it to determine similarity between concepts specified in $\mathcal{ALCHQ}$.

The framework consists of the following five steps. Their concrete implementation depends on the semantic similarity theory on the one hand and the underlying representation language on the other hand.

1. Selection of query (search) and target concepts.
2. Transformation of concepts to canonical form.
3. Definition of an alignment matrix for concept descriptors.
4. Application of constructor specific similarity functions to selected pairs.
5. Determination of normalized overall similarity.

For reasons of readability all equations forming the SIM-DL measure (steps 4 and 5) have been moved to the appendix.

3.1   Query and Target Concepts

Before measuring similarity it needs to be determined which concepts from the examined ontology should be compared. Depending on the application scenario and theory, the query (search) concept C_s can be part of the ontology or phrased using a shared vocabulary [5,19]. The target concepts $\{C_t\}$ are selected by hand or determined by the context of the query. Such a context specifies the domain of application either by explicitly selecting the compared-to concepts or implicitly by defining a context concept C_c. In the latter case the target concepts are all concepts subsumed by C_c. Same as for the matching distance similarity measure defined by Rodriguez and Egenhofer [2], SIM-DL defines the set of target concepts as $\{C_t | C_t \subseteq C_c\}$. All similarity functions (see section 3.4) are defined with respect to this context.

3.2   Canonical Normal Form

Before similarity can be computed, the compared concepts have to be rephrased to a canonical normal form to reduce potential syntactic influence. The procedure can be further distinguished into a normalization step and the application of rewriting rules. Both steps mostly depend on the underlying representation language and their importance increases with the expressivity of the used language.

In case of geometric representations a canonical normal form can be achieved through mappings between reference spaces if they approximate the same quality space (see [20]).

In case of model driven measures based on description logics, the procedure is more complex. For $\mathcal{ALCHQ}$ we have developed the following disjunctive normal form (DNF): A concept description C is in normal form iff $C = \top$, $C = \bot$ or $C = C_1 \sqcup ... \sqcup C_n$ and each $C_i (i = 1, ...n)$ is of the form:

$$C := \prod_{A \in primitive(C_i)} A \sqcap \prod_{R \in N_R} \left(\prod_{C' \in exists_R(C_i)} (\exists R.C') \sqcap \forall R.forall_R(C_i) \right.$$

$$\left. \sqcap \prod_{C' \in min_R(C_i)} (\geq |min_R(C_i)|R.C') \sqcap \prod_{C' \in max_R(C_i)} (\leq |max_R(C_i)|R.C') \right) \tag{1}$$

The set $primitive(C)$ represents all (negated) primitives (and $\bot$) at the top-level of C. N_R is the set of available roles, and $exists_R(C)$, $min_R(C_i)$ and $max_R(C_i)$ denote the sets of all C' for which there exists $\exists R.C'$ (respectively min/max restrictions) at the top-level of C. $forall_R(C_i)$ denotes the intersection of concepts $(C1 \sqcap ... \sqcap C_n)$ derived by merging all value restrictions for the role R ($\forall R.C_i$) on the top level of C. $|min_R(C_i)|$ and $|max_R(C_i)|$ represent the minimum and maximum cardinalities for the role R on the top-level of C. Note that the concepts $forall_R(C_i)$ and C' are again in $\mathcal{ALCHQ}$ normal form.

To ensure that the SIM-DL measure is not influenced by the syntactic form, rewriting rules (see also [21,22]) have to be applied in order to get a canonical representation of the compared concepts. On the one hand these rewriting rules map between equivalent expressions such as $(\forall R.\bot)$ and $(\leq 0R.\top)$. On the other hand they ensure that only such descriptions are used within concept specifications which (by definition) have an impact on the cardinality of the regarded sets. For instance $(\geq 1R.C) \sqcap (\geq 2R.C)$ is mapped to $(\geq 2R.C)$, while $(... \sqcap \top)$ can be skipped without changing the extension of the specified concept.

3.3   Alignment Matrix and Blocking

While section 3.1 describes how concepts $(C_s, C_{t_1}...C_{t_n})$ are selected, an alignment matrix [5,23] is necessary to determine which parts of their descriptions are compared. Most theories assume similarity to be a binary relation, hence the alignment matrix creates tuples $sim(X_s, Y_{t_n})$ for all possible combinations of the Cartesian product $C_s \times C_{t_n}$. While C_s and C_{t_n} denote two compared-to concepts, X_s and Y_{t_n} are parts of their descriptions (e.g. concrete number restrictions).

In case of feature based representations such as used for MDSM [2], the alignment matrix is reduced to a 0/1 matching. If two parts of compared concept descriptions have the same label, they count as common features, if not they are distinguishing features. The impact of these features on the overall similarity depends on sub/super relations between the compared concepts (see section 3.5). Note that MDSM distinguishes between three feature types: functions, parts and attributes. Features are only compared if they belong to the same feature type.

Geometric approaches which take relations into account, choose such tuples for later comparison where the target relation is a subtype of the source relation.

For SIM-DL the alignment matrix is defined as follows. If two concepts are compared, an alignment matrix M_1 with all possible combinations of their parts is created. Once similarity for each tuple is calculated (see section 3.4), those tuples with the highest similarity values are chosen for further computation. Note that each X_s respectively Y_{t_n} is only selected once and similarity can only be calculated if both elements of the tuple are based on the same constructor. For instance, a value restriction is never compared to a quantification. For each selected tuple the normalization factor (see section 3.5) is increased by 1.

To handle circular definitions[6] such as $C \equiv \ldots \sqcap (\forall R.C)$ the matrix (and the similarity functions) need to implement a blocking mechanism as known from tableaux algorithms for subsumption reasoning in DL. For instance, consider the tuple $sim(C, D)$ from the matrix M_1 used to compare a search and target concept (where C is defined as above and $D \equiv \ldots \sqcap (\forall R.D)$). In order to calculate the similarity between C and D, an alignment matrix M_2 that contains tuples for all possible combinations of the Cartesian product $C \times D$ is created. Since the definition of concept C (and D) is circular, all tuples from M_2 containing $(\forall R.C)$ (and $(\forall R.D)$) will end up in a loop (creating infinite alignment matrices). Instead such tuples are set as *blocked*. All similarity values for tuples in the matrix M_2 are calculated leaving the blocked tuples aside. The result is an approximated similarity between C and D. Using this value, the blocked tuples can now be computed and M_2 (and finally M_1) can be re-calculated without loops. This tuple-wise blocking often appears in case of negation. If only one part of the tuple is blocked (e.g. if $(\forall R.D)$ is replaced by $(\forall R.E)$) the process continues unfolding E and building matrices until no expression to be compared to $(\forall R.C)$ is left, or its filler is either $\top$ or primitive. As similarity can be computed for this tuple, the value is now used one level (matrix) higher and so on until $sim(C, D)$ can be determined. This kind of blocking is called expression-wise here.

3.4 Similarity Functions and Neighborhoods

After choosing the compared-to concepts and aligning their descriptions, similarity is measured for each selected tuple $sim(X_s, Y_{t_n})$. Depending on the constructors used for X_s and Y_t different similarity functions have to be applied.

In case of MDSM, features are distinguished into different types during the alignment process, however, the same similarity measure (a weighted and asymmetric feature ratio function) can be applied to all of them. Geometric approaches allow for several functions either based on different metrics (such as Euclidian or city-block) and, if they support relations, distinguish between similarity (inverse distance) within a conceptual space and network-based similarity measures for relations.

Because the $\mathcal{ALCHQ}$ knowledge representation language allows for more expressive conceptualizations, SIM-DL has to offer a similarity function for each

---

[6] The problem of circularity also affects other similarity measures, but was not taken into account so far.

constructor. The measurement process always starts at the union level (see $\mathcal{ALCHQ}$ canonical normal form; section 3.2) with the sim_u function. Each concept on this level is itself formed by intersection and similarity between such concepts is measured by sim_i[7]. Each concept of this intersection is either a primitive (sim_p), an existential quantification (sim_e), a value restriction (sim_f) or a qualified number restriction (sim_{min}, respectively sim_{max}). In addition to role hierarchies (sim_r) SIM-DL supports temporal and topological neighborhoods (sim_n) to calculate similarity between roles. This allows to determine the similarity between tuples such as ($\exists inside.Lake, \exists overlap.Lake$); see [5] for more details. All necessary similarity functions are listed in the appendix.

3.5 Overall Similarity

The overall similarity determines the similarity between compared concepts C_s and C_t based on the similarities for all considered tuples $sim(X_s, Y_{t_n})$. In most examined theories this step was a summation function, normalized to values between 0 and 1.

For MSDM the overall similarity is the weighted sum of the similarity determined between functions, parts and attributes. While the weighting indicates the relative importance of each feature type, at the same time it acts as the normalization factor ($\sum \omega = 1$)[2]. In case of geometric approaches the overall similarity is given by the normalized (via z-transformation) sum of compared dimensions.

For SIM-DL each similarity function discussed in section 3.4 takes care of its normalization using the number of compared tuples. Each similarity function returns a value between 0 and 1 to the function (on a higher level) it was called by.

4 Similarity Server and Interfaces

This section gives a brief overview of the architecture of the DIG-based semantic similarity server. A plug-in for the Protégé Ontology Editor will be described. The SIM-DL server and the plug-in are still under development, but already available as an open-source cross-platform project at Sourceforge.net. The current beta version[8] supports subsumption reasoning and similarity measurement up to $\mathcal{ALCHQ}$, support for more expressive description logics is under development.

4.1 Architecture

The SIM-DL server is based on an embedded Jetty HTTP server[9]. Incoming requests via XML-over-HTTP are processed by a request handler who interprets DIG operations and starts the similarity and reasoning engines. The reasoner implements a tableaux algorithm to determine TBox subsumption based

---

[7] Of course primitives, restrictions and quantifications can already appear on union level without violating the measurement process (see appendix).

[8] The current release can be downloaded at http://sim-dl.sourceforge.net/.

[9] http://jetty.mortbay.org/

on ABox satisfiability, while the similarity engine is based on the presented SIM-DL framework and theory. Both components implement their own normalization and blocking methods. Each similarity request involves interaction with the reasoning component to determine target concepts out of the context. The reasoner is also used for some similarity functions such as sim_p. In this paper, we propose the Protégé[10] plugin and gazetteer Web interface (see section 5) as clients; however, every DIG compatible client software can be used.

4.2 SIM-DIG Interface

A short introduction to the DIG interface was given in section 2.2. The interface has to be extended to enable similarity measurement between concepts. First, the *Ask* syntax has to be extended by a similarity query which defines a search concept (C_s) and a context concept (C_c). The search concept is compared to all subclasses of the context concept. Table 2 shows the supported queries as well as our extension.

Table 2. Supported Ask language, *similarity extensions* and query syntax

Request Category	Tag Syntax
Satisfiability	`<satisfiable>C</satisfiable>`
Concept Hierarchy	`<parents>C</parents>`
	`<children>C</children>`
	`<ancestors>C</ancestors>`
	`<descendants>C</descendants>`
	`<equivalents>C</equivalents>`
Similarity Queries	`<ccsimilarity>CS CC</ccsimilarity>`

The result of a similarity query contains a set of concepts where each concept has a value indicating the similarity to the source concept. Since the existing response operators do not allow for assigning a value to a concept, the response syntax has to be extended, too. Table 3 shows the supported response operators and, additionally, the syntax extension that permits similarity queries.

4.3 Protégé Plug-In

To enable the use of reasoning services there is a need for suitable graphical user interfaces. This holds for standard reasoning tasks, such as subsumption reasoning, as well as for similarity reasoning. Today's standard front-end for DL based reasoning is Protégé, a Java based open source ontology editor and knowledge base framework. It is built upon an extensible architecture that provides the possibility to add further functionality via plug-ins. The Protégé OWL plugin is one of the most popular plug-ins that have been developed for the Protégé framework. It enables users to create, explore and modify OWL ontologies supporting OWL-Lite, OWL-DL and OWL-Full [24]. Additionally, it provides DIG-based access to DL reasoners such as Pellet[11]. The combination of DL theory,

---

[10] http://protege.stanford.edu/
[11] http://pellet.owldl.com

Table 3. Supported Ask language, *similarity extensions* and response syntax

Response Category	Response Syntax	Request Category
Boolean	`<true/>` `<false/>`	Satisfiability

Concept Set	`<conceptSet>`
	  `<synonyms>S11...S1N</synonyms>`
	  `<synonyms>SM1...SMN</synonyms>`
	`</conceptSet>`
		Concept Hierarchy

Similarity Set	`<conceptSet>`
	  `<catom name=S1>`
	    `<simValue>s1</simValue>`
	  `</catom>`
	  `<catom name=SN>`
	    `<simValue>sN</simValue>`
	  `</catom>`
	`</conceptSet>`
		Similarity Query

reasoning services and Protégé as a graphical frontend was a prerequisite for establishing OWL as the standard for creating semantic web applications. A similar combination will be necessary to initiate the spread of DL based similarity measurement. The Protégé OWL API includes several extension points for implementing OWL specific plug-ins. To provide a graphical frontend for accessing the SIM-DL similarity server we developed the SIM-DL plug-in as a GUI-plugin based on Protégé OWL. The possibilty to view and explore the ontologies that are involved in the similarity measurement process is mandatory. This functionality is already provided by Protégé -OWL and reused for the SIM-DL plug-in. Due to the architecture of the similarity server the SIM-DL plug-in has to support the DIG interface. We reused the DIG implementation provided by Protégé OWL and added the SIM-DL specific DIG elements. Figure 1 shows a screenshot of the current state of the plugin.

5   Gazetteer Application Scenario

The use of similarity measurement in current gazetteers is hampered by a lack of formalism in the corresponding feature type thesauri. In the following, we show how subsumption and similarity based user interfaces can improve the gazetteers' functionality and usability, based on a transformation of feature type thesauri into ontologies.

5.1   From ADL FTT to Feature Type Ontology

Georeferencing is the core functionality of gazetteers as place name directories. The distinction between different place (feature) types is enabled by thesauri, which contain semi-formal descriptions of the feature types and can be queried via *type-lookup* functionality. To fully support subsumption and similarity-based reasoning, a transformation of these thesauri into formal ontologies is required. We use the example of the Alexandria Digital Library (ADL) Feature Type

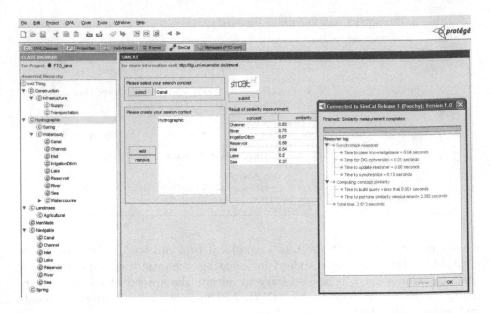

Fig. 1. SIM-DL Protégé Plug-in (beta version)

Thesaurus (FTT)[12] in the following to demonstrate the required steps and design decisions. The procedure can be transferred to other thesauri as well.

The ADL FTT contains textual definitions for *preferred* terms in the form of *scope notes* (SN); in addition, *non-preferred* terms are listed as pointers to preferred terms via the *Use* (USE) and *Used for* (UF) relations, e.g. *lakes UF lagoons*. Inheritance between preferred terms is marked by the *broader term* (BT) and *narrower term* (NT) relations, which are not directly comparable to the sub- and supertype relations in ontologies [25], so that transitivity cannot be taken for granted. Moreover, there is only one broader term for every term in the ADL FTT (despite the ANSI-NISO 39.19 standard allowing for multiple inheritance). This single inheritance structure forces every term to be a NT of only one of the six top terms; for example, *cities* are only classified as *administrative areas*, but not as *manmade features*. Beyond NT and BT, the *related term* (RT) relation is used to express diverse kinds of relations between terms, so that the semantics of RT remain ambiguous—for example, RT is used to describe the relation of *lakes* to *reservoirs*, i.e. a functional relation, but also to *wetlands*, i.e. a topological relation.

It must be pointed out that the structure of the ADL FTT is not wrong or badly designed, since thesauri are developed for different purposes than ontologies. However, there is a lack of formalism and explicit semantics from an ontological point of view, so that an automatic transformation into a feature type ontology is not possible. To manually transform the thesaurus and preserve the original naming and structure, a syntactic and semantic conversion as

---

[12] http://www.alexandria.ucsb.edu/gazetteer/FeatureTypes/ver070302/index.htm

described in [25] must be performed. The resulting ontology (see figure 1 for an extract) uses the top level concept *Feature*, subsumed by different classes such as *Manmade*, *Hydrographic* or *Transportation*; note that these classes are not disjoints, i.e. the concept *Canal*, for example, subsumes all these feature classes at the same time. Moreover, feature types (or concepts in ontology terminology) can be related to each other with an arbitrary number of hierarchically ordered properties which have to be extracted manually from the RT relations and the scope notes in the thesaurus. For example, we introduce the property *hasConnection*, with sub-properties *hasOrigin* and *hasDestination*, to specify that a canal connects (*hasDestination*) two hydrographic features. This brief insight into the conversion process shows that the generation of a feature type ontology requires a significant effort; in the following, we argue that such a conversion is worthwhile, as gazetteer Web interfaces can greatly benefit from a feature type ontology.

5.2   Towards a Distributed Gazetteer Infrastructure

The long term vision of current gazetteer research is focussing on the development of a distributed local-responsibility service infrastructure instead of a single world gazetteer. Such an infrastructure can be compared to the Domain Name Service (DNS) which maps hostnames on the internet to their IP addresses. Each gazetteer offers lookup for local places within its spatial and thematic scope. If the gazetteer cannot answer a request, it redirects the query to a higher level gazetteer which decides whether it or another gazetteer can resolve the query. The underlying idea is that gazetteers should contain and maintain data of interest for the community running the service. This ensures that the stored data is both accurate and up-to-date.

A distributed gazetteer infrastructure raises several challenges for both the georeferencing and the type-lookup function. For georeferencing, the main challenge is that several names may point to the same place using different footprints, which includes divergences between the referred-to coordinates, but especially between the type of footprint such as point versus polygon representation (see also [26]). In the case of type-lookup, one must ensure that all involved gazetteers share a common understanding of the feature types used. Gazetteers are developed for different thematic scopes and spatial scales, which may require different conceptualizations of the described features. Consequently, a common feature type specification needs to be generic enough to form a top level for all gazetteers and extensible to allow for local type definitions. Figure 2 illustrates the role of the SIM-DL server within the proposed gazetteer infrastructure.

5.3   Similarity-Based Gazetteer Web Interface

To efficiently use the ADL gazetteer's Web interface[13], the user needs detailed knowledge of the FTT hierarchy to select the adequate preferred term for what he

---

[13] http://www.alexandria.ucsb.edu/clients/gazetteer/

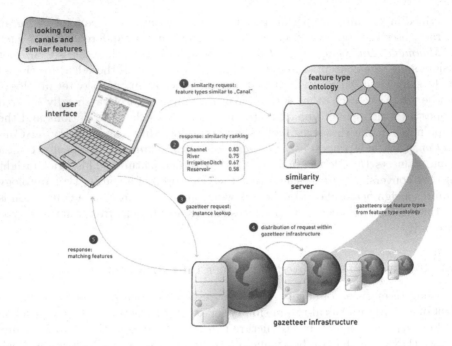

Fig. 2. Similarity-based feature type lookup within the proposed gazetteer infrastructure

is looking for. If the user is not aware of the FTT hierarchy, retrieving the desired information is complicated and tedious, as the user must first consult the FTT to find out about the preferred term for his query. To overcome these difficulties, we propose a subsumption and similarity based gazetteer Web interface based on a feature type ontology, as shown in figure 3.

The proposed interface utilizes AJAX technology in a *search-while-you-type* input field: as the user types in the place type he has in mind, results are automatically loaded in the background. The suggested types are based on a syntactic match of the letters already typed in by the user; next to every suggestion, its supertypes and the most similar other types from the ontology are presented, where the size and color of the type indicate its similarity to the suggested type in the leftmost column. This way, there is no need for the user to know about the underlying feature type hierarchy, as similar types are automatically suggested by the interface. All suggestions are hyperlinked and can be moved to the input field with a single click. Moreover, the interface also allows for spatial restriction by simply zooming the map to the desired extent. The proposed interface thus allows for an intuitive workflow that supports also novice users in the selection of the appropriate feature types for a query. Apart from up-to-date Web technology, this functionality is made possible by the feature type ontology in the background, and by the similarity server accessing it.

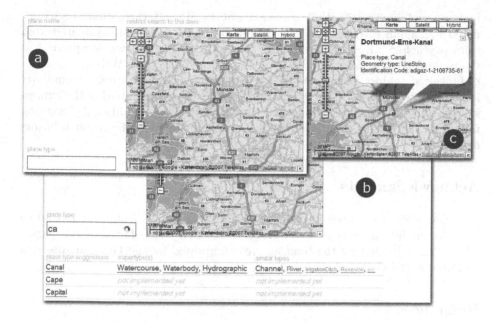

Fig. 3. Conceptual design for the gazetteer Web interface: search interface with input fields for place name and type, and map for spatial restriction (a); automatic suggestion of place types during user input (b); display of results as map overlays (c)

6   Conclusions and Further Work

Most existing similarity theories cannot be implemented as parts of semantically enabled information retrieval infrastructures because they do not support the current standards for knowledge representations (such as OWL). In this paper, we introduced an extended version of the SIM-DL theory [5] and its implementation within an open source similarity server. The server is based on an extended DIG interface and can hence interact with existing tools such as reasoners and editors. An application scenario from gazetteer research demonstrated how similarity measurement can be integrated into user interfaces and existing geo-services. In addition one may also think of the SIM-DL server as a web service within a geo-processing chain as realized in spatial data infrastructures. This would enable to query a Web Feature Service for all features of types similar to *Canal*. As (leaving the list view aside) our approach does not include a visualization component, the integration into ConceptVISTA may be an interesting further step. The presented Protégé plug-in allows ontology engineers to integrate similarity into their development process. For instance, similarity can be used to examine whether a constructed ontology reflects the users view (i.e. conceptualizations).

Further work has to focus on similarity measures for even more expressive description logics and especially for taking modal logics into account as discussed by Poole and Smyth [27]. While some parts of the SIM-DL theory have been evaluated by human subject tests (see [18]) or based on previously evaluated work from psychology or computer science, the evaluation of the whole approach is the next step to be done. In addition the proposed gazetteer Web interface has to be tested against existing interfaces to determine to which degree similarity improves interaction. Finally, while this work focuses on comparing the expressions forming the examined concepts, further work should additionally focus on the ABox. Similarity could then be measured in the style of current tableaux algorithms.

Acknowledgments

The comments from three anonymous reviewers as well as from Martin Raubal provided useful suggestions to improve the content and clarity of the paper. This work is funded by the SimCat project granted by the German Research Foundation (DFG Ra1062/2-1).

References

1. Goldstone, R., Son, J.: Similarity. In: Holyoak, K., Morrison, R. (eds.) Cambridge Handbook of Thinking and Reasoning, Cambridge University Press, Cambridge (2005)
2. Rodríguez, A., Egenhofer, M.: Comparing geospatial entity classes: an asymmetric and context-dependent similarity measure. International Journal of Geographical Information Science 18(3), 229–256 (2004)
3. Raubal, M.: Formalizing conceptual spaces. In: Varzi, A., Vieu, L. (eds.) Formal Ontology in Information Systems, Proceedings of the Third International Conference (FOIS 2004). Frontiers in Artificial Intelligence and Applications, vol. 114, pp. 153–164. IOS Press, Amsterdam, NL (2004)
4. Schwering, A., Raubal, M.: Spatial relations for semantic similarity measurement. In: Akoka, J., Liddle, S., Song, I.Y., Bertolotto, M., Comyn-Wattiau, I., van den Heuvel, W.-J., Kolp, M., Trujillo, J., Kop, C., Mayr, H.C. (eds.) Perspectives in Conceptual Modeling. LNCS, vol. 3770, pp. 259–269. Springer, Heidelberg (2005)
5. Janowicz, K.: Sim-dl: Towards a semantic similarity measurement theory for the description logic $\mathcal{ALCNR}$ in geographic information retrieval. In: Meersman, R., Tari, Z., Herrero, P. (eds.) SeBGIS 2006, OTM Workshops 2006. LNCS, vol. 4278, pp. 1681–1692. Springer, Heidelberg (2006)
6. Baader, F., Calvanese, D., McGuinness, D.L., Nardi, D., Patel-Schneider, P.F. (eds.) The Description Logic Handbook: Theory, Implementation, and Applications. Cambridge University Press, Cambridge (2003)
7. Medin, D., Goldstone, R., Gentner, D.: Respects for Similarity. Psychological Review 100(2), 254–278 (1993)
8. Goodman, N.: Seven Strictures on Similarity. In: Goodman, N. (ed.) Problems and projects, Bobbs-Merrill, pp. 437–447. New York (1972)
9. Keßler, C.: Similarity Measurement in Context. In: Sixth International and Interdisciplinary Conference on Modeling and Using Context. LNCS (LNAI), pp. 277–290. Springer, Heidelberg (2007)

10. Tversky, A.: Features of Similarity. Psychological Review 84(4), 327–352 (1977)
11. Gaerdenfors, P.: Conceptual Spaces - The Geometry of Thought. Bradford Books, MIT Press, Cambridge (2000)
12. Janowicz, K., Raubal, M.: Affordance-based similarity measurement for entity types. In: 5th Conference on Spatial Information Theory (COSIT 2007). LNCS, vol. 4736, pp. 133–151. Springer, Heidelberg (2007)
13. d'Amato, C., Fanizzi, N., Esposito, F.: A dissimilarity measure for ALC concept descriptions. In: Proceedings of the 2006 ACM Symposium on Applied Computing (SAC), Dijon, France, pp. 1695–1699 (2006)
14. Borgida, A., Walsh, T., Hirsh, H.: Towards measuring similarity in description logics. In: Proceedings of the 2005 International Workshop on Description Logics (DL2005). CEUR Workshop Proceedings, CEUR, Edinburgh, Scotland, UK, vol. 147 (2005)
15. Li, B., Fonseca, F.: Tdd - a comprehensive model for qualitative spatial similarity assessment. Spatial Cognition and Computation 6(1), 31–62 (2006)
16. Nedas, K., Egenhofer, M.: Spatial similarity queries with logical operators. In: Hadzilacos, T., Manolopoulos, Y., Roddick, J.F., Theodoridis, Y. (eds.) SSTD 2003. LNCS, vol. 2750, pp. 430–448. Springer, Heidelberg (2003)
17. Bechhofer, S.: The dig description logic interface: Dig/1.1. In: DL2003 Workshop, Rome (2003)
18. Janowicz, K.: Similarity-based retrieval for geospatial semantic web services specified using the web service modeling language (wsml-core). In: Scharl, A., Tochtermann, K. (eds.) The Geospatial Web - How Geo-Browsers, Social Software and the Web 2.0 are Shaping the Network Society. LNCS, Springer, Heidelberg (2007)
19. Lutz, M., Klien, E.: Ontology-based retrieval of geographic information. Journal of Geographical Information Science 20(3), 233–260 (2006)
20. Probst, F., Espeter, M.: Spatial dimensionality as classification criterion for qualities. In: International Conference on Formal Ontology in Information Systems, Baltimore, Maryland, IOS Press, Amsterdam, Trento, Italy (2006)
21. Brandt, S., Küsters, R., Turhan, A.Y.: Approximation and difference in description logics. In: Fensel, D., Giunchiglia, F., McGuiness, D., Williams, M.A. (eds.) International Conference on Principles of Knowledge Representation and Reasoning (KR2002), pp. 203–214. Morgan Kaufman, San Francisco, CA (2002)
22. Molitor, R.: Structural Subsumption for ALN. LTCS-Report LTCS-98-03, LuFG Theoretical Computer Science, RWTH Aachen, Germany (1998)
23. Markman, A.: Structural alignment, similarity, and the internal structure of category representations. In: Similarity and Categorization, pp. 109–130. Oxford University Press, Oxford, UK (2001)
24. Knublauch, H., Fergerson, R., Noy, N., Musen, M.: The Protégé OWL Plugin: An Open Development Environment for Semantic Web Applications. In: Third International Semantic Web Conference, pp. 229–243 (2004)
25. van Assem, M., Menken, M., Schreiber, G., Wielemaker, J., Wielinga, B.: A method for converting thesauri to rdf/owl. In: McIlraith, S.A., Plexousakis, D., van Harmelen, F. (eds.) ISWC 2004. LNCS, vol. 3298, Springer, Heidelberg (2004)
26. Janée, G.: Rethinking gazetteers and interoperability. In: International Workshop on Digital Gazetteer Research & Practice, Santa Barbara, California, (December 7-9, 2006) (2006)
27. Poole, D., Smyth, C.: Type uncertainty in ontologically-grounded qualitative probabilistic matching. In: Godo, L. (ed.) ECSQARU 2005. LNCS (LNAI), vol. 3571, pp. 763–774. Springer, Heidelberg (2005)

A   Appendix

The appendix gives an overview about the involved similarity functions described in section 3.4; for a detailed description see [5]. The sets of tuples selected by the alignment matrix are represented by the letter S followed by an abbreviation for the type of constructor. For instance, SI is the set of concepts on union level of C where each C_i is formed by intersection.

sim_u is the weighted sum of similarities for all tuples (C_i, D_j). The weighting ω ($\sum \omega_{ij} = 1$) can be either determined by the count of tuples or by analyzing the ontological structure [5]. If the similarity of a particular tuple is 1, $sim_u = 1$.

$$sim_u(C, D) = \sum_{(C_i, D_j) \in SI} \omega_{ij} * sim_i(C_i, D_j) \tag{2}$$

Following the $\mathcal{ALCHQ}$ canonical normal form (see section 3.2), each C_i (respectively D_j) is an intersection of primitives or concepts formed by restrictions or quantifications. sim_i is the function that determines similarity on this level as normalized sum derived from the similarity functions for the involved constructors. The normalization factor σ is defined as the sum of cardinalities derived from the sets of compared tuples (SP, SE, SF, $SMIN$ and $SMAX$).

$sim_i(C, D) =$

$$\frac{1}{\sigma} \Bigg(\sum_{(A,B) \in SP} sim_p(A, B) + \sum_{(R,S) \in SE} sim_e(exists_R(C), exists_S(D))$$

$$+ \sum_{(R,S) \in SF} sim_f(forall_R(C), forall_S(D)) + \sum_{(R,S) \in SMIN} sim_m(min_R(C), min_S(D))$$

$$+ \sum_{(R,S) \in SMAX} sim_m(max_R(C), max_S(D)) \Bigg) \tag{3}$$

Primitives have no description that can be compared, hence an information theoretic approach (comparable to the Jaccard coefficient) is used to determine their similarity. Primitives are the more similar, the more complex concepts (within the context) are subsumed by both.

$$sim_p(A, B) = \frac{|\{C \mid C \sqsubseteq A) \sqcap (C \sqsubseteq B)\}|}{|\{C \mid C \sqsubseteq A) \sqcup (C \sqsubseteq B)\}|} \tag{4}$$

sim_e compares concepts formed by existential quantifications. The similarity is the product of role and filler similarity. The second sum (see sim_i) is necessary as there may be more than one existential quantification for the same role.

$$sim_e(exists_R(C), exists_S(D)) = sim_r(R, S) * \sum_{(C'_i, D'_j) \in SE} sim_u(C'_i), D'_j)) \tag{5}$$

sim_f compares concepts formed by value restriction. The similarity is the product of role and filler similarity.

$$sim_f(forall_R(C), forall_S(D)) = sim_r(R, S) * sim_u(forall_R(C), forall_S(D)) \tag{6}$$

The similarity (sim_m) between concepts formed by quantified number restrictions is the product of the similarities determined for the involved roles, fillers and their maximal or minimal occurrence (cardinality). sim_m is used as an abbreviation here, in fact minimum and maximum restrictions are handled separately (i.e. m is replaced by min respectively max). The normalization $m_{RS}(total)$ is the highest maximum (respectively minimum) restriction for R or S within the context. If one cardinality is explicitly set to 0 (while the other is not), $sim_m = 0$.

$$sim_m(mR(C), m_S(D)) = sim_r(R, S) * \left(1 - \frac{\mid m_R(C) - m_S(D) \mid}{m_{RS}(total)} \right) * sim_u(C_i'), D_j')) \quad (7)$$

The similarity between roles (sim_r) is their normalized distance within the hierarchy. The normalization is depth-dependent to indicate that the distance from node to node decreases with increasing depth of R and S within the hierarchy.

$$sim_r(R, S) = \frac{depth(lub(R, S))}{depth(lub(R, S)) + edge\_distance(R, S)} \quad (8)$$

If roles are not organized within a hierarchy but within a neighborhood, sim_n is used for comparison.

$$sim_n(R, S) = \frac{max\_distance_n - edge\_distance(R, S)}{max\_distance_n} \quad (9)$$

A Location and Action-Based Model for Route Descriptions

David Brosset, Christophe Claramunt, and Eric Saux

Naval Academy Research Institute
BP 600, 29240, Brest Naval, France
{brosset,claramunt,saux}@ecole-navale.fr

Abstract. Representing human spatial knowledge has long been a challenging research area. The objective of this paper is to model a route description of human navigation where verbal descriptions constitute the inputs of the modeling approach. We introduce a structural and logical model that applies graph principles to the representation of verbal route descriptions. The main assumption of this approach is that a route can be modeled as a path made of locations and actions, both being labeled by landmarks and spatial entities. This assumption is supported by previous studies and an experimentation made in natural environment that confirm the role of actions, landmarks and spatial entities in route descriptions. The modeling approach derives a logical and formal representation of a route description that facilitates the comprehension and analysis of its structural properties. It is supported by a graphic language, and illustrated by a preliminary prototype implementation applied to natural environments.

1   Introduction

Over the past years the study of human navigation has been the object of considerable research efforts [1,2,3]. This reflects a trend in modern sciences where human behaviours in the environment are studied as internal and external processes whose analysis should help to conceptualize and understand their semantics in space and time. This implies several research domains from cognitive to computer sciences, and where the objective is to better understand how people conceptualize the environment, and the way they act in it. Human navigation relies on a cognitive interpretation of a dynamic environment, and how it is perceived and interpreted in space and time. External representations entail how human beings conceptualize space and displacements in their environment. It has been shown that cognitive representations of human navigation rely on topological and qualitative abstractions that share some similarities with map representations [4]. Cognitive maps form a set of concepts that formalize such knowledge. Differences with map representations result from the nature of human conceptualizations which are imprecise and volatile per nature. Cognitive collages model the way humans derive a logical structure of a navigational space [5]. These mental representations are mainly qualitative, based on relations, rather

F. Fonseca, M.A. Rodríguez, and S. Levashkin (Eds.): GeoS 2007, LNCS 4853, pp. 146–159, 2007.

than on quantitative geographical information. Route descriptions, either verbal or graphic, reflect the spatial knowledge of a human being acting in the environment [6,7].

Wayfinding processes generally consider two different aspects of human navigation. The first one is route planning whose aim is to facilitate target-oriented navigation [8]. The second one addresses route modeling and the understanding of how people navigate in their environment [2]. Several models of navigation processes have been proposed in large-scale urban environments [5,9,10,11]. Route descriptions constitute modeling references for favouring human navigations. They have been applied to application contexts that cover large-scale urban environments [12], built environments [13], and urban undergrounds [14].

The aim of our research is to contribute to the modeling and representation of the knowledge and linguistic terms involved in a navigation process. Our analysis is developed at the structural level, and searches for the linguistic constructs and words used by humans when navigating. The proposed model is based on the three main components of a route description previously identified by Michon and Denis [15]: action, landmark and spatial entities. An action represents the displacement behaviour of a human acting in the environment. A landmark is the most salient feature used in human navigation [16]. Spatial entities denote two-dimensional entities on which moves are executed (e.g., a street) or non-salient and non-punctual entities used in navigation (e.g., a forest) [15]. The approach is first experimented in the context of a natural environment. Our research is developed from the case study of a foot orienteering race, a sporting activity involving navigation in natural landscapes. In a related work, the peculiarities of route descriptions produced in the context of foot orienteering races have been studied and qualified, and particularly the respective roles of landmarks and actions [17]. Experimental data come from a set of verbal descriptions of several orienteers who were asked to describe a part of their itinerary. The approach developed relies on the extraction and representation of the verbal constructs derived from route descriptions. This paper goes further by introducing a formal and structural representation of route descriptions also supported by a prototype implementation. The aim of our research is to provide a graph-based support of route descriptions, and where locations, landmarks and spatial entities will act as privileged primitives to facilitate derivation of GIS-based representations.

The remainder of this paper is organized as follows. Section 2 presents the motivation of our research. Section 3 introduces the principles of the modeling approach. Section 4 develops the prototype implementation developed so far, and an application to a case study in a natural environment. Finally, section 5 concludes the paper and outlines further work.

2   Research Background

2.1   Urban Environments

A better understanding and search for formal representations of human processes have led to several recent attempts where the objective is to characterise the linguistic or graphic constructs used in route descriptions [18]. Several approaches have been proposed, either based on visual, graphic or verbal constructs [19]. They tend to identify the basic primitives and constructs encompassed in route descriptions. In particular, these studies show that landmark, spatial entity, action and direction terms are amongst the predominant verbal forms identified in route descriptions [20,15].

Itinerary descriptions have been studied from a linguistic point of view where terms, verbs and basic constructs are identified as the main components [10,21,22]. Similarities and differences between verbal and schematic descriptions have been also studied [23]. Verbal structures show the predominant role of landmarks, spatial entities and actions [20,15]. Elementary scenes categorize the way landmarks and actions interact [18]. Elementary scenes make a difference between directed actions, located actions, integrated actions, referenced landmark, identified landmarks, located landmarks, and landmark descriptions. Landmarks are referenced by nouns, and most of the time qualified by adjectives, representing the most salient features of the environment along the route described [24], [25]. Landmarks are closely related to a kind of environment and depend on the perception and judgment of human beings. Landmarks can be classified into three categories: visual, cognitive, and structural [26]. The prominent role of landmarks in wayfinding and route descriptions have been already emphasized in urban systems [27,16].

2.2 Natural Environments

Although considerable attention has been given to the modeling of human navigation in urban environments, little work has been oriented, to the best of our knowledge, to natural contexts. In order to study to which degree the prominent roles of landmarks and actions also apply to natural environment, we have conducted an experimental study [17]. Foot orienteering has been chosen as an experimental context to support the analysis of wayfinding descriptions in natural environments. Foot orienteers have to visit a set of control places in a given order, and in a minimum of time. Control places are placed on features which are prominent in the environment, and specified on a "control description sheet" given to the orienteers. An orienteer generally has an accurate and detailed map in hands, and a compass to identify control places in the landscape. Our experiment was setup with fifteen experienced orienteers (12 men and 3 women) who were asked to remember and communicate their route at the end of their race.

This experiment confirms the role of landmarks used in orienteering races compared to orientation constructs and other spatial and temporal metrics. Actions are also significantly present, although in a smaller proportion than landmarks. It also appears that two-dimensional constructs are far more represented than three-dimensional constructs (76 % are two-dimensional constructs), and with a high proportion of landmarks (65 % of two-dimensional constructs). This provides a good example of the complementary roles of landmarks and actions in route descriptions, and supports the observed fact that two-dimensional constructs are easier to describe and memorize than three-dimensional constructs.

The figures of our experiment also show that orienteers mainly employ relative references. While absolute constructs are occasionally used in two-dimensional terms, they are never used in three-dimension terms.

3   Modeling Approach

3.1   Model Principles

The term model is hereafter used as a conceptual representation of a phenomenon. A conceptual model is a theoretical construct composed of a set of variables, and a set of relationships between them. It provides a framework for applying logic and mathematics that can be used for representing and reasoning over complex information systems. Such representations can also serve as a basis for simulation.

Actions, landmarks and spatial entities are the main primitives considered as the core elements of our modeling approach. When combined, actions, landmarks and spatial entities generate elementary navigation expressions, where landmarks and spatial entities are often associated with an action [15,20]. A landmark is commonly defined in navigation as a decision point, or assimilated to a decision point, and where decisions are taken [26,28]. A spatial entity models a two-dimensional entity on which moves are executed or a non-salient and non-punctual entity used in navigation. Our objective is to identify and integrate within our modeling approach the features that can be geo-referenced. Spatial entities and landmarks belong to this category. Actions expressed by verbs convey the dynamic component of a human navigation. They describe elementary displacements and can be schematized by a directed path between two locations. A navigation process can be modeled as a path in the sense of graph principles, where the nodes of the path represent locations, edges of the path actions between these locations. A primitive displacement is defined by an origin and an arrival and materializes a route segment. This allows us to model a route by an ordered sequence of route segments.

Actions, spatial entities and landmarks interact in different ways. Let us consider the following route description: "From the forest go to the bridge". This action is terminated by a landmark (i.e., the bridge) and started by a spatial entity (i.e., the forest) that cannot be considered as a landmark. Actions can be also associated to landmarks or spatial entities (e.g., "cross the bridge", "follow the watercourse", respectively).

In order to develop our modeling approach, we then consider the fact that a location can be a landmark or a spatial entity, and similarly that an action can be associated with a landmark or a spatial entity. In order to characterize these categories, we introduce two Boolean functions f, and g, that are true when respectively a location, or an action, are related to either a landmark or a spatial entity.

$$f : N \rightarrow \{0, 1\}$$
$$g : E \rightarrow \{0, 1\}$$

More formally, a route is modeled as an oriented graph $G(N, E, l, d)$ where N denotes the set of nodes, E the set of edges, l a function that associates a location to a node, and d a function that associates an action to an edge. Let L be the set of locations and A the set of actions.

A route description is modeled by an ordered set r of connected 3-tuples (p_i, a_i, p_{i+1}), named route segments, where p_i, $p_{i+1} \in L$ and $a_i \in A$. An elementary route segment is characterized and refined by the outputs of the functions f an g, applied respectively to p_i, p_{i+1} and a_i. Let S be the set of route segments, and h a function that characterizes a route segment, then:

$$h : \left| \begin{array}{l} S \qquad\qquad\qquad \to \{0,1\}^3 \\ h((p_i, a_i, p_{i+1})) \mapsto (f(p_i), g(a_i), f(p_{i+1})) \end{array} \right.$$

An action starts and terminates at a location, that might be either a landmark or a spatial entity. Similarly, an action might interact with either a landmark or a spatial entity during its execution. The instances of $h(S)$ are the elementary and orthogonal cases illustrated in table 1.

In order to give a visual component to the formalism that will be used at the interface level of our prototype, we provide a schematic representation of a path. This representation considers an elementary part of a route description, and supplies a Boolean-based and schematic view of a directed path between two locations (table 1). In order to derive a visual representation of a given path, modeled as an ordered sequence of location - action - location, the schematic language maps every 3-tuples of $h(S)$ to an equivalent graphic symbol. These elementary cases provide a complete set of orthogonal configurations. They outline the respective roles of landmarks, spatial entities and actions in route descriptions. Their sequential description can be used to exhaustively represent a route description.

3.2   Model Refinement

At a finer level of granularity, and this provides a multi-scale component to the approach, actions in the environment can be also qualitatively described by orientation and three-dimensional terms.

An action can be qualified by its cardinal or relative directions (e.g., go to the *north*, turn to the *right*), and its three-dimensional component whatever the way it is reflected by its linguistic representation (e.g., *climb* the hill, go to the *top* of the knoll). This information qualitatively refines the description of a given action, and integrates additional spatial relationships that characterize edges. When an orientation qualifies a displacement action, this corresponds to a displacement vector. More formally, let Or_R and Or_C respectively denote the set of relative orientation terms, and the set of cardinal terms. The union of these two sets gives the set of orientation directions Or. Note that the value *null* denotes here and in the following notations the absence of information, that is, no orientation information.

Table 1. Actions-landmark elementary cases

Id	Definition	Boolean representation	Graphic representation
α_0	An action that starts at a location and terminates at a location	$[0,0,0]$	○———▶○
α_1	An action that starts at a location and terminates at a landmark or a spatial entity	$[0,0,1]$	○———▶□
α_2	An action that starts at a location, qualified by a landmark or a spatial entity and terminates at a location	$[0,1,0]$	○—□▶○
α_3	An action that starts at a location, qualified by a landmark or a spatial entity and terminates at a landmark or a spatial entity	$[0,1,1]$	○—□▶□
α_4	An action that starts at a landmark or a spatial entity and terminates at a location	$[1,0,0]$	□———▶○
α_5	An action that starts at a landmark or a spatial entity and terminates at a landmark or a spatial entity	$[1,0,1]$	□———▶□
α_6	An action that starts at a landmark or a spatial entity, qualified by a landmark or a spatial entity and terminates at a location	$[1,1,0]$	□—□▶○
α_7	An action that starts at a landmark or a spatial entity, qualified by a landmark or a spatial entity and terminates at a landmark or a spatial entity	$[1,1,1]$	□—□▶□

$$Or_R = \{R, L, F, B\}$$
$$Or_C = \{N, S, E, W, NE, NW, SE, SW\}$$
$$Or\ = Or_R \cup Or_C \cup \{null\}$$

Although, pedestrian navigation is mostly considered as a two-dimensional process, human beings acting in natural environments integrate the third spatial dimension in the description of their displacement [17]. This information gives additional information on the landscape, and therefore improves the precision of route descriptions. Let V denote the set of three-dimensional constructs:

$V = \{+, -, =, null\}$ where the symbol + denotes an action upward and the symbol - an action downward whereas the symbol = an horizontal action.

Similarly, two functions are introduced to integrate orientation and three-dimensional constructs into the model. Let g^+ be a function that characterizes an edge as a landmark or not, and by its cardinal or relative directions and elevation value. Let h^+ be a function that characterizes a 3-tuple $[n_i, e_i, n_{i+1}]$ by the f values as applied to n_i, and n_{i+1} and the g^+ value as applied to e_i :

$g^+ : N \rightarrow g(E) \times Or \times V$ where E, Or, and V denote respectively the sets of edges, orientation terms and three-dimensional terms

$h^+ : N \times E \times N \rightarrow f(N) \times g^+(E) \times f(N)$

Orientation and three-dimensional constructs are integrated within the formal component of our modeling approach, and as additional symbols of the graphic language (cf. table 2).

Table 2. Orientation and elevation symbols

Orientation	Symbol	Definition
Cardinal orientation		North
		South
		North West
		North East
		South West
		South East
		West
		East
Relative orientation		Forward
		Backward
		Left
		Right
Elevation		Up
		Down

While qualitative terms give an additional component to route descriptions, landmark and spatial entity categories complement route descriptions. For each kind of environment and navigation, a set of landmark and spatial entity categories should be identified and ideally derived from an ontological or standard reference that apply to the environment considered. For instance, the vegetation category is important in natural navigation but not in urban navigation. These categories are modeled as follows. Let L_K be the set of spatial entity categories derived from the environment where the navigation occurs. Let k_l and k_a be two functions that respectively characterise a location and an action given by the true value of the functions f and g, that is, by the corresponding value of L_K:

$$k_l : N \rightarrow P(L_K)$$
$$k_a : E \rightarrow P(L_K)$$

4   Computational and Experimental Validations

The formal model and the graphic language have been implemented by an experimental prototype. Verbal descriptions give the input of the prototype interface, and support an interactive modeling of the routes provided by the orienteers. In the current version of the prototype, the translation task is processed manually as an automatic execution is far beyond the objective of our research. This section introduces the main principles and properties of the prototype, and some of the analysis supported by the model.

4.1   Prototype Principles

The Java language has been chosen as the software environment for the prototype development due to its portability and web compliance. Three frames form the prototype interface (cf. fig. 1). The main frame supports integration of route descriptions and derivation of model representations. The top part of this interface displays the textual route descriptions. Route descriptions support derivation of the model constructs at different levels of granularity, this being left to the user according to its objectives. Additional frames are used to label the actions (relative, cardinal, and elevation constructs) and to categorize each landmark.

Let us take the example of an illustrative path: "I started from the meadow to the north, climbed up the knoll, turn to right, crossed the river and went south to the forest". The first elementary route segment ("I started from the meadow to the north") is composed of an action that starts by a landmark ("the meadow") and ends with no precise data on the destination. The resulting route description at the node level gives an elementary path that contains six nodes and five edges, and where four landmarks are identified ("meadow","knoll", "river", "forest"). Figure 1 illustrates the representation of such a route description at the interface level of the prototype.

Within the context of foot orienteering, landmarks and spatial entities are qualified according to the symbols derived from the International Orienteering Federation (IOF). These symbols characterise the land features that are likely to have a specific role in wayfinding processes. The identified symbols are classified into five main categories: landforms, rock and boulders, water and marsh, vegetation and man-made features.

However, and although the IOF classification identifies different types of landmark and spatial entities in natural environments, these landmarks and spatial

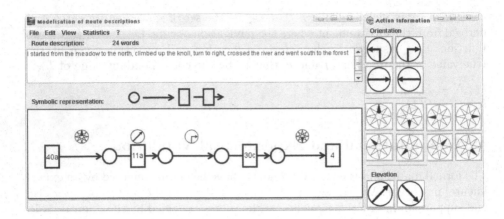

Fig. 1. From verbal to graph- and graphic-based descriptions

entities have been refined for the purpose of our study. For instance, the landmark referenced by the term "forest" corresponds to several landmarks in the IOF classification. This implies an integration of additional landmark types (e.g., 40a for "meadow"). A second problem comes from the fact that a a given category can be characterized by more than one symbol. For instance, the term "meadow" corresponds to either the spatial entity "open land" or "open land with scattered trees". Additional symbols are then defined in order to complement IOF categories.

4.2   Route Description Analysis

The modeling approach outlines the core structure of a given route that does not immediately emerge readily with verbal descriptions. Table 3 shows some examples of route descriptions that describe and model the same itinerary (d_0, d_1, d_2, d_3). These graph representations provide the main logical view of every itinerary description, although some ambiguities due to the interpretation of natural language expressions might remain. The first route description d_0 is derived from the example presented in figure 1 without direction terms. In contrast to the second verbal description d_1 which is relatively similar, the description d_0 starts by a spatial entity. This difference is important as the route description d_1 is not entirely bounded. Consequently, the starting point of the d_1 description is not precisely described, and the two routes are likely to have different lengths when interpreted. It is also worth to note that a given node in a route description is qualified by a landmark or a spatial entity when either its terminating or starting edge is qualified by a landmark or a spatial entity.

Another significant pattern to study concerns a quantitative comparison of graph structures. The descriptions d_0 and d_3 concern the same itinerary, however their model transcriptions have not the same number of nodes (five nodes for d_0 whereas d_3 has only three nodes). This clearly denotes the fact that d_0 has

Table 3. Descriptions and models: logical differences

Id	Description and model
d_0	I started from the meadow, climbed up the knoll, crossed the river and went to the forest.
d_1	I crossed the meadow, climbed up the knoll, crossed the river and went to the forest.
d_2	After climbing up the knoll, I went to the forest.
d_3	I climbed up the knoll after the meadow. Then I passed over the river to the forest.

a richer semantic description. On the contrary, the route descriptions d_2 and d_3 have a same number of nodes whereas the itinerary d_2 is shorter.

At a finer level of granularity, the orientation and elevation constructs used reveal several characteristics (cf. table 4). The first route description (d_{0+}) is the one presented in the figure 1. The structures of the model transcriptions d_{0+} and d_4 are equals, and their route descriptions are very similar. Nevertheless, the orientation and elevation terms used in these itinerary descriptions are largely different. On the contrary, d_{0+} and d_5 have several structural differences, and orientation and elevation similarities. This shows that some of the orienteers are relatively precise in describing the structure of their route (d_4), while others are more likely to qualify their actions using spatial relationships (d_5), some of them being precise in both respects (d_{0+}).

Table 4. Route descriptions: orientation and elevation constructs

Id	Description and model
d_{0+}	I started from the meadow to the north, climbed up the knoll, turn to the right, crossed the river and went South to the forest. $h^+(d_{0+}) = [\ [1, (0, N, null), 0], [0, (1, null, +), 0], [0, (0, R, null), 0],$ $[0, (1, null, null), 0], [0, (0, S, null), 1]\ ]$
d_4	I started from the meadow, went down the knoll, went straight, passed over the river and went North to the forest. $h^+(d_4) = [\ [1, (0, null, null), 0], [0, (1, null, -)0,], [0, (0, F, null), 0],$ $[0, (1, null, null), 0], [0, (0, N, null), 1]\ ]$
d_5	I went to the north, then I went up the knoll, turned to the right and went to the South. $h^+(d_5) = [\ [0, (0, N, null), 0], [0, (1, null, +), 0],$ $[0, (0, R, null), 0], [0, (0, S, null), 0]\ ]$

Additional statistics on route descriptions can be easily inferred from the model descriptions and prototype implementation. Figure 2 illustrates some emerging properties computed from our experimental study. First, the distribution of the different elements of the modeling approach (location, location with either a landmark or a spatial entity, action, and action interacting with either a landmark or a spatial entity) characterize relatively well the structure of a given route. Secondly, the relative importance of the landmarks and spatial entity identified shows the semantic richness of the route descriptions. The landmarks and spatial entities distribution within the path description also gives additional information on the homogeneity of the route descriptions. Finally, the different landmark and spatial entity categories present in the description characterize the environments as they appear in the route descriptions.

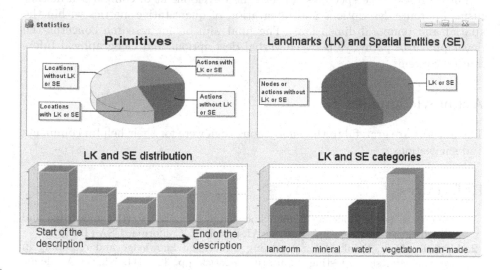

Fig. 2. Route descriptions analysis

Quantitative evaluations might be derived from additional comparisons of the length of route descriptions, and frequencies of the modeling primitives used. Qualitative and structural aspects can be derived from the study of different properties and patterns. Although left to further work, we plan to develop a range of graph-based similarity and structural measures that can support cross-comparison of route descriptions at the local (i.e., route segment) and global levels (i.e., itinerary).

5 Conclusions

The objective of the research presented in this paper was to extract and represent route knowledge and constructs provided by human verbal descriptions, and to develop a logical representation of route descriptions based on actions, landmarks and spatial entities. This is supported by the assumption that locations are often characterized as landmarks or spatial entities, and similarly that an action can be associated with a landmark. Routes are modeled as oriented graphs where nodes denote locations, and edges actions. A logical language and a schematic representation support the modeling approach. Route descriptions are represented at different levels of abstraction. At the lowest level, orientation constructs and spatial relationships complement the formal approach. These representations provide interactive descriptions, categorization and comparison of route descriptions where landmarks, spatial entities, actions, orientation and three-dimensional constructs provide different modeling levels of granularity. A preliminary prototype implementation illustrates the potential of the approach and its application to navigation in natural environment.

Future research perspectives concern the development of comparison mechanisms for route descriptions and integration of quantitative metrics within the spatial and temporal dimensions. The final aim of our research concerns the development of pathways between verbal descriptions and Geographical Information Systems.

Acknowledgments

The authors are grateful to the anonymous reviewers for their helpful comments and suggestions.

References

1. Kuipers, B. (ed.): Modeling human knowledge of routes: Partial knowledge and individual variation, pp. 216–219. AAAI, Stanford, California, USA (1983)
2. Montello, D.R., Frank, A.U.: Modeling directional knowledge and reasoning in environmental space: testing qualitative metrics, pp. 321–344. Kluwer Academic Publishers, Dordrecht (1996)
3. Allen, G.: Spatial Abilities, Cognitive Maps, and Wayfinding - Bases for individual Differences in Spatial Cognition and Behavior, pp. 45–80. Johns Hopkins University Press, Baltimore (1999)
4. Kuipers, B.: The map in the head metaphor. Environment and Behavior 14(2), 202–220 (1982)
5. Tversky, B.: Cognitive maps, cognitive collages, and spatial mental models. In: Campari, I., Frank, A.U. (eds.) COSIT 1993. LNCS, vol. 716, pp. 14–24. Springer, Heidelberg (1993)
6. Kuipers, B.: Modeling spatial knowledge. Cognitive Science 2, 129–153 (1978)
7. Golledge, R.G.: Path selection and route preference in human navigation: A progress report. In: Kuhn, W., Frank, A.U. (eds.) COSIT 1995. LNCS, vol. 988, pp. 207–222. Springer, Heidelberg (1995)
8. Allen, G.L.: From knowledge to words to wayfinding: Issues in the production and comprehension of route directions. In: Hirtle, S.C., Frank, A.U. (eds.) COSIT 1997. LNCS, vol. 1329, pp. 363–372. Springer, Heidelberg (1997)
9. Claramunt, C., Parent, C., Thériault, M.: An entity-relationship model for spatio-temporal processes. Data Mining and Reverse Eng., 455–475 (1997)
10. Couclelis, H.: Verbal directions for Wayfinding: Space, Cognition and Language, pp. 133–153. Kluwer Academic Publishers, Dordrecht (1996)
11. Fontaine, S., Edwards, G., Tversky, B., Denis, M.: Expert and non-expert knowledge of loosely structured environments. In: Cohn, A.G., Mark, D.M. (eds.) COSIT 2005. LNCS, vol. 3693, pp. 363–378. Springer, Heidelberg (2005)
12. Denis, M., Pazzaglia, F., Cornoldi, C., Bertolo, L.: Spatial discourse and navigation: An analysis of route directions in the city of Venice. Applied Cognitive Psychology 13, 145–174 (1999)

13. Raubal, M., Worboys, M.: A formal model of the process of wayfinding in built environments. [29] 381–399
14. Fontaine, S.: Spatial cognition and the processing of verticality in underground environments. [30] 387–399
15. Michon, P.E., Denis, M.: When and why are visual landmarks used in giving directions? [30] 292–305
16. Weissensteiner, E., Winter, S.: Landmarks in the Communication of Route Directions. In: Egenhofer, M.J., Freksa, C., Miller, H.J. (eds.) GIScience 2004. LNCS, vol. 3234, pp. 313–326. Springer, Heidelberg (2004)
17. Brosset, D., Claramunt, C., Saux, E.: Wayfinding in natural and urban environments: a comparative study, accepted, Cartographica, Toronto University Press (2006)
18. Przytula-Machrouh, E., Ligozat, G., Denis, M.: Vers des ontologies transmodales pour la description d'itinéraires. le concept de scène élémentaire. Revue internationale de Géomatique, Hermès, Paris 14(2), 285–302 (2004)
19. Tversky, B., Lee, P.U.: Pictorial and verbal tools for conveying routes. [29] 51–64
20. Denis, M.: The description of routes: A cognitive approach to the production of spatial discourse. Current Psychology of Cognition 16, 409–458 (1997)
21. Gryl, A.: Analyse et modélisation des processus discursifs mis en oeuvre dans la description d'itinéraires. Unpublished PhD report, Université de Paris XI Orsay, France (1995)
22. Fraczak, L.: Description d'itinéraire: de la référence au texte. Unpublished PhD report, Université de Paris-Sud, France (1998)
23. Przytula-Machrouh, E.: Information verbale et information graphique pour la description d'itinéraires. Unpublished PhD report, Université René Descartes, France (2004)
24. Yeh, E., Kriegman, D.: Toward selecting and recognizing natural landmarks (1995)
25. Nothegger, C.: Automatic selection of landmarks. Unpublished PhD report, Vienna Technical University, Austria (2003)
26. Sorrows, M.E., Hirtle, S.C.: The nature of landmarks for real and electronic spaces. [29] 37–50
27. Raubal, M., Winter, S.: Enriching wayfinding instructions with local landmarks. In: Egenhofer, M.J., Mark, D.M. (eds.) GIScience 2002. LNCS, vol. 2478, pp. 243–259. Springer, Heidelberg (2002)
28. Golledge, R.G.: In: Human Wayfinding and Cognitive Maps, pp. 5–45. Johns Hopkins University Press, Baltimore (1999)
29. Freksa, C., Mark, D.M. (eds.): COSIT 1999. LNCS, vol. 1661. Springer, Heidelberg (1999)
30. Montello, D.R. (ed.): COSIT 2001. LNCS, vol. 2205. Springer, Heidelberg (2001)

Spatio-temporal Conceptual Schema Development for Wide-Area Sensor Networks

Mallikarjun Shankar[1], Alexandre Sorokine[1], Budhendra Bhaduri[1],
David Resseguie[1], Shashi Shekhar[2], and Jin Soung Yoo[2]

[1] Oak Ridge National Laboratory
PO Box 2008 MS 6017 Oak Ridge, TN 37831-6017
{shankarm,sorokina,bhaduribl,resseguiedr}@ornl.gov
[2] Department of Computer Science and Engineering, University of Minnesota
200 Union ST SE 4192S Minneapolis, MN 55414
{shekhar,jyoo}@cs.umn.edu

Abstract. A Wide-Area Sensor Network (WASN) is a collection of heterogeneous sensor networks and data repositories spread over a wide geographic area. The diversity of sensor types and the regional differences over which WASNs operate result in semantic interoperability mismatches among sensor data, and a difficulty in agreeing on methods for sensor data access and exchange. We assume that sensors and their associated data have an explicit spatio-temporal basis (or tagging) in their representation. In this paper, we describe a spatio-temporal loosely-coupled federated database model for the WASN data storage problem - that of unifying query and data representation given a heterogeneous WASN - and propose a conceptual schema to ease the problem of integration of sensor data representations. This is a continuing and critical challenge as sensor networks become more ubiquitous and data interoperation becomes increasing vital for a variety of applications (such as homeland security, transportation, environmental monitoring, etc.). We employ a top-down ontology-driven software development methodology. We use the SNAP/SPAN ontology as a sample framework for the conceptual schema. We compare our methodology of conceptual schema development with a bottom-up entity-oriented schema construction and discuss the differences in the two approaches. A unique contribution is the discussion of deployment experiences to evaluate proposed approaches in the context of a concrete WASN testbed.

1 Introduction

We address the problem of unifying sensor data representation (e.g., for queries and storage) in Wide-Area Sensor Networks (WASNs). A WASN is a collection of heterogeneous sensor networks and data repositories spread over a wide geographic area (e.g., a deployment across one or more states). We aim to find commonality in the representation of the data records and their semantic description given the heterogeneity of the constituent components. We observe that a majority of WASN-related data is explicitly or implicitly spatio-temporally tagged.

F. Fonseca, M.A. Rodríguez, and S. Levashkin (Eds.): GeoS 2007, LNCS 4853, pp. 160–176, 2007.

In many cases, it is easy to argue for an explicit spatial and temporal context for the WASN data, since the location and time-instant of data measurements directly affect the command and control operations within the WASN. Example domains for such sensor networks include transportation networks and CBRNE (Chemical-Biological-Radiation-Nuclear-Explosive, [1]) detection networks. Sensor networks have been a focus of much research in recent years [2,3]. Much of the research targets resource-constrained wireless sensor networks. WASNs, by contrast, are often not resource constrained. Although the concerns of power efficient computing, networking, and data access are relevant to WASNs near the edges of the network, we address the complementary requirement for WASNs: that they cover significantly larger geographic areas and incorporate a wide range of sensor and application types.

The systematic representation and storage of the spatio-temporal data related to WASNs presents difficulties because such networks typically support a large number of heterogeneous sensors (types and manufacturers) and consist of several autonomous regional domains carrying out sensor data collection and sensor actuation. The diversity of types and the distances over which the network operates results in semantic interoperability mismatches among sensor data and a difficulty in agreeing upon sensor data access and exchange formats and protocols. For example, data from the same type of sensor but from different manufacturers may disagree in data formats, or may be implemented in a way that the database stores the records differently. The difficulty arises in building common structures for data transport, storage, and access that can apply to all or the majority of the individual domains. The problem is exacerbated by the spatio-temporal representation of the data-records which present challenges in treating characteristics (like location and movement) systematically. While we recognize that it will not generally be reasonable to fix or mandate one schema across geographically disparate deployments or sensor modalities, we believe that it is still a reasonable expectation to have a common baseline framework to represent and serve sensor data. Experiences constructing components of a wide-area sensor network for the real-time collection and integration of sensor data [4,5] show that a critical prerequisite to offering sensor data to a wide variety of applications (a model or software component that is a customer of the sensor data as well as a potential actuator of the sensor) is uniformity of representation and storage. Analogous to the interoperable Internet (for accepted access modes such as http-based communication), data providers can participate in the sensor network operation, regardless of the nature of the data source, or of the application model, if they comply with recommended guidelines for data representation and access. Here, we employ a methodology for sensor data characterization that uses ontological principles to guide the definition of the entity categories. Specifically, we use an ontology-driven software development methodology [6] building upon the SNAP/SPAN [7] upper-level spatio-temporal ontology as a foundation for the basic conceptual schema. The ontology-driven design methodology results in a desired common framework, and enables us to construct the sensor network's data plane generically. For purposes of this paper we refer to the ontology-driven approach as *top-down* in contrast to an entity-oriented approach which we call *bottom-up*.

1.1 Contributions

We make the following contributions. First, we discuss integration concerns in WASNs from the position of the well-studied area of federated databases, and propose a spatio-temporal loosely-coupled federated database model for WASN sensor data storage. Second, we design a high-level conceptual schema for achieving flexible integration of sensor data representations. The conceptual schema we offer, based on a geospatial ontology, is capable of addressing the spatio-temporal aspects of WASN data collection. Third, to evaluate our ontology-driven methodology of conceptual schema design, we qualitatively compare two conceptual schemas developed over the course of the SensorNet architecture design process, and discuss their strengths and weaknesses. One approach is a bottom-up entity-oriented modeling approach that takes advantage of GML [8] concepts and the other is a top-down approach derived from the ontology-driven approach. We find that although the ontology-driven conceptual schema offers the richer semantics that one may need, we do not yet address the implementation gap that conventional entity-oriented methodologies fill effectively.

We consider our deployment experiences in SensorNet [5], a WASN testbed to enable plug-n-play sensors and applications. SensorNet incorporates several standardized techniques for ingesting and disseminating data including standards from the Open Geospatial Consortium (OGC) suite of specifications [8,9,10,11] and the IEEE [12]. These specifications give us a unique and concrete foundation to investigate the state-of-the-art and state-of-the-practice in creating interoperable WASN data models.

1.2 Scope and Outline

This paper addresses the problem of schema integration in the context of WASN data interoperation. We relate our work to conceptual schema construction and work in ontology-driven software development(e.g., [6]) - our goal here is to compare two approaches to the schema construction in the specific domain of WASN. The connection between generic schema construction for wide-area sensor networks has not received very much attention in the literature. While they do not directly address conceptual schema construction, IrisNet [13] is an earlier effort where the focus is a querying mechanism acting on an Extensible Markup Language (XML) based distributed data collection and transmission framework, and, more recently, the Global Sensor Network effort [14] aims to use a virtual sensor abstraction to make sensors accessible uniformly in a middleware. We make references through the paper to related work and approaches we build upon in conceptual schema and ontology-driven software development. More general questions of schema integration are beyond the scope of the paper as are other practical challenges that are rooted in the distributed nature of WASN data such as, for example, the security aspects of access. The remainder of the paper is organized as follows. Section 2 details general WASN architecture and discusses the problem of data integration in WASN. Section 3 presents our ontology-driven approach to conceptual schema design. In Section 4, we qualitatively compare the developed top-down and the bottom-up schemas. Section 5 contains a summary and avenues for future work.

2 WASN Architecture and Data Integration

In this section, we present general WASN architecture considerations and discuss the data interoperation problem.

2.1 Architecture and Operation

The WASN architecture [5] accomplishes distributed data dissemination from a large variety of sensors using different types of connectivity and data acquisition modes. A regional instantiation of a WASN consists of four primary entities: sensors (and actuators), aggregation nodes, regional data centers, and user applications (Figure 1).

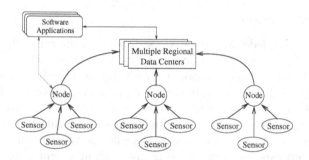

Fig. 1. WASN Architecture

The SensorNet Node is a device which collects data from various kinds of sensors and actuators, provides intermediate storage for the collected data, and transmits the data either to a regional center for permanent storage or a user application for analysis. The Node is built around an embedded PC-compatible computer that runs Linux OS and a Java-based server that implements the Open Geospatial Consortium (OGC) Web Feature Service (WFS) protocol [10]. (WFS was chosen because it was an accepted standard that supported spatio-temporal data transport and querying. Until recently, WFS alone had these characteristics.) Communication with the sensors takes place through a variety of means such as RS232 ports, USB or IEEE 1451 enabled devices [15]. The node is equipped with multiple standard wired and wireless communication devices such as modems and network ports. Nodes can be configured either as stationary platforms installed in the environment or compact transportable devices [16].

The WASN node supports several types of sensors (chemical, radiological, biological, weather and etc.). The Node data processing functions include:

- connecting to sensors,
- collecting data from sensors,
- issuing commands to sensor and actuators,
- preprocessing collected data to perform validation and calibration of the sensor readings,
- tagging the data with location coordinates and time stamps, and
- issuing alerts when a threshold is exceeded.

Applications access data from the regional data centers that act as hubs of the regional federate. Regional data centers thus provide permanent storage and retrieval capabilities for the collected data. Communications between software applications, nodes and regional centers on the WASN are performed using the WFS protocol for data insertions and retrievals. Entities within the message are described as OGC features. An OGC feature is a set of property-value pairs, one of whose properties is an OGC geometry with a geospatial reference [17]. Other feature properties carry information such as time stamps and measurements. Each feature has a feature type description associated with it.

2.2   Data Integration

A WASN imposes a federated database structure in the sense that, on the one hand, it provides unified access to a multitude of sensor data sources while, on the other hand, it allows regional autonomy on the data. For example, a WASN does not define a concrete set of regional centers or specific models of sensor devices that are needed to participate in the WASN. Such an approach allows for great flexibility and extensibility of the information system. It also minimizes the cost of connecting new and legacy sensor networks to a growing WASN. However, this approach brings many challenges typical to any federated and ad-hoc expanding data system.

To address the problems of unifying query and record structure and preserving the semantic integrity of the data, we first discuss database schema transformation and mapping in the context of a WASN. The database schema is a description of the data managed by the database system. Database schemata are composed of descriptions of the databases objects (tables, classes, types, etc.) and the relations between these objects. Schemata may be associated with vocabularies and ontologies to maintain relevant semantic information. In federated databases, compatibility between record structures and query languages is achieved by transforming or mapping component database schemata into a single federated schema and by converting the corresponding data as shown in Figure 2. In loosely-coupled federated databases, component database schemata typically develop independently of the federated schema and the data has to be converted according to several database schemata that play different roles in the federation. Component databases of a federated database system (FDBS) typically store their data using internal schemata that are optimized for the particular

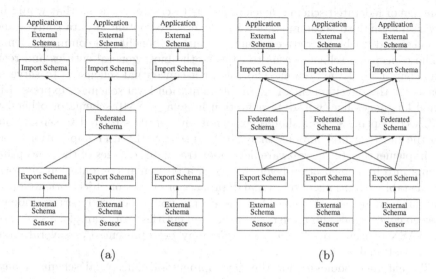

(a)   (b)

Fig. 2. Schema Translation in WASN

purpose a component database was developed for. Component databases expose their data using external schemata that are a subset of the internal schema and that contain the elements that are relevant for the FDBS.

In a WASN, the role of component databases on the data production side is played by sensors and partly by the WASN nodes. The internal schema of a sensor is either based upon a sensor standard (e.g., the IEEE 1451 Transducer Electronic Data Sheet (TEDS)) or defined during the WASN node setup. A sensor schema may contain data that is not necessarily pertinent to the overall operation of the WASN such as the calibration or sensor health information. Such data is omitted from the external sensor-level schema. On the WASN node, the data is tagged with a spatio-temporal reference that also becomes a part of the external sensor schema. An external schema of a sensor may be specific to that sensor. For a federation of a large number of sensors, the sensor data has to be translated into some common federated schema ("Federated Schema" in Figure 2(a)). We do not preclude multifederation with several federated schemata (Figure 2(b)) but do not address it here for space reasons.

For an application to be able to retrieve WASN data, it must be able to translate the WASN federated schema into its own external schema. In most cases a user will need to develop an import schema that is compatible with the application external schema which represents a subset of the WASN federated schema as shown in Figure 2(a).

3   Conceptual Schema Design

Due to the potentially unlimited variety of applications that can utilize WASN data and the many types of sensors accessible through the WASN, the main challenge of federated schema development is to create a schema that would be universal enough to accommodate a large variety of potential data sources and data uses and yet be restrictive enough to gain the benefit of a common schema.

The conceptual schema needs to capture the most general entities and relations contained in the system. The role of a conceptual schema is to serve as a basis for the development of the implementation-level schemata expressed in formal languages like SQL data definition language, XML schema or others. A WASN conceptual schema should cover not only entities related to sensors but also the most general notions that can be used by the user applications. Before implementation, the software developer translates entities of the conceptual schema into implementation-level objects of a particular computation platform such as table definitions of relational databases or classes of object-oriented programming languages. We note that there may not be a direct correspondence between conceptual- and implementation-level entities. For example, a sensor on the WASN conceptual schema (Figure 5) may be represented as several inter-linked relational tables.

Different methodologies for the development of conceptual schemata have been discussed in the literature [18,19]. In this study we present two approaches for WASN conceptual schema development in the SensorNet project. One approach is a bottom-up approach based on entities, and the other is a top-down approach using the upper-level SNAP/SPAN ontology.

3.1   Bottom-Up Conceptual Schema

Figure 3 shows the structure for the bottom-up schema [20]. The data interoperation is accomplished through WFS clients located on the WASN nodes and WFS servers at the data centers. For the communications inside the WASN that we consider here, the data is encoded in a GML application schema and each record is represented as an OGC feature tagged with a spatial location and a time stamp.

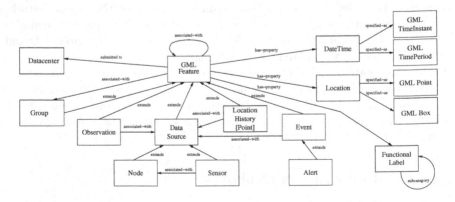

Fig. 3. Bottom-Up SensorNet Concept Map

The central entity of the SensorNet schema is the GML Feature inherited from the GML schema that is a set of property-value pairs, some of whose values represent geometry. Each Feature and its descendants has properties to characterize its spatial and temporal locations. To represent locations, the schema incorporates a limited subset of GML geometries including GML Points, TimeInstants and bounding boxes. Although this set of geometries can potentially be extended to other GML geometry types, the types used satisfy most of the requirements for storage and retrieval of the data in the WASN and limiting to these types significantly simplifies implementation.

Data centers, Observations, Events, Alerts and Data Source all extend the GML Feature and inherit its properties. An Observation is used to store measurements performed by the Sensors. Nodes and Sensors represent Data Sources that can be queried or instructed to perform Observations. Each Observation is always associated with a Data Source. In the WASN architecture, locations are associated with the Nodes which are equipped with global positioning system (GPS) devices (a Node inherits its Location property from the GML feature). After a measurement has been taken, its geographic coordinates are stored in the Node database as a GML Point. Sensor measurements are stored within Observations as XML records whose structure is defined by sensor-specific Data Type Definitions (DTDs). Events and Alerts are intended to communicate information such as sensor observations exceeding certain values. Events represent changes notable from the SensorNet operation point of view, e.g., low battery reading, communication channel failures. Alerts are a narrower category of Events intended for SensorNet users. Alerts are issued in cases of, for example, radiation sensor readings that exceed a certain threshold. Each Alert and Event is associated with a Node and Sensor that it has originated from. Location History is used to represent movements of Sensors and Nodes. Each record of the Location History type describes a point and associated time stamp for the location of a Sensor or a Node. It can be retrieved from a WFS just like any other Feature.

The approach for developing the schema shown in Figure 3 is based on the GML specification and accumulated practical knowledge of developing sensors and sensor platforms. The schema evolved in parallel with the architecture prototype being built. Additional entities (for example, Location History) are added to the schema in order to accommodate the needs of the prototype development and continually translated to an XML structure using standard ER-Modeling based translation.

3.2   Ontology-Driven, Top-Down Approach

The breadth of the potential application area of the conceptual schema necessitates the use of a systematic methodology for the study of conceptualization. To develop such a schema we investigate the conceptualization of the WASN domain area. Our goal is to produce a conceptual schema that later would serve as a foundation for implementing a WASN federated schema following the approach called "ontology-driven development" as proposed in [6]. We build a conceptual schema using a formal ontology[1] as a tool for comprehending the conceptualization of the WASN domain.

We express our domain conceptualization (or domain ontology) as a repertoire of entities and relations that are necessary to communicate the concepts (as opposed to the artifacts). Examples of such entities within the WASN domain are sensors, measurements, classes of sensors, etc. Relations are defined between entities. Examples of relations in the WASN context would be an association between a measurement and a sensor, or a "kind-of" relation between a specific sensor and a class of the sensor device that it belongs to. To create the catalog of entities, we use a methodology similar to that described in [19]. We study the various narratives describing sensor data acquisition and use. To create a list of entities, we identify nouns and noun phrases specific to the sensor domain. Then we clean the list of synonyms and entities with low importance in the context of a WASN. We obtain the list of relations by analyzing association between the entities and also using general expertise in the domains of sensors and geographic data. After collecting a catalog of entities and relations, the next step we take in the development of the domain ontology is to classify the domain entities according to the broadest top-level ontological categories. Development of an exhaustive but meaningful classification of all entities of some domain would be a significant challenge, and consequently we rely on a sample top-level ontology[2] discussed in the philosophical literature and described in [7].

SNAP/SPAN Ontology. Here we use the top-level categories of the SNAP/SPAN ontology (Figure 4, [7, page 74, modified and highly simplified]) to build a conceptual schema for the WASN - shown in in Figure 5. This ontology subdivides all entities along class/individual and SNAP/SPAN dimensions (which we describe below) and has advantages such as independence from any particular application domain in the WASN and thus has an intrinsic ability to account for the dynamic nature of the reality. It also has an extensive formalization in first-order predicate logic [31]. The choice of SNAP/SPAN ontology was largely influenced by its explicit support of dynamic entities.

Support for dynamics is critical for WASNs because monitoring change and detection of processes is a goal of any WASN operation. The SNAP/SPAN ontology framework that we adopt here has previously been successfully applied to the design of spatio-temporal applications [32,33,34], analysis of the conceptualization in hydrologic models [26], and the solving of semantic interoperability problems for CAD/GIS data models [35].

---

[1] Formal ontology is a study of the basic constituents of reality and is deeply rooted in philosophy [21]. Here we use the notion of ontology as an explicit specification of the conceptualization [22]. Ontological methodologies have proved their efficiency in other domains including medical information systems [23,24] and biological classifications [25].

[2] The use of top-level classifications is common in ontology development [26,27,28]. Examples of systems of top-level categories of entities are IEEE SUMO ontology or John Sowa's upper-level ontology [29,30].

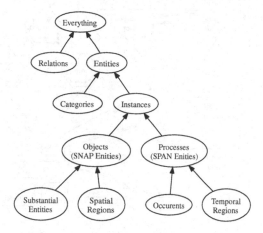

Fig. 4. Top-level Ontological Categories

Top-level ontological categories are systematized as a proper tree with each domain entity instantiating one and only one terminal (leaf-level) category (Figure 4). At the root of the category hierarchy, all entities are subdivided into classes and individuals. Classes represent concepts, while individuals are entities that have specific spatial and/or temporal locations [36]. For example, "temperature sensors" is a class and a particular temperature sensor outside a building is an example of an individual.

The central dichotomy of the SNAP/SPAN ontology in the branch of individuals is the subdivision of all individuals into two non-overlapping classes of entities — SNAP (or object-like) entities and SPAN (or process-like entities) as shown in Figure 4. Occasionally, in the literature, these SNAP and SPAN entities are referred to as endurants and perdurants respectively. SNAP entities are capable of preserving their identity through time. Examples of SNAP entities are humans, planets, cities, crevices, etc. Examples of SNAP entities in the WASN domain are sensors, actuators, nodes or service personnel. By contrast, SPAN entities unfold in time and only a part of a SPAN entity exists at each moment of time. (A perdurant is always dependent upon an endurant, for example, in the pair $(you, your\_life)$ you is an endurant and $your\_life$ is a perdurant. $your\_life$ has such parts as childhood, youth, adulthood and old age. At any point of time only one part of your life exists. $your\_life$ is dependent on you in the sense that it does not make sense to discuss $your\_life$ without referring to you and there is nothing like $your\_life$ if you does not exist.) Processes and events (instant occurrences) are kinds of SPAN entities. Other examples of SPAN entities are communication sessions, an alert event (as opposed to an alert message) and many others.

WASN Ontology-Driven Conceptual Schema. The central notion (or entity or concept) of the WASN conceptual model is a $sensor$[3], which is a particular

---

[3] $Sensors$ can be equipped with $actuators$ that control $sensor$ position, orientation, etc.

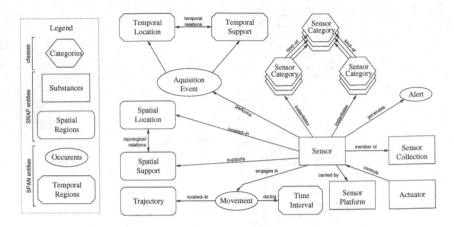

Fig. 5. WASN Ontology-driven Conceptual Schema

device capable of performing measurements of some properties of the environment. To house *sensors*, we require *sensor platforms*. *Sensor platforms* can be stationary, like poles or wall brackets, or mobile. A single *sensor platform* can carry several *sensors*. In this case, *sensors* are said to be arranged into *sensor collections* that share the same *platform*. An example of a *sensor collection* is a WASN node. Individual entities of the conceptual schema correspond to the substantial entities of the top-level ontological categories in Figure 4. Each *sensor, sensor collection, platform* and *actuator* has a corresponding record in the database of a regional center it belongs to. Each of these entities is first registered with a WASN node. Then the node transmits the descriptive data to a regional center from where it becomes available to the WASN users. The main purpose of these records is to support inventory and discovery of the available equipment that can be used for data collection.

Sensor categories are another facility for WASN resource discovery. They provide users with a class view over available sensors and other resources. *Sensor categories* represent types and classes of sensors and are organized strictly hierarchically. There are several hierarchies intended to capture various dimensions of the category space of *sensors*. For instance, there can be hierarchies created by the type of measured property or measurement technology. A measured property hierarchy is used to query the different kinds of measurements available on the WASN. Another hierarchy can be set by the manufacturer to be used for hardware maintenance tasks. Each sensor belongs to at most one category in a single hierarchy but it can be represented in multiple hierarchies. *Sensor categories* also play another important role by supporting automatic aggregation of the data collected from *sensors*. For example, sensors capable of measuring temperature can be made by different manufacturers. The WASN cannot impose the requirement to use a specific sensor model of a sensor produced by a specific manufacturer. However, the data collected by the models produced by different manufacturers are likely to be comparable and should be visible to the end-user application as a single aggregated dataset of temperature measurements. Such aggregations can be performed along the specialized sensor category hierarchy. *Sensor categories* are Classes in the top-level ontological categories hierarchy (Figure 5). The system of categories can be extended to entities other than *sensors*.

A *sensor* position within the geographic space is represented through two kinds of locational entities: *spatial location* and *spatial support*. The *spatial location* of a *sensor* is a point on the Earth surface where a *sensor platform* is found at the moment when a *sensor* is active and performs measurements. Static *platforms* have fixed *locations* while *locations* for movable and mobile *platforms* can change. In the WASN, a *sensor location* is determined on the basis of the GPS coordinate of a WASN node. A *sensor* is said to have a *support*, that is, a spatial region that is characterized by the sensor. In most cases a *support* is different from a sensor *location*. For example, a pan-tilt-zoom camera can be installed on a poll with known geographic coordinates. The area that this camera captures is the *support* of this *sensor*. Another example is a temperature sensor whose *support location* is limited to the room in which the sensor is installed. Sensor location and sensor support can be deduced from the properties of the *sensor* itself and the *sensor platform*.

Spatial location and *spatial support* belong to the category of "spatial regions" of the top-level hierarchy (Figure 4). In the WASN database they are recorded as GML Feature Geometry objects [8]. For automatic inferencing, relations between location and support can be expressed using an Egenhofer nine intersection model or region connectedness calculus [37,38].

Each occurrence of a measurement performed by a *sensor* is called an *acquisition event*. An *acquisition event* is a SPAN entity in the top-level hierarchy in Figure 4. Records of *acquisition events* comprise the bulk of data stored in the WASN and are the major target of WASN queries. Each *acquisition event* is associated with the *sensor location* and *support* at the moment that the measurement was performed. As a SPAN entity, an *acquisition event* has a *temporal location* and a *temporal support*, both representing temporal regions from the top-level category hierarchy (Figure 4). *Temporal support* plays the same role in relation to *temporal location* as *spatial support* in relation to *spatial location*, that is, a temporal region representing the time period for which the result of the measurement remains valid. An *acquisition event temporal location* is not necessarily a time instant and may have more than a zero time length.

A *temporal support* is typically confined within a *temporal location*. For example, a camera usually needs some time to adjust to lighting conditions after being switched on. The first few frames typically contain erratic data and will be a part of the *acquisition event temporal location* but not of the *temporal support*.

Movement of the sensor is represented by a special SPAN entity called *Movement*. *Movement* is intended to capture the constantly or intermittently changing location of a *sensor*. *Movement* has two entities associated with it: *Trajectory* is a spatial region to represent trajectory of the movement, and *Time Interval* represents a temporal region during which the movement has occurred.

4   Comparison of Conceptual Schema Constructions

The schemata in Figure 3 and Figure 5 were developed for the same domain but on different premises and they exhibit different sets of entities and relations.

The biggest difference between the schemata is the direct use of the well-defined GML Feature in the bottom-up schema (Figure 3). Most other entities of the schema extend (i.e., inherit their properties from) the GML Feature that enables them to have such properties as Functional Labels and spatial and temporal locations. In terms of object-oriented programming (OOP), the GML Feature represents a superclass of other classes in the schema. The benefit of structuring the schema in this bottom-up way allows implementors to translate the schema to a WFS engine or a database mapping directly. The ontology-driven schema (Figure 5) does not use the extend relation in its OOP sense, leaving that to more implementation-level schemata, thus gaining semantic flexibility for some ambiguity in implementation. As may be expected, the goal of allowing a common semantic structure to describe diverse application concepts comes at the cost of not providing specific mappings for implementors. If WASN architects recommend a specific implementation (e.g., WFS or a web-service implementation) along with an advertised vocabulary, the ontology-driven approach provides semantic commonality (and a top-down systematic structure to build upon) and enables implementation interoperability.

4.1   Representation of WASN Infrastructure

Representation of the WASN infrastructure in the two schemata have important differences stemming from the central component used for the modeling. The GML Feature element provides a common glue that does not easily enable the model to capture the semantic interdependence of the other entities. For the edges in the graph in Figure 3, the named relationships start resembling a part of the notions of a Sensor and Observation (Acquisition Event in Figure 5). The benefit of the latter of course is that the modeler gravitates to these representations at the outset instead of in retrospect. As an example, the SensorNet Node present in the bottom-up schema corresponds to the Sensor Collection in the ontology-driven schema. However, the notion of Sensor Collection is a more general concept and easily extends to other groupings of sensors such as actuators. While an expert designer would inject a systematic structure at the beginning in Figure 3, our claim is that the approach illustrated in the top-down ontology-driven approach orients the designer's mindset to allow systematic model construction.

4.2   Representation of Space and Time

There are significant differences in the models of space and time used in the schemata. The subset of GML geometric primitives that includes Point, Box, TimeInstant and TimePeriod are specified first as building blocks. It takes designer ingenuity to collect and apply them for a WASN use case. In contrast, the ontology-driven schema orients itself to apply to the domain of interest and the intended meaning. In addition to Sensor Spatial and Temporal Location (Acquisition Time), it introduces the notion of Support, i.e., a spatial or temporal region that is characterized by the sensor. Emphasis on user-level queries and applications in the ontology-driven schema necessitates introduction of such entities as Spatial and Temporal Support. The top-down representation of space and time allows designers to easily include more complex notions of temporal phenomena and their relationship to the spatial components. For example, topological configurations may be described (and discussed) as first-class objects. It is harder to instantiate this structure in the bottom-up model because aggregations of constituent components requires additional entities (e.g., a trajectory as a collection of points).

4.3   Representation of Motion and Dynamics

The schemata take different approaches towards representation of movement and dynamics. In the bottom-up schema, Location History is an add-on entity that stores coordinates and time stamps of sensor locations. The ontology-driven schema uses a domain-aware, and therefore directly applicable entity called Movement that is associated with spatial and temporal regions. Relating *Movement* to an implementation helps a designer in bringing to the surface the true intention of the entity, as opposed to using a construction such as Location History.

5   Summary and Future Work

This study addresses the problem of semantic heterogeneity and semantic interoperability of data collected in WASNs. The study presents an entity-oriented (bottom-up) and an ontology-driven (top-down) methodology for conceptual schema construction and illustrates a set of differences. The top-down conceptual schema offers the potential for better interoperability with user-level software applications because user semantics are taken into account early in the development process. The interoperability advantage comes with the higher level of independence of the conceptual schema from a computing platform because it relies on a very general conceptual framework. The schema can ultimately be implemented using the technologies that the bottom-up approach uses. The bottom-up approach has the benefit of being closer to the implementation and thus allows rapid translation to an implementation structure. An overall conclusion of the study is that ontology-driven methodologies can benefit extensibility and cross-domain interoperability over time but require larger efforts at the early design and development stages.

Several research activities can build upon our work. A direction for future development of the ontology-driven conceptual schema proposed in this paper is to follow through with the prime motivator for ontologies in the first place - formal models in order to identify internal inconsistencies and ambiguities. This issue becomes particularly relevant as the research community starts formalizing the systematic mapping of semantics (from top-down conceptupal schema design) into the design/development of the database schema implementations (the latter remains a gap in the state-of-the-art). Another clear direction for research is evaluating quantitatively the performance differences in the implementations that derive from the two schema constructions.

Acknowledgments

Authors thank Ranga Raju Vatsavai from the Oak Ridge National Laboratory for helpful advice and comments concerning the paper.

Research sponsored by SensorNet Program, Office of Naval Research, and Oak Ridge National Laboratory (ORNL) managed by UT-Battelle, LLC for the U. S. Department of Energy. The submitted manuscript has been authored by a contractor of the U.S. Government under contract DE-AC05-96OR22464. Accordingly, the U.S. Government retains a nonexclusive, royalty-free license to publish or reproduce the published form of this contribution, or allow others to do so, for U.S. Government purposes.

References

1. Heyer, R.J.: Introduction to CBRNE Terrorism: An Awareness Primer and Preparedness Guide for Emergency Responders. The Disaster Preparedness and Emergency Response Association. 3 edn. (2006), http://www.disasters.org
2. Akyildiz, I.F., Su, W., Sankarasubramaniam, Y., Cayirci, E.: Wireless sensor networks: a survey. Computer Networks 38(4), 393–422 (2002)
3. Karl, H., Willig, A.: A short survey of wireless sensor networks. Technical report, Technical University Berlin (2003)
4. Shankar, M., Liu, C., Bhaduri, B.: From static to dynamic models: Enabling real-time geocomputation infrastructures. In: GeoComputation 2005 (2005), http://igre.emich.edu/geocomputation2005/
5. Gorman, B.L., Shankar, M., Smith, C.M.: Advancing sensor web interoperability. Sensors Online (4) (2005), http://sensorsmag.com/articles/0405/14/
6. Frank, A.U.: Ontology for spatio-temporal databases. In: Sellis, T., Koubarakis, M., Frank, A., Grumbach, S., Güting, R.H., Jensen, C., Lorentzos, N.A., Manolopoulos, Y., Nardelli, E., Pernici, B., Theodoulidis, B., Tryfona, N., Schek, H.-J., Scholl, M.O. (eds.) Spatio-Temporal Databases. LNCS, vol. 2520, pp. 9–77. Springer, Heidelberg (2003)
7. Grenon, P., Smith, B.: SNAP and SPAN: Towards dynamic spatial ontology. Spatial Cognition and Computation 4(1), 69–104 (2004)
8. Open GIS, Consortium 35 Main Street, Suite 5, Wayland, MA 01778, USA: Geography Markup Language (GML). v3.1 edn (2004) http://portal.opengeospatial.org/files/
9. Open, G.I.S.: Consortium: OpenGIS® Sensor Model Language (SensorML) Implementation Specification. 1.0 edn (2006), http://vast.nsstc.uah.edu/SensorML/
10. Vretanos, P.A.: Web Feature Service Implementation Specification. Open GIS Consortium Inc., 35 Main Street, Suite 5, Wayland, MA 01778, USA. Version: 1.0.0 edn. (2002) OGC 02-058

11. Consortium, O.G.: Sensor web enablement initiative. web page, (2006)
    http://www.opengeospatial.org/projects/groups/sensorweb
12. IEEE Standards Board: Draft Standard for A Smart Transducer Interface for Sensors and Actuators - Common Functions, Communications Protocols and Transducer Electronic Data Sheets (TEDS) Formats (2007)
13. Gibbons, P.B., Karp, B., Ke, Y., Nath, S., Seshan, S.: Irisnet: An architecture for a world-wide sensor web. IEEE Pervasive Computing 4 (2003)
14. Aberer, K., Hauswirth, M., Salehi, A.: A middleware for fast and flexible sensor network deployment. ACM Very Large Databases (VLDB), 1199–1202 (2006)
15. IEEE Standards Board: IEEE Std 1451.2-19974: IEEE Standard for a Smart Transducer Interface for Sensors and Actuators (1997)
16. Chin, J.C., Hou, I.H., Hou, J.C., Ma, C., Rao, N.S., Saxena, M., Shankar, M., Yang, Y., Yau, D.K.Y.: Sensornet platforms: Sensor-cyber network testbed for plume detection, identification, and tracking. In: Information Processing in Sensor Networks (2007)
17. Open GIS, Consortium 35 Main Street, Suite 5, Wayland, MA 01778 USA: The OpenGISTM Abstract Specification. Topic 1: Feature Geometry. Version 4 edn. (1999)
18. Daniels, J.D., Cook, S.: Designing Object Systems: Object-Oriented Modelling with Syntropy. Prentice Hall, NY (1994)
19. Mannino, M.V.: Database Application Development and Design. Irwin McGraw-Hill, Boston (2001)
20. Resseguie, D.: Sensornet data center services design document. web page (2006),
    http://www.us.sensornet.gov/SensorNetDocs/
21. Smith, B.: Ontology and information systems. Technical report, NCGIA (2000),
    http://ontology.buffalo.edu/smith/articles/ontologies.htm
22. Gruber, T.R.: A translation approach to portable ontologies. Knowledge Acquisition 5(2), 199–220 (1993)
23. Rector, A., Rogers, J.: Ontological issues in using a description logic to represent medical concepts: Experience from GALEN. In: IMIA WG6 Workshop: Terminology and Natural Language in Medicine, January 1997, Phoenix, Arizona, International Medical Informatics Association (1999)
24. Smith, B., Rosse, C.: The role of foundational relations in the alignment of biomedical ontologies. In: Proceedings of Medinfo (March, 22–24) San Francisco (2004)
25. Smith, B.: The logic of biological classification and the foundations of biomedical ontology. In: Invited Papers from the 10th International Conference in Logic Methodology and Philosophy of Science, Oviedo, Spain, 19–25 August, Elsevier-North-Holland (2004)
26. Feng, C.C., Bittner, T., Flewelling, D.: Modeling surface hydrology concepts with endurance and perdurance. In: Egenhofer, M.J., Freksa, C., Miller, H.J. (eds.) GIScience 2004. LNCS, vol. 3234, pp. 67–80. Springer, Heidelberg (2004)
27. Russomanno, D.J., Kothari, C.R., Thomas, O.A.: Building a sensor ontology: A practical approach leveraging ISO and OGC models. In: Arabnia, H.R., Joshua, R. (eds.) IC-AI, CSREA Press, pp. 637–643. CSREA Press (2005)
28. Keet, C.M.: Factors affecting ontology development in ecology. In: Ludäscher, B., Raschid, L. (eds.) DILS 2005. LNCS (LNBI), vol. 3615, pp. 46–62. Springer, Heidelberg (2005)
29. Niles, I., Pease, A.: Towards a standard upper ontology. In: FOIS 2001: Proceedings of the international conference on Formal Ontology in Information Systems, pp. 2–9. ACM Press, New York (2001)

30. Sowa, J.F.: Knowledge Representation: Logical, Philosophical, and Computational Foundations. Brooks Cole Publishing Co., Pacific Grove, CA (2000)
31. Bittner, T., Smith, B.: Formal ontologies for space and time (2003), http://ontology.buffalo.edu/geo/sto.pdf
32. Worboys, M., Hornsby, K.: From objects to events: GEM, the geospatial event model. In: Egenhofer, M.J., Freksa, C., Miller, H. (eds.) Third International Conference on GIScience. LNCS, vol. 2334, pp. 327–345. Springer, Heidelberg (2004)
33. Worboys, M.F.: Event-oriented approaches to geographic phenomena. International Journal of Geographical Information Science 19(1), 1–28 (2005)
34. Worboys, M.: Knowledge discovery using geosensor networks. Technical report, University Consortium for Geographic Information Science (2003), http://www.ucgis.org/Visualization/whitepapers/Worboys%20paper.pdf
35. Bittner, T., Donnelly, M., Winter, S.: Ontology and semantic interoperability. In: Prosperi, D., Zlatanova, S. (eds.) Large-Scale 3D Data Integration, pp. 139–160. CRC Press, London (2004)
36. Lowe, E.J.: A Survey of Metaphysics. Oxford University Press, Oxford (2002)
37. Randell, D.A., Cui, Z., Cohn, A.G.: A spatial logic based on regions and connection. In: Kaufmann, M. (ed.) 3rd International Conference on Knowledge Representation and Reasoning (1992)
38. Egenhofer, M.J.: A formal definition of binary topological relationships. In: Litwin, W., Schek, H.-J. (eds.) FODO 1989. LNCS, vol. 367, pp. 457–472. Springer, Heidelberg (1989)

Modeling Spatio-temporal Network Computations: A Summary of Results*

Betsy George** and Shashi Shekhar

Department of Computer Science and Engineering
University of Minnesota
200 Union St SE, Minneapolis, MN 55455, USA
{bgeorge,shekhar}@cs.umn.edu

Abstract. Spatio-temporal network is defined by a set of nodes, and
a set of edges, where the properties of nodes and edges may vary over
time. Such networks are encountered in a variety of domains ranging from
transportation science to sensor data analysis. Given a spatio-temporal
network, the aim is to develop a model that is simple, expressive and
storage efficient. The model must also provide support for the design
of algorithms to process frequent queries that need to be answered in
the application domains. This problem is challenging due to potentially
conflicting requirements of model simplicity and support for efficient algorithms. Time expanded networks which have been used to model dynamic networks employ replication of the network across time instants,
resulting in high storage overhead and algorithms that are computationally expensive. This model is generally used to represent time-dependent
flow networks and tends to be application-specific in nature. In contrast,
the proposed time-aggregated graphs do not replicate nodes and edges
across time; rather they allow the properties of edges and nodes to be
modeled as a time series. Our approach achieves physical data independence and also addresses the issue of modeling spatio-temporal networks
that do not involve flow parameters. In this paper, we describe the model
at the conceptual, logical and physical levels. We also present case studies
from various application domains.

1 Introduction

Given a spatial network and its variations (e.g., travel times in road networks over
time) the aim is to develop a model that can represent the temporal changes of
the network. This problem has application in several domains such as crime analysis and transportation networks. In transportation networks, travelers might be
interested in finding the best time to start their travel so that they spend the

* This work was supported by the NSF/SEI grant 0431141, US Army Corps of Engineers (Topographic Engineering Center) grant, and Minnesota Department of Transportation. The content does not necessarily reflect the position or policy of the
government and no official endorsement should be inferred.
** Corresponding author.

F. Fonseca, M.A. Rodríguez, and S. Levashkin (Eds.): GeoS 2007, LNCS 4853, pp. 177–194, 2007.
© Springer-Verlag Berlin Heidelberg 2007

least time on the road. Crime data analysts might be interested in finding temporal patterns of crimes at certain locations or the routes in the network that show significantly high crime rates. In these application domains, it is often necessary to develop a model that captures the time dependence of the data and the underlying connectivity of the locations. There are significant challenges in developing a model for spatio-temporal networks. The model needs to balance storage efficiency and expressive power and provide adequate support for the algorithms that process the data. Second, the proposed model should ensure physical data dependence without compromising on the representational power. Third, the time series of spatial data could be infinite. The data model should be able to add data to the existing information and efficiently compute results based on the dynamic data.

Related research in the area of databases falls into the categories of graph databases, and spatio-temporal databases. Graph databases [1, 2, 3, 4, 5, 6] primarily deal with spatial networks that do not vary with time. Research in graph databases that accounts for temporal variations perform computations over a snapshot of the network [7,8,9], and does not consider the interplay between the edge travel times and the existence of edges. Chorochronos [10] studied various aspects of spatio-temporal databases including ontology, modeling, and implementation. However, the researchers have yet to study spatio-temporal networks in this framework.

Operations Research uses a model called the time expanded network [11, 12, 13, 14, 15, 16, 17]. This model duplicates the original network for each discrete time unit $t = 0, 1, \ldots, T$ where T represents the extent of the time horizon. The expanded network has edges connecting a node and its copy at the next instant in addition to the edges in the original network, replicated for every time instant. This significantly increases the network size and is very expensive with respect to memory. Because of the increased problem size due to replication of the network, the computations become expensive. In addition, time expanded graphs have representational issues when modeling non-flow networks as described in Section 2.1.1. Time expanded graphs require a prior knowledge of the length of the time period and hence might lead to a semantic mismatch while handling infinite time series. This model incorporates the time dependent edge attributes into the graph in the process of graph expansion making it more application-dependent, thus making physical data independence harder to achieve.

Various temporally enhanced entity relationship models have been proposed [18]. Some of these models capture the temporal properties of relationships in terms of their existence and validity periods; these do not explicitly capture the changes in relationship types. Other models such as TERC+ [19] capture the temporal nature of relationship types by expressing the relationship changes in terms of entity transformations. This model basically uses entity subtypes to represent temporal evolution of entities as well as relationships and hence might not be able to represent evolving relationships between entities without subtypes.

Our Contribution: The paper describes a model for spatio-temporal networks called the time aggregated graph, which uses a time series to represent time-varying attributes. We illustrate the representational capability of the model through various application domains such as transportation science and emergency planning. We compare this model with the existing graph-based model, the time expanded graph, in the context of various application domains. Preliminary analysis [20, 21] has shown that time aggregated graphs are more storage efficient and support the path computations encountered in transportation networks.

A comparative study on storage and computational efficiency of the time aggregated model has been done previously and this paper presents a comparison of the model with time expanded graphs in the context of representational power. Analysis shows that the model offers better precision of expression and reduces the potential for inconsistent updates since it avoids replication.

1.1   Illustrative Application Domains

Modeling spatio-temporal networks has significant applications in a number of scientific domains. Transportation networks are the kernel framework of many advanced transportation systems such as the Advanced Traveler Information System and Intelligent Vehicle Highway Systems. Transportation networks are spatio-temporal in nature and require significant database support to handle the storage of their large amounts of multi-dimensional data. Many important applications based on transportation networks, including travelers' trip planning, consumer business logistics, and evacuation planning need to be built upon spatio-temporal network databases. For example, commuters try to find a suitable time to start their commute so that they spend the least time in traffic. Varying levels of congestion on road networks during a day can result in changes to the shortest route travel times at different times of the day. With the increasing use of sensor networks to monitor traffic data on spatial networks and the subsequent availability of time-varying traffic data, it becomes important to incorporate this data into the models and algorithms related to transportation networks. As an example, Figure 1 shows a layout of traffic sensors in the Twin Cities of Minneapolis-St Paul, Minnesota and the congestion measurements at two different times of a day.

In crime analysis and prevention, identifying the areas of increasing criminal activity is a key step. Computing the routes that show significantly high crime rates can improve the efficiency of the patrol operations. Crime data usually consists of the geographical location of the crime, type of crime and its time of occurrence [22]. To compute the routes of high criminal activity, a model is required to represent the underlying transportation network along with the time dependent crime data associated with its edges and nodes. For example, the crime rates can vary with the time of the day and the interesting routes can change. With the availability of time-varying data, it becomes important to incorporate this data in the models and analysis of crime data.

Another interesting area of exploration is the effect of temporal dimension on conceptual models such as Entity-Relationship (ER) model [23] and more specifically on the Pictogram-Enhanced Entity-Relationship (PEER) diagram [24]. A simple example is shown in Figure 2. It illustrates a scenario where a moving

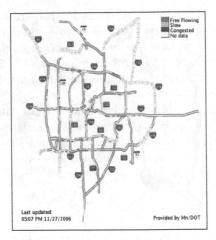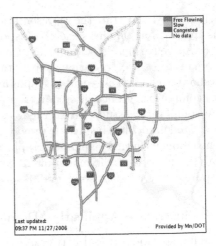

Sensors on Twin Cities, MN Road Network

Fig. 1. Sensors periodically report time-variant traffic volumes on Minneapolis-St Paul highways (Best viewed in color, Source: Mn/DOT)

sensor B crosses a geographic area A. Figure 2(a) shows the locations of B at discrete time instants ($t = t_1, t_2, t_3, t_4, t_5, t_6, t_7, t_8, t_9$). The relationship of object B with object A changes with time. This has been represented in Figure 2(b) using a series of PEER diagrams. Each diagram represents the relationship at an instant. For example, the first diagram represents the time instant $t = t_1$ when the relationship between the objects is 'disjoint'. The figure shows the representations for the first four instants; the rest are modeled in a similar manner.

1.2   Problem Definition

Spatio-temporal networks serve as the underlying networks for many applications. They can be broadly classified into flow networks and non-flow networks, based on the physical scenarios they represent. Popular examples of flow networks would be transportation networks and communication networks. Networks that represent scenarios where the connectivity between the entities is based on physical relationships other than a flow, (e.g., geographical proximity) would fall under the category of non-flow networks. Models of these networks need to capture the possible changes in topology and values of network parameters with time and provide the basis for the formulation of computationally efficient and correct algorithms for the frequent computations. We formulate this as the following problem:

Given: A spatial network and the temporal changes in network topology and parameters.

Output: A model which supports efficient and correct algorithms for computating the query results.

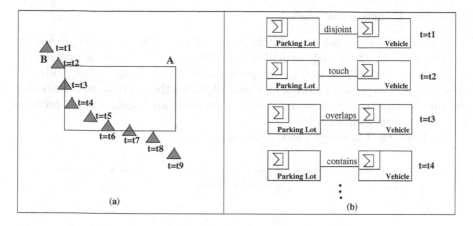

Fig. 2. Illustration of a dynamic relationship between two objects and its representation

Objective: Minimize the storage and computational costs.
Constraints: (1) Edge travel times are positive integers.

1.3   Scope and Outline of the Paper

The paper describes a model called the time aggregated graph for the representation of spatio-temporal networks. It presents the conceptual, logical, and physical models in the representation and provides some case studies that involve transportation networks and emergency planning. The paper also presents some initial steps towards implementing sliding windows in the representation of time series. However, the paper does not specify algorithms facilitated by time aggregated graph due to the nature of the forum and to reduce redundancy with respect to [20]. The paper does not provide a complete formal specification of the model for application domains such as PEER diagrams.

The rest of the paper is organized as follows. Section 2 presents basic concepts related to time aggregated graphs. Section 3.3 presents several case studies and proposes an extension to handle infinite time series data. Section 4 concludes this paper and discusses the direction of future work.

2   Time Aggregated Graph

Spatio-temporal networks have wide applications in domains such as crime analysis, sensor networks, and transportation science. Models of these networks need to capture the possible changes in topology and values of network parameters with time and provide the basis for the formulation of computationally efficient and correct algorithms. In this section we discuss the basics of the model used to represent time dependent spatial networks called "Time Aggregated Graphs" [21].

2.1   The Conceptual Model

A graph $G = (N, E)$ consists of a finite set of nodes N and edges E between the nodes in N. If the pair of nodes that determines the edge is ordered, the graph is directed; if it is not, the graph is undirected. In most cases, additional information is attached to the nodes and edges. In this section, we discuss how the time dependence of these edge/node parameters are handled in the proposed time-aggregated graph model.

We define the time-aggregated graph as follows.

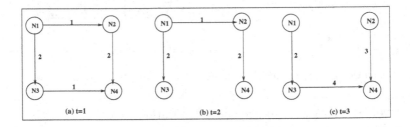

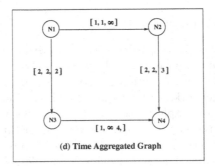

(d) **Time Aggregated Graph**

Fig. 3. Network at various time instants and the Time Aggregated Graph

$$taG = (N, E, TF, f_1 \ldots f_k, g_1 \ldots g_l, w_1 \ldots w_p | f_i : N \to \mathbb{R}^{TF}; g_i : E \to \mathbb{R}^{TF}; w_i : E \to \mathbb{R}^{TF})$$

where N is the set of nodes, E is the set of edges, TF is the length of the entire time interval, $f_1 \ldots f_k$ are the mappings from nodes to the time-series associated with the nodes, $g_1 \ldots g_l$ are mappings from edges to the time series associated with the edges, and $w_1 \ldots w_p$ indicate the time dependent weights (eg. travel times) on the edges.

Each edge has an attribute, called an edge time series that represents the time instants for which the edge is present. This enables the time aggregated graph to model the topological changes of the network with time. We assume that each edge travel time has a positive minimum and the presence of an edge at time instant t is valid for the closed interval $[t, t + \sigma]$.

Figure 3(a,b,c) shows a network at three time instants. The network topology and parameters change over time. For example, edge N3-N4 is present at time instants $t = 1, 3$, and absent at $t = 2$, and its weight changes from 1 at $t = 1$ to 4 at $t = 3$. The time aggregated graph that represents this dynamic network is shown in Figure 3(d). In this figure, edge N3-N4 has an attribute, $[1, \infty, 4]$, which is its weight time series, indicating the weight of the edge at instants $t = 1, 2, 3$. This model can include spatial properties at nodes and edges.

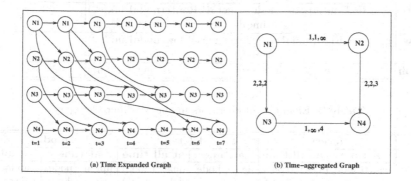

(a) Time Expanded Graph   (b) Time-aggregated Graph

Fig. 4. Time-ggregated Graph vs. Time Expanded Graph

Figure 4(a) shows the time aggregated graph (corresponding to Figure 3(a), (b),(c)) and a time expanded graph that represents the same scenario. Edge weights in a time expanded graph are not explicitly shown as edge attributes; instead they are represented by edges that connect the copies of the nodes at various time instants.

2.2   The Logical Data Model

Basic Graph Operations
The logical model is based on the most commonly used graph model, which is further extended to incorporate the time dependence of the network. The framework of the model consists of two dimensions (1) graph elements, namely node, edge, route and graph and (2) the operator categories which consist of accessors, modifiers and predicates. A representative set of operators for each operator category is provided in Tables 1, 2 and 3. Table 1 lists a representative set of 'access' operators. For example, the operator *getEdge(node1,node2,time)* returns the edge properties of the edge from node 1 to node 2, such as the edge identifier (if any) and associated parameters at the specified time instant. For example operator *getEdge(N1,N2,1)* on the time-aggregated graph shown in Figure 3 would return the travel time of the edge N1-N2 at $t = 1$, that is 1. Similarly, *get_edge(node1,node2)* returns the edge properties for the entire time interval. In Figure 3, the operator *get_edge(N1,N2)* would result in $(1, 1, \infty)$. *get_edge_earliest(N3,N4,2)* returns the earliest time instant at which the edge

Table 1. Examples of operators in the Accessor Category

	at_time	at_all_time	at_earliest
Node	get(node,time)	get_node(node)	get_node_earliest (node,time)
Edge	getEdge(node1,node2,time)	get_edge(node1,node2)	get_edge_earliest (node1,node2,time)
Route	getRoute(node1,node2,time)	getRoute(node1,node2)	get_route_earliest (node1,node2,time)
Route	getSP_Route(node1,node2, time)	getSP_Route(node1,node2)	
Flow	get_max_Flow(node1,node2, time)	get_max_Flow(node1,node2)	
Graph	get_Graph(time)	get_Graph()	—

Table 2. Examples of operators in the Modifier Category

	insert		delete		modify	
	at_time	at_all_time	at_time	at_all_time	at_time	at_all_t
Node	insert(node, time,value)	insert(node, valueseries)	delete(node, time)	delete(node) delete(node)	update(node, time,value)	update(node,series)
Edge	insert(node1, node2, time,value)	insert(node1, node2, valueseries)	delete(node1, node2, ,time)	delete(node1, ,node2) ,node2)	update(node1, node2,time value)	update(edge,series)
Route	insert(node1, node2,time)	insert(node1, ,node2)	delete(node1 ,node2,time)	delete(node1, node2)		
Graph	insert(graph time)	insert(graph)	delete(graph, time)	delete(graph)	update(graph, ,time)	update(graph)

Table 3. Predicate operators in Time-aggregated Graphs

	exists_at_time_t	exists_after_time_t
Node	exists(node u,at_time_t)	exists(node u,after_time_t)
Edge	exists(node u,node v, at_time_t)	exists(node u,node v, after_time_t)
Route	exists(node u,node v,a_route r at_time_t)	exists(node u,node v,a_route r, after_time_t)
Flow	exists(node u,node v,a_flow r at_time_t)	exists(node u,node v,a_flow r, after_time_t)

N3-N4 is present after $t = 2$ (that is $t = 3$). Table 2 shows a set of modifier operators that can be applied to the time aggregated graphs. We also define two predicates on the time-aggregated graph.

exists_at_time_t: This predicate checks whether the entity exists at the start time instant t.

exists_after_time_t: This predicate checks whether the entity exists at a time instant after t.

Table 3 illustrates these operators. For example, node v is adjacent to node u at any time t if and only if the edge (u, v) exists at time t as shown in the table. $exists(N1,N2,1)$ on the time aggregated graph in Figure 3 returns a "true" since the edge N1-N2 exists at $t = 1$.

2.3   Physical Data Model

A static graph $G = (V, E)$ can be represented using an adjacency matrix A, a $|V| \times |V|$ matrix, such that the element a_{ij} is defined as $a_{ij} = w_{ij}$ if $ij \in E$, and w_{ij} is the weight of the edge ij and $a_{ij} = 0$, otherwise. This representation requires $O(N^2)$ memory. It can be seen that the storage required for this representation is independent of the number of edges in the graph, in relation to the number of nodes. In other words, there is no saving in memory even when the graphs are sparse. A representation that can exploit this sparsity is adjacency list representation.

The adjacency list representation of a graph $G = (V, E)$ consists of an array of lists, one for each vertex $v \in V$. The list corresponding to a vertex v contains all vertices that are adjacent to v in G. For a directed graph, the space requirement for the lists is $O(m)$ where $m = |E|$. The total memory reuirement is $O(n + m)$ where $n = |V|$. The weight of each edge uv is stored with the vertex v in u's adjacency list. This representation is specially suitable for sparse graphs.

2.3.1   Data Structures

Time aggregated graphs can be represented by either adjacency list of adjacency matrix representation, with the necessary modifications. These representations need to be extended to include the time series representations on edges (corresponding to time dependent edge costs) and nodes. Adjacency list representation is extended by adding a list to each vertex in the adjacency list. Adjacency list representation uses an array of pointers, one pointer for each node. The pointer for each node points to a list of immediate neighbors. At each neighbor node, attribute time series for the edge starting from the first node to this neighbor are stored. Since the length of the time series is T where T is the length of the time period, the adjacency list representation would require $O(m + n + mT)$ where n is the number of nodes and m is the number of edges if every edge has a time series of length T. In reality, not all time series would be of length T and assuming an average length α, the storage would be $O(n + m + \alpha m)$. The time series store a single value if the value of the attribute remains constant, indicated by the character 'F'. If the value of the attribute changes over time, it is indicated by the character 'V'.

To extend the adjacency matrix to represent the time aggregated graph, a third dimension can be added. The new matrix A would be $n \times n \times T$, requiring $O(n^2 T)$ memory. Figure 5 (a) and (b) show the adjacency list and adjacency matrix representations for the time aggregated graph shown in Figure 3. For example, the edge N1-N2 in the graph at $t = 1$ is represented by the pointer from N1 to N2 in the adjacency list. The array $(1, 2, \infty)$ is stored at N2 to represent the travel times at $t = 1, 2, 3$ for the edge N1N2. In adjacency matrix

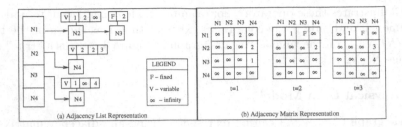

Fig. 5. Storage structures for Time Aggregated Graph

the presence of edge N1N2 at a time instant $t = 1$ is represented by $A[1, 2, 1] = 1$, since the travel time for the edge is 1 unit at $t = 1$. Since the edge is absent at an instant $t = 3$, $A[1, 2, 3] = \infty$. Note that the start node, end node and the time instant are represented by the first, second and third dimensions of the matrix. Though the adjacency matrix has been illustrated as three separate snapshots in Figure 5(b) for the sake of clarity, the entire matrix is stored as a single three-dimensional matrix.

Logical operations on a time-aggregated graph can be classified as

1. Topology first operators (graph dominated operations). Examples include get_route(n1,n2) and get_edge(n1,n2).
2. Time-first operators (Time dominated queries).
   Some examples are get_Graph(time t) and get_edge_at_t(n1,n2,t).

Both representations are equally capable of handling graph dominated queries. To compute time first operations (snapshot queries such as to find the graph at a given time instant), adjacency matrix representation is more suitable. In this representation, these queries represent the time slices of the matrix at the given time instants.

Graphs representing spatio-temporal networks such as transportation networks are generally sparse and hence adjacency list representation is more likely to be storage efficient compared to adjacency matrix representations. The choice is hence a tradeoff between the storage cost and the frequency of the time dominated queries. We expect route queries (which are topology first queries) to be more frequent and since adjacency list representation is capable of handling these, based on storage costs, we used adjacency lists in our implementations. Moreover, most databases use adjacency list representation.

2.3.2 Towards Handling Infinite Time Series

In most domains that involve spatio-temporal networks such as transportation networks, crime data analysis, and sensor networks data is continuously collected at discrete instants of time. For example, sensors on urban highways measure congestion levels every 30 seconds and crime data is appended with every time a crime occurs.. Conceptually, the time aggregated graph can be viewed as a time series of graphs. Each graph represents the attribute values and the topological structure of the network at the given instant of time. Based on the periodicity of data collection, the application domains can be broadly classified into 1) applications where data is measured periodically and 2) applications such as crime analysis where data is recorded when an event occurs.

When data is measured periodically, the underlying model should be able to capture the changes that take place in the spatio-temporal network at every instant. Time aggregated graphs represent this as a time series of graphs, each graph in the series modeling the state of the network. For example, the state of a road network at $t = t_1$ would be represented as a graph corresponding to this instant. The state of a sensor network, which would include the measurements at an instant would also be modeled in a similar manner.

In application domains where the network state changes due to an event, the time aggregated graph stores the tuples of time stamp and the event.

Implementation

The time series of graphs would be implemented as a graph where the node and edge attributes are time series. Most application domains deal with 'infinite' streams of data, and the edge and node attributes are possibly infinite time series. One implementation uses sliding windows implemented through circular buffers. Figure 6 shows a possible implementation of the time aggregated graph shown in Figure 3(d). Figure 6(a) shows a time aggregated graph with time series attributes on its edges. Figure 6(b) shows the modified adjacency list representation that implements an infinite time series. Each time series is stored in a circular buffer.

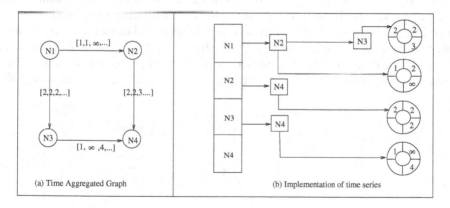

Fig. 6. Representation of Sliding Windows in Time Aggregated Graph

3   Evaluation and Validation

3.1   Representational Comparison: Time Aggregated Graphs vs. Existing Models

A time-expanded network has one copy of the set of nodes for each discrete time instant. Corresponding to each edge with transit time t in the original network, there is a copy of an edge (called the cross edge) between each pair of copies of nodes separated by the transit time t [12,25,15]. Thus, a time-dependent flow in a dynamic network can be interpreted as a static flow in a time expanded network. This allows application of static algorithms on such networks to solve dynamic flow problems. Apart from the "enormous increase in the size of the underlying network" [15] the suitability of the model in some application domains needs further exploration.

A time expanded graph assumes that the edge weight represents a flow parameter, and it represents the time taken by the flow to travel from the source node to the end node. This is represented by the cross edges between the copies of the graph. Since the cross edges in a time expanded graph represent a flow across the nodes, the representation of non-flow networks using this model is not obvious. By contrast, the time aggregated graph model does not impose such a restriction because the attributes are collected into a time series. This difference can be illustrated through the example of the possible extension of the PEER diagram explained in Section 1.1. While time aggregated graph would model the time-dependent relationships as a time series on the edge connecting the nodes (that represent the entitites), the representation of the same scenario is not obvious when time expanded graphs are used. An illustration of the representation of time-dependent relationships using time-aggregated graph representation for the scenario depicted in Figure 2 is shown in Figure 7. Figure 2 shows the locations of B at discrete time instants ($t = t_1, t_2, t_3, t_4, t_5, t_6, t_7, t_8, t_9$). The relationship of object B with object A changes with time. This has been represented in Figure 7(a) using an aggregated representation. The line segment that represents the relationship has an attribute which is an ordered set, each element indicating the current relationship of object B with A. For example, the second entry 'o' indicates that the object B touches A at $t = t_2$ and overlaps A at $t = t_3$. In the domain of crime analysis, the number of crimes reported on a road segment (represented by an edge) at a given time might not be meaningfully

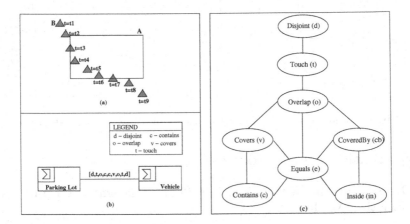

Fig. 7. Illustration of a dynamic relationship between two objects and its representation

represented by an edge in the time expanded graph. The time aggregated graph would represent this as an element in its time series attribute.

In most spatio-temporal networks, the length of the time period (indicated by T in this paper) might not be known in advance since data arrives as a sequence at discrete time instants. For example, sensors in transportation networks collect data at a rate of about once every 30 seconds. Crimes are reported whenever an incident occurs. In addition to being able to represent these attributes, the model must be capable of handling infinite sequences of data. Since time expanded networks require a prior estimate of the length of the time period T, handling of infinite time series might not be easy and obvious. Also, the necessity for the prior knowledge of T might lead to problems in the algorithms based on time expanded networks since an underestimation of T can result in failure of finding a solution. On the other hand, an over-estimated T will result in an over-expanded network and hence lead to unnecessary storage and run-time and would adversely affect the scalability of the algorithms.

Time expanded graphs model the time-dependence of edge parameters through the cross edges that connect the copies of the nodes. This representation, thus, does not provide the means to separate data (for example, an edge attribute series) from its physical representation and hence can adversely affect physical data independence.

The temporal conceptual model TERC+ [19] models dynamic relationships between entities using evolutions of the entities involved. The temporal nature is captured through representing transitions of objects. An example is shown in Figure 8. It represents a dynamic relationship between a person and a University. The relationship changes from an applicant to a donor after graduation. The change in the relationship is represented through various classes of the same entity as shown in Figure 8(a). An aggregated model of the same scenario is shown in Figure 8(b). Though at the finest level, the representations would be the same, the aggregated model facilitates a better high level summarization. This model might not be sufficient to represent cases where entity subtypes cannot be used to model evolving relationships. For example, Figure 2, represents a scenario where the entities

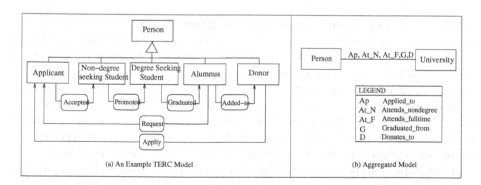

Fig. 8. Representations of dynamic relationships in TERC and Aggregated Graph. (Figure (a) adapted from [19]).

(a sensor and a geograhic area) involved in the dynamic relationship do not have subtypes and hence might not yield itself to this model.

Using Resource Description Framework (RDF) in PEER Diagram: Resource Description Framework (RDF) [26] has been extensively used in representing information about resources in the world wide web. There has been an increasing suuport provided in Databases such as Oracle to query RDF data [27]. Since RDF can be used to capture domain semantics, one possible area of application would be in the temporal extension of PEER diagrams. For example, while representing a dynamic relationship between two spatial objects (as depicted in Figure 2), we need to ensure that the transitions of relationships follow the topological neighborhood graph [4] (Figure 7(b)). Since RDF has the ability to search an arbitrary pattern against a graph structure, the validity of the relationship time series can be checked against the neighborhood graph represented as RDF.

3.2   Comparison of Storage Costs with Time Expanded Networks

According to the analysis in [28], the memory requirement for a time expanded network is $O(nT)+O(n+mT)$, where n is the number of nodes, m is the number of edges in the original graph, and T is the length of the travel time series. The framework of a time aggregated graph would require a memory of $O(n + m)$, where n is the number of nodes and m is the number of edges. Each edge that has a time-varying attribute has an attribute time series associated with it. If the average length of the time series is $\alpha(\leq T)$, the memory required is $O(\alpha m)$, assuming an adjacency list representation. The total memory requirement for a time aggregated graph is $O(n + m + \alpha m)$. This comparison shows that the memory usage of time-aggregated graphs is less than that of time expanded graphs $nT > n$ and $\alpha \leq T$.

3.3   Case Studies

This section discusses time aggregated graph in the context of two application domains, namely, transportation networks and emergency traffic management.

3.3.1   Transportation Networks: Best Start Time

Since the network paramaters and topology can change over time in a transportation network, connectivity and the shortest paths between nodes can be time-dependent. For example, the shortest path travel time from node N1 to node N4 is 3 units if the travel starts at $t = 1$; a commute on the same route would take 4 units if the start time is moved to $t = 2$. The fact that the shortest paths in a time-dependent network vary with time adds an interesting dimension to shortest path computation. A path that takes the smallest travel time for source-destination traversal over the entire time horizon (called 'Best Start Time shortest Path') can be computed. The potential waits at intermediate nodes can increase the total journey time even if an initial part of the path turns out to be optimal. It is significant to note that the prefix journeys of the best start time shortest path journey are not always optimal since some optimal prefix journeys can lead to longer waits at intermediate nodes. An algorithm to compute the best start time in a network was proposed in [20]. For the sake of completeness the key ideas of the algorithms are provided here.

The algorithm that computes the best start time is based on a node-cost time series [20]. The route-finding in the graph is based on the the updation of this node cost time series. The algorithm uses the time aggregated network model to represent a time dependent spatial network.

While computing the best start time, each node needs to keep track of the travel times to the destination for every start time instant. The algorithm attributes each node with a time series. The ith entry in the series represents the current, least travel time to the destination node for the start time t_i. Due to the lack of optimality of prefix paths and lack of ordering of nodes based on the costs (ie. travel times), nodes cannot be selected and "closed" based on a minimum scalar cost. The algorithm uses an iterative, label correcting approach [29] and each entry in a node time series is modified according to the following condition.

$$C_u[t] = minimum\{C_u[t], \sigma_{uv}(t) + C_v[t + \sigma_{uv}(t)]\} \tag{1}$$

where

  $uv \in E$
  $C_n[t]$ - Travel time from $u \in N$ to the destination for the start time t.
  $\sigma_{uv}(t)$ - Travel time of the edge uv at time t.

The algorithm maintains a list of all nodes that change the costs according to the condition and terminates when there is no further improvement indicated by an empty list.

3.3.2 Emergency Traffic Management

A key step in emergency management is the evacuation of a population from areas affected by disasters to safe locations. One significant challenge in this step comes from the time-dependence of the transportation network. Travel times on the road segments and the available capacities of the roads are time-dependent. The dynamic nature of the networks raises some interesting questions (as given in Table 4) and the model for the transportation networks should provide support for such queries.

Table 4. Example queries in time-varying networks

Static	Time-Variant
Which is the shortest travel time path from downtown Minneapolis to airport?	Which is the shortest travel time path from downtown Minneapolis to airport at different times of a work day?
What is the capacity of Twin-Cities freeway network to evacuate downtown Minneapolis ?	What is the capacity of Twin-Cities freeway network to evacuate downtown Minneapolis at different times in a work day?

The proposed time aggregated graph model can model time-varying capacities and travel times and hence would be able to support algorithms to process queries that arise in emergency planning.

4 Conclusion and Future Work

Spatio-temporal networks are a key component of critical applications such as transportation networks, sensor data analysis, and crime analysis. The paper describes a model to represent a spatio-temporal network and presents case studies to illustrate the applicability of the model in various domains. Existing approaches mostly rely on time expanded networks, which leads to high storage overhead and computationally expensive algorithms. Time-aggregated graphs model the time dependence using an aggregation of network parameters across the time horizon without the need to replicate the entire network. Our case studies and related analysis show that this model is less memory expensive. Experiments show that the algorithms based on time aggregated graphs significantly reduce the computational cost compared to similar algorithms based on time expanded networks [20, 21].

The extension of the model to incorporate infinite time series and sliding windows needs to be developed. Extensions of various techniques used in time series indexing or spatial graph indexing need to be explored. We are currently working on the possibility of using this model in the context of mining information from sensor data and we plan to evaluate the performance using real data in the near future. We feel that this model would be applicable in application domains not mentioned in this paper and we plan to explore such domains in the future.

Acknowledgments

We are grateful to the members of the Spatial Database Research Group at the University of Minnesota for their helpful comments and valuable suggestions. We would also like to express our thanks to Kim Koffolt for improving the readability of this paper.

This work was supported by the NSF SEI grant (grant number IIS-0431141), US Army Corps of Engineers (Topographic Engineering Center) grant, and Minnesota Department of Transportation. The content does not necessarily reflect the position or the policy of the government and no official endorsement should be inferred.

References

1. Erwig, M.: Graphs in Spatial Databases. PhD thesis, Fern Universität Hagen (1994)
2. Erwig, M., Guting, R.: Explicit Graphs in a Functional Model for Spatial Databases. IEEE Transactions on Knowledge and Data Engineering 6(5), 787–804 (1994)
3. ESRI: ArcGIS Network Analyst (2006),
   http://www.esri.com/software/arcgis/extensions/

4. S., S., S., C.: Spatial Databases: Tour. Prentice Hall, Englewood Cliffs (2003)
5. Shekhar, S., Liu, D.: CCAM: A Connectivity-Clustered Access Method for Networks and Networks Computations. IEEE Transactions on Knowledge and Data Engineering 9 (1997)
6. Stephens, S., Rung, J., Lopez, X.: Graph Data Representation in Oracle Databese 10g: Case Studies in Life Sciences. IEEE Data Engineering Bulletin 27(4), 61–66 (2004)
7. Ding, Z., Guting, R.: Modeling Temporally Variable Transportation Networks. In: Proc. 16th Intl. Conf. on Database Systems for Advanced Applications, pp. 154–168 (2004)
8. Hamre, T.: Development of Semantic Spatio-temporal Data Models for Integration of Remote Sensing and in situ Data in Marine Information System. PhD thesis, University of Bergen, Norway (1995)
9. Rasinmäki, J.: Modelling Spatio-temporal Environmental Data. In: 5th AGILE Conference on Geographic Information Science, Palma, Balearic Islands, Spain (2002)
10. Koubarakis, M., Sellis, T.K., Frank, A.U., Grumbach, S., Güting, R.H., Jensen, C.S., Lorentzos, N.A., Manolopoulos, Y., Nardelli, E., Pernici, B., Schek, H., Scholl, M., Theodoulidis, B., Tryfona, N.: Spatio-Temporal Databases. LNCS, vol. 2520. Springer, Heidelberg (2003)
11. Dreyfus, S.: An Appraisal of Some Shortest Path Algorithms. Operations Research 17, 395–412 (1969)
12. Ford, L., Fulkerson, D.: Constructing maximal Dynamic Flows from Static Flows. Operations Research 6 (1958)
13. Ford, L., fulkerson, D.: Flows in Networks. Princeton University Press, Princeton, NJ (1962)
14. Kaufman, D., Smith, R.: Fastest Paths in Time-Dependent Networks for Intelligent Vehicle Highway Systems Applications. IVHS Journal 1(1), 1–11 (1993)
15. Kohler, E., Langtau, K.: Time-Expanded Graphs for Flow-Dependent Transit Times. In: Proc. 10th Annual European Symposium on Algorithms, pp. 599–611 (2002)
16. Orda, A., Rom, R.: Minimum Weight Paths in Time-dependent Networks. Networks 21, 295–319 (1991)
17. Pallottino, S., Scuttella, M.G.: Shortest Path Algorithms in Tranportation Models: Classical and Innovative Aspects. Equilibrium and Advanced transportation Modelling (Kluwer), 245–281 (1998)
18. Gregerson, H., Jensen, C.: Temporal Entity Relationship Models - A Survey. IEEE Transactions on Knowledge and Data Engineering 11(3), 464–497 (1999)
19. Zimayi, E., Parent, C., Spaccapietra, S.: TERC+: A Temporal Conceptual Model. In: Proceedings of International Symposium on Digital Media Information Base (1997)
20. George, B., Kim, S., Shekhar, S.: Spatio-temporal Network Databases and Routing Algorithms: A Summary of Results. In: Proceedings of International Symposium on Spatial and Temporal Databases (SSTD 2007). LNCS, vol. 4605, pp. 460–477. Springer, Berlin (2007)
21. George, B., Shekhar, S.: Time-aggregated Graphs for Modeling Spatio-Temporal Networks - An Extended Abstract. In: Proceedings of Workshops at International Conference on Conceptual Modeling (2006)
22. Levine, N.: CrimeStat 3.0: A Spatial Statistics Program for the Analysis of Crime Incident Locations. Ned Levine & Associatiates: Houston, TX / National Institute of Justice: Washington, DC (2004)

23. Chen, P.: The Entity-Relationship Model - Towards a Unified View of Data. ACM Transactions on Database Systems 1(1), 9–36 (1976)
24. Shekhar, S., Vatsavai, R., Chawla, S., Burk, T.: Spatial Pictorgram Enhanced Conceptual Data Models and Their Translation to Logical Data Models. In: Agouris, P., Stefanidis, A. (eds.) ISD 1999. LNCS, vol. 1737, Springer, Heidelberg (1999)
25. Hamacher, H., Tjandra, S.: Mathematical Modeling of Evacuation Problems: A state of the art. Pedestrian and Evacuation Dynamics, 227–266 (2002)
26. W3C: RDF Primer: W3C Recommendation (2004),
    http://www.w3.org/TR/rdf-primer/
27. Chong, E., Das, S., Eadon, G., Srinivasan, J.: An Efficient SQL-based RDF Queryin Scheme. In: Proceedings of the 31st International Conference on Very Large Databases (2005)
28. Sawitzki, D.: Implicit Maximization of Flows over Time. Technical report, University of Dortmund (2004)
29. Cherkassky, B., Goldberg, A., Radzik, T.: Shortest Paths Algorithms: Theory and Experimental Evaluation. Mathematical Programming 73, 129–174 (1996)

Building Geospatial Ontologies from Geographical Databases

Miriam Baglioni[1], Maria Vittoria Masserotti[2], Chiara Renso[2], and Laura Spinsanti[2]

[1] KDDLab Computer Science Department – University of Pisa
baglioni@di.unipi.it
[2] KDDLab ISTI, CNR, Pisa
{masserotti,renso,spinsanti}@isti.cnr.it

Abstract. The last few years have seen a growing interest in approaches that define methodologies to automatically extract semantics from databases by using ontologies. Geographic data are very rarely collected in a well organized way, quite often they lack both metadata and conceptual schema. Extracting semantic information from data stored in a geodatabase is complex and an extension of the existing methodologies is needed. We describe an approach to extracting a geospatial ontology from geographical data stored in spatial databases. To provide geospatial semantics we introduce new relations which define geospatial ontology that can serve as a basis for an advanced user querying system. Some examples of use of the methodology in the urban domain are presented.

1 Introduction

Historically, Geographical Information Systems (GIS) evolve from numeric cartography putting together remote sensing and digital images, typically skipping any design and modeling phase. Therefore, quite often they lack both metadata and the conceptual schema, thus losing part of the semantic geographical information.

In the last few years, ontologies [13] have gained increasing interest in the GIS community [17], because they are essential to create and use data standards as well as human computer interfaces and to solve heterogeneity/interoperation problems [6]. The use of ontologies as a middle layer between the user and the database, adds a conceptual level over the data and allows the user to query the system on semantic concepts without having any specific information about the database at hand [21]. This ontology should be capable to represent both high level semantic concepts as well as concepts that have a correspondence to database tables. This allows to build a mapping between ontological concepts and data.

Such an ontology can be constructed manually from data analysing the structure of the database and the contents of tables. However, this is a complex, expensive and time consuming task, and it could also lead to mistakes and missing information. Recently, the literature has seen a growing interest in approaches that define methodologies to automatically extract semantics from databases. Most of these approaches represent knowledge by means of ontologies. When dealing with geographic information, this automatic extraction becomes more complex, due to the complex semantics of spatial data.

F. Fonseca, M.A. Rodríguez, and S. Levashkin (Eds.): GeoS 2007, LNCS 4853, pp. 195–209, 2007.
© Springer-Verlag Berlin Heidelberg 2007

Our work is a first step in the direction of defining a methodology for automatically building a geospatial ontology as a semantic view of data stored in a geospatial database in order to support a user-oriented querying system. In this context, the use of ontologies has already been experimented (see for example [4, 25]). Indeed, having a conceptual and taxonomical representation of spatial data provides the query system with a semantic representation of spatial concepts. This enables the user to pose "semantic geospatial queries" instead of the classical "geospatial queries" provided by the query language of the DBMS.

For example, let us suppose that a spatial database contains information about hospitals, museums, shopping malls, and archaeological areas. The user can easily ask the database *Which are the museums in Garibaldi Street?* by means of a spatial SQL query. However, he/she is not allowed to ask which are, for example, the *buildings* in the same street. Here, *buildings* is a concept, not explicitly represented in the database, that subsumes, for example, museums, hospitals, shopping malls. Intuitively, exploiting the taxonomy allows the system to answer this kind of queries since the abstract concept is replaced by all its subclasses.

Another example of a "semantic geospatial query" is: *Where is IperMarket?* Here we refer to the concept of location of an object, in this case of a specific shopping mall. In spatial databases, geographic locations can be represented in different ways, as we will see in detail in Section 3. In order to answer this kind of queries, we need to define a geospatial ontology capable of representing these different types of locations that can be directly represented as an attribute of an object or they can be inferred by the relationship with another spatial object.

The contribution of our work is twofold. On the one hand, we define a geospatial ontology where new relations are introduced to provide geospatial semantics. On the other hand, we describe an approach for the automatic definition of this geospatial ontology from geographical data stored in spatial databases. This approach proposes the extraction of an application ontology from spatial database tables. This ontology is then enriched by means of a domain ontology in order to add semantics. In particular, we define both the new extraction rules for the case of spatial relations and a method to enrich the application ontology with the domain ontology. The extracted enriched ontology can serve as a basis for an advanced user querying system as shown, for example, in [4] where a natural language user interface has been built ·on top of a spatial database.

This paper is organized as follows. Section 2 presents the related work, Section 3 shows the system architecture. Then, Section 4 regards the definition of the *Geospatial Ontology* and the spatial properties we introduced. Section 5 describes *Application Ontology* and *Extraction Module* with the definition of new extraction rules. Section 6 introduces the *Domain Ontology*, whereas Section 7 illustrates the technique to build an *Enriched Ontology*. Finally, Section 8 and Section 9 show an application example, and conclusions and future work, respectively.

2 Related Work

To the best of our knowledge this is the first attempt to design a methodology to automatically build a geospatial ontology from data stored in a spatial database.

However, the field of automatic ontology building from relational database is active in Semantic Web research. These approaches (see for example [2,15,16,20,26]), propose extraction rules to build an ontology that represents the relational data schema in OWL formalism. In particular, [16] proposes, besides the definition of the extraction rules, an enrichment of the extracted ontology with the domain ontology having the purpose of adding semantics. We exploit and extend this idea for the case of spatial databases.

The proposal in [25] has some similarities with ours, since authors propose an ontological semantic layer to query a geographical database. In particular, this approach allows different community users to access the same geographic database. However, compared with our approach, it focuses on the representation of spatial relationships such as the topological ones (i.e. *touches*) and does not consider specifically the problem of representing the location of a geographical object. Again, the proposal is not interested in defining an automatic ontology extraction procedure directly from data.

There are approaches that define new ontology formalisms to represent spatial information, like [23], whereas other approaches define geospatial ontologies as a first step to build geospatial database/GIS ([1, 3, 5, 9, 10, 11, 12, 14]).

The work presented in this paper is an evolution of a project illustrated in [4] where the geospatial ontology was built manually from a spatial database. Initially, a Conceptual Model was constructed from data and then translated in the ontology formalism OWL [19]. The query system was composed by a natural language module and an ontology-based query interpreter capable of translating queries in OpenGIS spatial-SQL [18]

3 System Architecture

The methodology proposed is based on the architecture shown in Fig. 1. Here, starting from a spatial database, an *Application Ontology* is built by means of the *Extraction Module*. This ontology represents, by means of concepts and relations, the structure of the database where spatial properties between objects are explicitly represented. This ontology is then enriched (via the *Enriching Module*) with a *Domain Ontology* in order to provide domain semantics. The resulting *Enriched Ontology* represents a semantic and taxonomical view of the spatial data stored in the spatial database. Finally, the *Mapping Module* allows to link (some of the) concepts of the *Enriched Ontology* back to the database to enable the translation of user queries to spatial SQL [4]. In this paper, we focus on the ontology construction phase, therefore we are going to describe the *Extraction* and *Enriching Modules* omitting details on the *Mapping Module*.

In this architecture we assume that the spatial database is based on an OpenGIS spatial data model [18] such as the ones used in PostGIS or MySQL. This model defines a data type *geometry* (stored in the attribute called *the_geom*) that contains the geometry of the object and its coordinates. When the object is located on the Earth surface we say this object is *georeferenced* with respect to a coordinate system. In this case, the spatial database is called *geodatabase*.

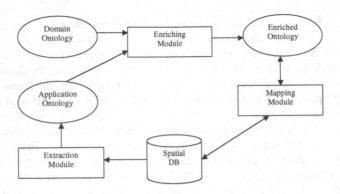

Fig. 1. System Architecture

Each object in a geodatabase may contain both thematic attributes and geographic information. We distinguish geographic information as:

- Location of the object (where is the object on the Earth surface)
- Geometric information (which geometry represents the spatial object, i.e. line, point, polygon)

Location is denoted as *direct* when the object has a *the_geom* attribute, that contains both a geometric and a location information. Location is *indirect* when the object itself does not have the *the_geom* attribute, but its location is implicitly contained in the thematic attributes. The following examples are aimed at showing direct and indirect location, respectively.

Examples
The following tables contain information about hospitals, museums and street numbers. The first table is an example of a direct location. Notice that the the_geom attribute contains the geometry (POLYGON), the coordinates of the point on the Earth surface (10.589,47.779 and all the polygon vertices) and the reference system used (WGS84).

Hospital

ID	Name	DayBeds	The_geom
3456	Santa Chiara	200	POLYGON(10.589,47.779,...,WGS84)

Example 1. An example of direct location

In this second example indirect location is used. Notice that the Museum table does not have a the_geom attribute, but it refers to another table (StreetNumber) with direct location. We can infer, in this case, that the location of "Arsenale" is at coordinates 10.590,47.756 in the WGS84 reference system.

Museum

ID	Name	Topics	ID_StreetNumber
3456	Arsenale	50	84

StreetNumber

ID	The_geom
84	POINT(10.590,47.756,WGS84)

Example 2. An example of indirect location

The methodology introduced in this paper aim to build the Enriched Ontology starting from spatial tables, where the geographic location can be either direct or indirect. Therefore, the next paragraphs outline how geographic information is represented in the ontologies and show the extraction rules that automatically translate tables with direct/indirect locations into ontology classes and properties.

It is worth noticing that all the ontologies built by this methodology must be of type *geospatial*, thus must be capable to represent spatial information in terms of direct/indirect location and geometry. For this reason, we need to define the notion of *geospatial ontology* as an ontology provided with these spatial concepts and relations. In the next section we give a formal definition of a geospatial ontology.

4   The Geospatial Ontology

Since the ontology must be able to represent abstraction of spatial data, we need to explicitly introduce special relations to express spatial properties. Here, we focus on *geographic location* and *geometry*. Other spatial relations, such as topological ones, will be considered in the future. To represent these properties, we introduce special high level concepts (see Fig. 2): *GeographicObject* as the root class to represent all objects that are geographic. This class is a parent of *GeoRef* that indicates the class of all objects that have a geographic location, and *Geometry* that represents all objects that have a geometry and subsumes all the geometry classes *Point, Line, Polygon*. Furthermore, new relations, that indicate the geometric and geographic properties of objects, are introduced: *is_at, has_geometry, has_georef*, formalized below.

Formally, an ontology is a 5-tuple $O:=\{C, R, HC, rel, A_0\}$, where C is a set of concepts, which represent the entities in the ontology domain; R is a set of relations defined among concepts; HC is a taxonomy or concept-hierarchy, which defines the *is_a* relations among concepts ($HC(C1, C2)$ means that $C1$ is a sub-concept of $C2$, or in other words $C2$ is a parent of $C1$), *rel*: $R \rightarrow C \times C$ is a function that specifies the relations on R ($rel(R)=(C1, C2)$ is also written as $R(C1, C2)$). Finally, A_0 is the set of axioms expressed in a logical language, such as first order logic.

We instantiate this definition by including the spatial concepts and relations. Therefore the *geospatial* ontology is a 5-tuple $Os:=\{C', R'\ HC, rel, A_0\}$, where $C'=C \cup Cs$ and $R'=R \cup Rs$. Cs contains the base spatial concepts {*GeographicObject*,

GeoRef, Geometry, Point, Line, Polygon} whereas Rs contains the new relations {has_georef, is_at, has_geometry} defined as follows:

- has_georef(GeographicObject, GeoRef). A has_georef B indicates that A has the coordinates location B.
- is_at(GeographicObject, GeographicObject). A is_at B means that if B has_georef C then A has_georef C, therefore A has the geographic location of object B. Intuitively, this means that in our system the query "where is A?", has as answer the location of B.[1]
- has_geometry(GeographicObject, Geometry). A has_geometry B indicates that object A has as geometry B.

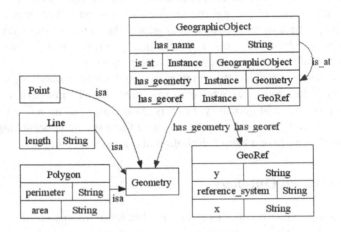

Fig. 2. Classes and relations of a geospatial ontology

5 Application Ontology and Extraction Module

Since the application ontology is derived from the geodatabase, and strictly depends on the structure of tables and their relations, it is not modeled *a-priori*.

The objective of the *Extraction Rules* module is to automatically build the Application Ontology starting from the database schema. There are approaches in the literature that define rules to automatically extract ontologies from relational database such as discussed in [15, 16]. Typically, these rules produce concepts and relations from tables depending on the schema of the database, such as the structure of the tables and the features of the primary and foreign keys. When dealing with geodatabase, new rules have to be defined in order to manage direct and indirect location and connect them with the geospatial ontology. Notice that we start the extraction phase considering each concept depicted in Figure 2 as enclosed in the Application Ontology. Each concept of the Application Ontology is considered a subclass (*is_a* relation) of *GeographicObject*.

---

[1] It is important to explicit the spatial information about geometry and location into two distinct relations. Indeed, the *is_at* definition for indirect location refers to *has_georef* and not to *has_geometry*.

Now, we are going to define the new extraction rules that produce the spatial relations in the application ontology. These rules can be added to a rule extraction module for relational database, such as [15, 16].

Formally, a relational schema S is an 8-tuple $S=(R,A,I,T,att, pkey,fkey,type)$ where R is a finite set of relations, A is a finite set of attributes, T is a set of atomic data type, att is a function that defines the set of attributes of the relations in R, $pkey$ is a function that defines the set of attributes that compose the primary key of a relation in R, $fkey$ is a function that defines the set of foreign key attributes of a relation in R, $type$ is a function $type: a_i \in A \rightarrow T$ that defines the type of the attribute. We also indicate with $value(att_i)$ the actual value of the attribute att_i in the table R_i stored in the database. I is the set that defines the inclusion dependency [16].

Let us focus on the definition of the two new extraction rules for direct/indirect location. It is worth recalling that these rules are aimed at building the spatial relations between concepts. We are assuming that the extraction module has already built the concepts relative to tables using, for example, rules illustrated in [15, 16]. Once all the concepts have been created, the location rules are triggered.

5.1 Extraction Rule for Direct Location

Direct location is represented in the spatial database by means of the *the_geom* attribute. The stored value of this attribute indicates the geometry. Therefore, when a table has a *the_geom* attribute, this means that it contains both the geometry and the coordinates, and the value of the attribute as split into these two components. This produces a relation *has_geometry* with the class representing the geometry and a *has_georef* property with the *GeoRef* class.

More formally,

> *Given the relation Ri, in the database schema, if the_geom $\in$ att(R_i) and Gi = value(the_geom) and Gi=(Cj,GeoRef) and Ri has produced concept Ci, then the relations: has_geometry(Ci,Cj) and has_georef(Ci,Georef) are added to the ontology.*

The extracted ontology fragment for the *Hospital* example (example 1) is:

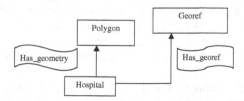

Fig. 3. The ontology fragment related to the *Hospital* table

Notice that for the implementation of this rule, an explicit query to the geodatabase is needed to capture the content of the *the_geom* attribute.

5.2 Extraction Rule for Indirect Location

Indirect location is represented in the geodatabase by means of a foreign key to a georeferenced entity. Therefore, if a table has a foreign key that refers to another table that has a direct location (*the_geom* attribute), a *is_at* property between the two classes is produced.

More formally:

> *Given the relations R_i, R_j in the database schema, if fkey(R_i) = pkey(R_j) and the_geom $\in$ att(R_j) and Rj produces a concept Cj, Ri produces a concept Ci, then the relation is_at(Ci,Cj) is added to the ontology.*

The intuitive meaning is that the first object has the location of the referenced object. Notice that the geometry is not inherited. It is worth noticing also, that we are not dealing here with the general case where the referred table has itself an indirect location, that produces a recursive definition.

The extracted ontology fragment for the *Museum* example (example 2) is:

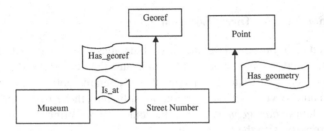

Fig. 4. The ontology fragment related to the Museum table

6 Domain Ontology

The Domain Ontology is not a specific ontology but a class of ontologies that represent the perspective of a given community about a predefined domain. Typically, it can be defined from public shared ontologies or can be built from domain experts knowledge. The primary purpose of the domain ontology is to represent concepts on which the user can query. For this reason, in our architecture we propose to enrich the application ontology, obtained from the geodatabase, with a domain ontology. Furthermore, since the spatial relations are explicitly represented in the ontology, the query system becomes capable to answer location queries also in presence of an indirect location.

Here, we give an example of a domain ontology defined in a "mereological fashion" in that we consider as main individuals the relationships between parts and wholes not taking care of any particular instance. This means that it describes both the geometry shapes and the physical entities expressed as the geo-referenced object.

As an example of a Domain Ontology, we describe here a simplified fragment of the Urban Ontology that was developed in [4]. The extensive domain ontology covers many urban objects and consists of 132 classes, 46 object properties and 112 data type properties. The type of objects considered is very different: from streets and buildings, to archeological areas and parks. A fragment is shown in Fig. 5.

GeographicalObject is the main concept, it subsumes *UrbanObject* that represents all the entities in a city such as the transportation system and the buildings.

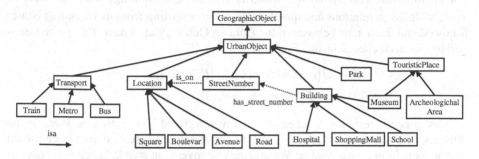

Fig. 5. A fragment of the Domain Ontology

The Domain Ontology is aimed at enriching the Application Ontology extracted from the geodatabase, thus it is characterized by a taxonomy of concepts that must be in the same domain of the data stored in the geodatabase. This allows to perform the enriching phase.

7 Enriching Module

The *Enriching Module* builds a new semantically-enriched ontology from the *Domain* and *Application Ontology*.

Let us define the Application Ontology as $O_A=(C_A,R_A,HC_A,rel_A)^2$, the Domain Ontology as $O_D=(C_D,R_D,HC_D,rel_D)$ and the Enriched Ontology as $O_G=(C_G,R_G,HC_G,rel_G)$.

The first part of the enriching process takes all the classes of the Application Ontology and searches for class correspondence in the Domain Ontology. The symbol $\cong$ indicates a linguistic correspondence, that is any lexical, semantics or structural correspondence between classes (can be obtained, for example, using a WordNet synset [26]).

Let us define the following two sets of concepts:

$$C_{A'} = \{c \in C_A \mid \exists c_i \in C_D \wedge c_i \cong c\} \qquad (1)$$

$$C_{A''} = C_A \setminus C_{A'} \qquad (2)$$

Where $C_{A'}$ represents all the classes belonging to C_A that have a linguistic correspondence class in C_D, and $C_{A''}$ is the set of classes of the Application Ontology for which there is no linguistic correspondence with the Domain Ontology.

---

[2] Notice that in this approach, axioms are not used and for the sake of readability we are omitting here A_O. However, this procedure is extendible to axioms.

Let C_{G_0} be the set of concepts in C_A'' plus those in C_D that have a linguistic correspondence with a class in C_A

$$C_{G_0} = C_A'' \cup \{c \in C_D \mid \exists c_1 \in C_A \wedge c_1 \cong c\} \tag{3}$$

Notice that C_A (or its linguistic equivalent classes) is completely included in C_{G_0}. For all linguistic correspondence found in the Domain Ontology with one class in C_A', all the *is_a* relations and the *object properties* starting from that class(es) is(are) followed and the reached classes of the Domain Ontology are taken. This procedure is iterated for each class selected. Formally

$$C_{G_n} = C_{G_{n-1}} \cup \{c \in C_D \mid \exists c_1 \in C_{G_{n-1}}.HC_D(c_1,c)\} \cup$$
$$\{c \in C_D \mid \exists c_1 \in C_{G_{n-1}} \wedge r \in R_D.r(c_1,c)\} \tag{4}$$

This process terminates when we reached a fixed point, that is there is no difference between the set at step n and the set at step $n+1$. Notice that, since the set of classes is finite, we converge. We assume to converge at step k, hence $C_G = C_{G_k}$.

In the second part, all the properties belonging to the domain ontology that are defined between two classes in C_G are taken

$$R_G = \{r \in R_D \mid \exists c_1, c_2 \in C_G.r(c_1,c_2)\} \tag{5}$$

Then we add to the set R_G all the properties defined in R_A between a pair of concepts from $C_{A''}$ and $C_{A'}$ and vice verse.

$$R_G = R_G \cup \{r \in R_A \mid \exists c_1, c_2 \in C_A.r(c_1,c_2) \wedge ((c_1 \in C_{A'}, c_2 \in C_{A''}) \vee (c_1 \in C_{A''}, c_2 \in C_{A'}))\} \tag{6}$$

Finally, all the properties belonging to R_A defined among concepts not present in C_G are added to the property set:

$$R_G = R_G \cup \{r \in R_A \mid \exists c_1, c_2 \in C_{A''}.r(c_1,c_2)\} \tag{7}$$

Notice that, since we add only the name of the properties, no redefinition is needed.

In the third part the HC_G set has to be defined. All the *is_a* relations of the Domain Ontology involving two classes of the Enriched Ontology are taken.

$$HC_G = \{h \in HC_D \mid \exists c_1, c_2 \in C_G.h(c_1,c_2)\} \tag{8}$$

Then we add all the *is_a* among concepts of C_A', and all the *is_a* relations from concepts in C_A' and C_A''.

$$HC_G = HC_G \cup \{h(c_1,c_2) \mid \exists h' \in HC_A, c_3, c_4 \in C_{A'}, c_1, c_2 \in C_G.h'(c_3,c_4) \wedge c_1 \cong c_3, c_2 \cong c_4\} \cup$$
$$\{h(c_1,c_2) \mid \exists h' \in HC_A, c_1 \in C_{A'}, c_2 \in C_G, c_3 \in C_{A'}.h'(c_1,c_3) \wedge c_2 \cong c_3\} \cup \tag{9}$$
$$\{h(c_1,c_2) \mid \exists h' \in HC_A, c_3 \in C_{A'}, c_1 \in C_G, c_2 \in C_{A''}.h'(c_3,c_2) \wedge c_1 \cong c_3\}$$

Notice that all the relations involving classes belonging to C_A' are redefined considering the correspondent class in C_G. Finally, we add all the *is_a* relations involving two concepts from C_A''.

$$HC_G = HC_G \bigcup \{h \in HC_A \mid \exists c_1, c_2 \in C_{A^*}. h(c_1, c_2)\} \tag{10}$$

The last part defines the rel_G set. Initially all the relations in rel_D that refer to properties enclosed in R_G for which the concepts representing the domain and co-domain of the property are enclosed in C_G are added to the set. Finally, all the relations in rel_A are taken, provided redefinition when it is needed. Formally:

$$rel_G = \{rel \in rel_D \mid \exists r \in R_G \wedge c_1, c_2 \in C_G . rel(r) = (c_1, c_2)\} \tag{11}$$

$$rel_G = rel_G \cup \{rel \in rel_A \mid \exists r \in R_A \wedge c_1, c_2 \in C_{A^*} . rel(r) = (c_1, c_2)\} \tag{12}$$

$$rel_G = rel_G \cup \{rel(r) = (c_1, c_2) \mid \exists r \in R_A, c_3 \in C_{A'}, c_2 \in C_{A^*}, c_1 \in C_G . r(c_3, c_2) \wedge c_1 \cong c_3\} \tag{13}$$

$$rel_G = rel_G \cup \{rel(r) = (c_1, c_2) \mid \exists r \in R_A, c_3 \in C_{A'}, c_1 \in C_{A^*}, c_2 \in C_G . r(c_1, c_3) \wedge c_2 \cong c_3\} \tag{14}$$

$$rel_G = rel_G \cup \{rel(r) = (c_1, c_2) \mid \exists r \in R_A \setminus R_G, c_3, c_4 \in C_{A'}, c_1, c_2 \in C_{A^*} . r(c_3, c_4) \wedge c_1 \cong c_3, c_2 \cong c_4\} \tag{15}$$

8 Application Example

In this section we present an application example to give the flavour of the approach. Consider the fragment of the geodatabase containing information about urban entities expressed in the following tables:

Hospital

ID	Name	DayBeds	The_geom
3456	Santa Chiara	200	POLYGON(10.589,47.779,...,WGS84)
3457	Ospedaletto	300	POLYGON(10.589,47.781,...,WGS84)

Museum

ID	Name	Topic	ID_StreetNumber	ID_Street
1	San Matteo	Art	3	12
2	Arsenale	Historical Ships	84	45

School

ID	Name	Type	ID_StreetNumber	ID_Street
34	Santa Caterina	High School	45	45
45	Fibonacci	Primary School	32	14

Street

ID	Name	The_geom
45	Via G. Garibaldi	LINE(10.509,47.708,...,WGS84)
12	Via G. Mazzini	LINE(10.523,47.746,..., WGS84)

StreetNumber

ID	The_geom
84	POINT(10.590,47.756,wgs84)

ShoppingMall

ID	Name	Parking	The_geom	ID_chain_st
2345	IperMarket	200	POLYGON($x_1,y_1,\ldots$, WGS84)	345
2346	MegaStore	100	POLYGON($x_2,y_2,\ldots$, WGS84)	567
2347	ShopCenter	234	POLYGON($x_3,y_3,\ldots$, WGS84)	432
2348	IperDrugStore	123	POLYGON($x_4,y_4,\ldots$, WGS84)	567

ChainStore

ID	Name	The_geom
345	Coop	POINT(10.509,47.708,…,WGS84)

An informal representation of the Application Ontology built by the Extraction Module is shown in Figure 6. Notice that Hospital has a direct location and therefore a *has_georef* relation has been created with the *GeoRef* class. Furthermore, a *has_geometry* relation has been created with *Polygon* class. Since the Museum table presents two indirect locations, two *is_at* relations has been created, one with *StreetNumber* and the other with *Street*. ShoppingMall table presents both direct and indirect locations. Indeed, the ShoppingMall class has relations *has_geometry has_georef* and *is_at*.

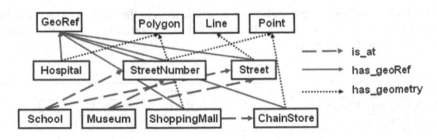

Fig. 6. The informal representation of the Application Ontology

Applying the Enriching Module to the above Application Ontology and the Domain Ontology shown in Fig. 5, we obtain:

- from the Application Ontology shown in Fig. 6, the classes extracted and matched to the DomainOntology are: *GeoRef, Polygon, Line, Point, Street* (equivalent to Road), *StreetNumber, School, ShoppingMall, Museum,* and *Hospital* (Step1 in white in Fig. 7)
- following the *is_a* relations and the *properties* defined among these classes in the Domain Ontology, *Geometry, UrbanObject, Location, Building, TouristicPlace* classes are added in Step 2, whereas *GeographicObject* and *TerritorialDivision* are added in Step 3.
- the class *ChainStore,* that has no matching class in the Domain Ontology, is taken (in gray in the figure 7).
- at the end, the taxonomy is reconstructed by *is_a* relations (dotted arrows in figure 7), relation are considered and instantiated (plain arrows in figure 7).

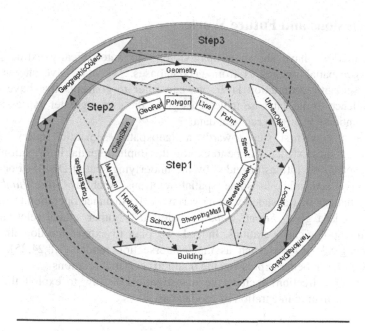

Fig. 7. Example of the enriching steps

The obtained Enriched Ontology makes it possible to answer "semantic geospatial queries". For example, consider the following queries, here expressed, for the sake of readability, in a natural language style:

Query: *Which are the buildings in Garibaldi street?*
This query could have not be answered by a standard (spatial) SQL query language since the geodatabase does not contain any "building" table. However, by using the enriched ontology as a middle layer between the query system and the geodatabase, we can abstract from the specific building (i.e. *hospital or school*) and refer to the concept of *building*. In a query execution phase, the *building* concept is expanded in the set of its subclasses, therefore, the original query is transformed into a set of spatial SQL-queries, one for each subclass that corresponds to a database table. In this example the answer of the query is Museum *Arsenale*, and School *Santa Caterina*.

Query: *Where is IperMarket?*
Here we have the two kinds of locations of *IperMarket*: direct and indirect. The direct location is the coordinates coming from *the_geom* attribute, whereas the indirect one refers to the chain central office, which in turn has a direct location. The answer is the coordinates of *IperMarket* and the coordinates of the chain central office. In this case the street name of *IperMarket* is not explicitly represented in the database. A way to find the street name of *IperMarket* could be to exploit the spatial analysis functionalities of the spatial DBMS such as buffer or distance functions to obtain the name of the street closest to the IperMarket coordinates.

9 Conclusions and Future Work

In this paper we have shown a methodology to automatically extract semantic enriched geospatial ontologies from geodatabases. We also have shown that, by adding a semantic abstraction layer, we can refer to concepts that have no direct correspondence to the database table. This gives the user a greater expressive power and a semantic view of geographical data.

This is a very first step towards a "geospatial semantic" query system to geodatabase. Our next step will be to exploit the implicit spatial information by using primitive spatial operations provided by the underlying spatial DBMS. For example, we can give semantics to some spatial relations, such as *StreetNumber is_on Location*, that can be translated with a overlay/buffer/distance operation to find out on which street is a given *StreetNumber*. Analogously, other spatial relations can be mapped to topological operations. Indeed, we plan to extend the domain ontology with topological relations such as the 9-intersection model [7,8,24,25]. This can support the user in better expressing the qualitative spatial relations.

Other future directions include the investigation of how to exploit the enriched ontology for semantic integration of geodatabases.

Acknowledgments

This work has been partially supported by GeoPKDD EU Project IST-6FP-014915, http://www.geopkdd.eu/. We would like to thank Alessandra Raffaetà for careful reading and helpful comments.

References

1. Abdelmoty, A.I., Smart, P.D., Jones, C.B., Gaihua, Fu., Finch, D.: A Critical Evaluation of Ontology Languages for Geographic Information Retrieval on the Internet. J. Visual Language and Computing 16(4), 331–358 (2005)
2. An, Y., Borgida, A., Mylopoulos, J.: Discovering the Semantics of Relational Tables through Mappings. AAAI, Stanford, California, USA (2006)
3. Arara, A., Laurini, R.: Towards a Formalization of Urban Ontologies with Multiple Perspectives. In: Proceedings of 12th Annual Conf. on GIS research UK, Norwich, pp. 168–174. University of East Anglia (2004)
4. Bartolini, R., Caracciolo, C., Giovannetti, E., Lenci, A., Marchi, S., Pirrelli, V., Renso, C., Spinsanti, L.: Creation and Use of Lexicons and Ontologies for NL Interfaces to Databases. In: LREC Conference, Genova (2006)
5. Bateman, J., Farrar, S.: Spatial Ontologies Baseline. ONTOSPACE Project Report, Bremen, Germany (2006)
6. Bolchini, C., Curino, C., Schreiber, F.A., Tanca, L.: Context integration for mobile data tailoring. In: Proc. IEEE/ACM ICDM, Nara, Japan, IEEE, ACM (2006)
7. Egenhofer, M.J.: Reasoning about Binary Topological Relations. In: Günther, O., Schek, H.-J. (eds.) SSD 1991. LNCS, vol. 525, pp. 143–160. Springer, Heidelberg (1991)

8. Egenhofer, M.J., Herring, J.: Categorizing Binary Topological Relations between Regions, Lines, and Points in Geographic Databases. Technical report, NCGIA, University of California, Santa Barbara (1990)
9. Fonseca, F.T., Egenhofer, M.J.: Ontology-Driven Geographic Information Systems. In: ACM-GIS, ACM Press, New York (1999)
10. Fonseca, F.T., Egenhofer, M.J., Davis, C.A.: Ontologies and Knowledge Sharing. Urban GIS. Computer, Environment and Urban Systems 24(3), 251–272 (2000)
11. Fonseca, F.T., Egenhofer, M.J., Agouris, P., Camara, C.: Using Ontologies for Integrated Geographic Information Systems. Transact. GIS 6(3), 231–257 (2002)
12. Frank, A.U.: Ontology for Spatio-temporal Databases. In: Koubarakis, M., et al. (eds.) Spatio-Temporal Databases. LNCS, vol. 2520, pp. 9–78. Springer, Heidelberg (2003)
13. Guarino, N.: Formal Ontology and Information Systems. FOIS 1998 (1998)
14. Klien, E., Probst, F.: Requirements for Geospatial Ontology Engineering. In: 8th Conference on Geographic Information Science (AGILE 2005), Estoril, Portugal (2005)
15. Li, M., Du, X., Wang, S.: Learning Ontology from Relational Database. In: Proceedings of the Fourth International Conference on Machine Learning and Cybernetics, Guangzhou (August 2005)
16. Macagnino, L.: Estrazione di Ontologie da Basi di Dati relazionali basata sulla Semantica. Master Thesis, Politecnico di Milano (in italian) (2006), http://poseidon.elet.polimi.it/ca/
17. Mark, D.M., Egenhofer, M.J., Hirtle, S., Smith, B.: UCGIS Emerging Research Theme: Ontological Foundations for Geographical Information Science (2006)
18. OpenGIS Simple Feature Access, http://www.opengeospatial.org/standards/sfa
19. OWL Web Ontology Language, http://www.w3.org/TR/owl-features/
20. de Perez, C.L., Conrad, S.: Relational.OWL A Data and Schema Representation Format Based on OWL. In: Second Asia-Pacific Conference on Conceptual Modelling (APCCM2005) (2005)
21. Peuquet, D.: Representations of Space and Time. The Guilford Press, New York (2002)
22. Protégé-OWL editor. http://protege.stanford.edu/overview/protege-owl.html
23. Spaccapietra, S., Cullot, N., Parent, C., Vangenot, C.: On Spatial Ontologies. In: GeoInfo (2004)
24. Torres, M., Quintero, R., Moreno, M., Fonseca, F.T.: Ontology-Driven Description of Spatial Data for Their Semantic Processing. In: Rodríguez, M.A., Cruz, I., Levashkin, S., Egenhofer, M.J. (eds.) GeoS 2005. LNCS, vol. 3799, pp. 242–249. Springer, Heidelberg (2005)
25. Viegas, R., Soares, V.: Querying a Geographic Database using an Ontology-Based Methodology. In: GEOINFO 2006 - VIII Brazilian Symposium on GeoInformatics, Campos do Jordão, Brazil (2006)
26. Volz, R., Oberle, D., Staab, S., Studer, R.: Ontolift demonstrator (2004), http://wonderweb.semanticweb.org
27. WordNet: A Lexical Database fort he English Language. http://wordnet.princeton.edu/

Applying Spatial Reasoning to Topographical Data with a Grounded Geographical Ontology

David Mallenby and Brandon Bennett

School of Computing
University of Leeds, Leeds
LS2 9JT, UK
{davidm,brandon}@comp.leeds.ac.uk

Abstract. Grounding an ontology upon geographical data has been proposed as a method of handling the vagueness in the domain more effectively. In order to do this, we require methods of reasoning about the spatial relations between the regions within the data. This stage can be computationally expensive, as we require information on the location of points in relation to each other. This paper illustrates how using knowledge about regions allows us to reduce the computation required in an efficient and easy to understand manner. Further, we show how this system can be implemented in co-ordination with segmented data to reason about features within the data.

1 Introduction

Geographic Information Systems are becoming increasingly popular methods of representing and reasoning with geographical data. In order to do this, we require methods of reasoning logically about geographical features and the relations that hold between them, including spatially. Ontologies have been cited as a method to perform this reasoning [1,2,3], but existing methodologies do not handle the inherent vagueness adequately. Features are often dependant on the context in which they are made, with local knowledge affecting definitions. Geographical objects are often not a clearly demarcated entity but part of another object [1,4]. The individuation of entities is therefore more important to geographical domains than to others.

One approach proposed to improve the handling of vagueness is to ground the ontology upon the data [5,6],making an explicit link between the ontology and the data, thus allowing reasoning to be made within the context of the particular data. So we require approaches that will allow spatial reasoning such as Region Connection Calculus (RCC) [7] to be used. RCC is a powerful representation of the principal relations between regions, but it can also be computationally expensive.

In this paper we examine developing the system introduced in [6], which takes topographical data as input and segments into polygons with attached attributes.

F. Fonseca, M.A. Rodríguez, and S. Levashkin (Eds.): GeoS 2007, LNCS 4853, pp. 210–227, 2007.

The data to be looked at is of the Hull Estuary[1], with the aim being to obtain a method of reasoning about the hydrological features implicit in the data. We examine how this segmented data can be stored effectively, and what is required in order to reason about the RCC relations between given polygons. Finally, we look at how these can be expanded to allow first order logic definitions of inland water features to be entered, with the appropriate regions returned. We do this by applying an example definition to see what results are returned.

2 Motivation

Vagueness is inherent to the geographical domain, with many features being context dependant, as well as lacking precise definitions and boundaries. Vagueness is not a defect of our language but rather a useful and integral part. It is a key research area of the Ordnance Survey[2], where it has been noted that GIS does not handle multiple possible interpretations well. Rather than attempting to remove vagueness, we should allow the user to make decisions about vague features. So rather than segmenting or labelling the image in advance, we require a mechanism for entering logical queries that may incorporate vagueness and can segment accordingly.

With GIS, we need to deal with several layers. We have our initial data level, which represents the points and polygons that make up a topographical map for example. An additional layer is the ontology level, whereby we define features and relations between the data. The ontology level is usually seen as separate to the data level; we reason within the ontology, and return the data that matches our queries. Thus the ontology is devoid of the data context. This has a clear impact upon handling vagueness, where context is important. A proposed improvement to this is to ground the ontology upon the data [5]. By grounding the ontology, we make an explicit link between the ontology and the data, thus allowing reasoning to be made within the context of the particular data.

The symbol grounding problem as proposed in [8] suggests that computers do not actually understand knowledge they are provided, as meanings are merely symbols we attach to objects. There have been no adequate solutions to this problem as yet and it remains an open problem [9] . Ontology grounding does not solve the problem. Rather, it allows the user to decide the meaning of concepts to some extent.

Grounding the ontology upon the data allows reasoning with the data in particular context, thus achieving our previously mentioned requirement of allowing the user control over the features generated. To ground the ontology upon the data, we need to work at both the data level and the ontology level. In [6]

---

[1] Landsat ETM+ imagery. Downloaded from the Global Landcover Facility (GLCF). Image segmented into water and land then vectorised.
http://glcfapp.umiacs.umd.ed:8080/esdi/index.jsp
[2] Ordnance Survey Research Labs: Modelling fuzzy and uncertain features
http://www.ordnancesurvey.co.uk/oswebsite/partnerships/research/research/data_fuzzy.html

linearity was shown as an example of such an attribute, and it was shown the work required on both levels to use such an attribute. To expand the system, we are required to implement approaches to generate polygons based upon the spatial relations between regions, such as if they are connected or disconnected.

Spatial reasoning can be computationally expensive, as we require information on the location of all points in relation to a given region. Previous work has looked at the problem at an abstract level [10]. By looking at how the relations are calculated, we can determine methods of reducing the calculations required based upon simpler observations. So instead of explicitly requiring every point location be determined, we could use other information to infer what relations are possible and reduce down our scope until we have our solution.

By implementing an RCC based system, we allow quantitative data to be reasoned with qualitatively. This significantly improves the expressiveness of GIS. This also allows for the individuation of features.

3   The Region Connection Calculus

The Region Connection Calculus (RCC) was introduced in [7]. RCC assumes an initial primitive relation C(x,y), which is true if x and y share a common point). From this initial connected relation, we can derive other relations that hold between two regions. A list of the basic key relations as listed in [11] follows:

$$DC(x,y) \equiv_{df} \neg C(x,y) \tag{1}$$

$$P(x,y) \equiv_{df} \forall z[C(z,x) \to C(z,y)] \tag{2}$$

$$PP(x,y) \equiv_{df} [P(x,y) \land \neg P(y,x)] \tag{3}$$

$$EQ(x,y) \equiv_{df} [P(x,y) \land P(y,x)] \tag{4}$$

$$O(x,y) \equiv_{df} \exists z[P(z,x) \land P(z,y)] \tag{5}$$

$$DR(x,y) \equiv_{df} \neg O(x,y) \tag{6}$$

$$PO(x,y) \equiv_{df} [O(x,y) \land \neg P(x,y) \land \neg P(y,x)] \tag{7}$$

$$EC(x,y) \equiv_{df} [C(x,y) \land \neg O(x,y)] \tag{8}$$

$$TPP(x,y) \equiv_{df} PP(x,y) \land \exists z[EC(x,z) \land EC(y,z)] \tag{9}$$

$$NTPP(x,y) \equiv_{df} PP(x,y) \land \neg \exists z[EC(x,z) \land EC(y,z)] \tag{10}$$

RCC-8 consists of eight of these relations: DC, EQ, PO, EC, TPP, TPPi, NTPP, NTPPi, where TPPi and NTPPi are the inverses of TPP and NTPP respectively. Fig. 1 shows graphically the RCC-8 set. This set is both jointly exhaustive and a pairwise disjoint set of base relations, such that only one can ever hold between two given regions [7]. RCC has previously been proposed as a method of spatial reasoning that could be applicable to GIS, for example in [12] where it was noted that the same set of relations have independently been identified as significant for GIS [13,14].

An additional property that we would like to express is the notion of self-connected regions, such that a region is self-connected if it is not divided into a

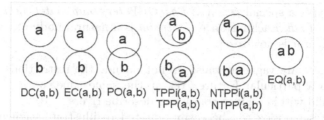

Fig. 1. The RCC-8 relations

number of DC parts. This is important, as in our system we will start with an initial set of segmented polygons, and wish to connect them to form larger regions that satisfy given properties. To do this, we first define a formula sum(x,y) which represents the spatial sum or union of two regions. From this we can define self-connectedness to be equal to the sum of a set of connected regions [11]:

$$\forall xyz[C(z, sum(x, y)) \leftrightarrow [C(z, x) \vee C(z, y)]] \tag{11}$$
$$CON(x) \equiv_{df} \forall yz[EQ(x, sum(y, z)) \rightarrow C(y, z)] \tag{12}$$

So equation 11 states that z represents the spatial sum of regions x and y f all parts of z are either connected to either x or y. This spatial sum is then used in 12 to define self-connectedness (CON); if x is self-connected, any two regions whose spatial sum is equal to x must be connected to each other. Thus x is a single connected region; if we imagine standing in any part of x it would be possible to travel to any other part of x without actually leaving the region.

4   Vagueness in Geography

Vagueness is ubiquitous in geographical concepts [15]. Both the boundaries and definitions of geographical concepts are usually vague, as well as resistant to attempts to give more precise definitions. For example, the definition of a river as given by the Oxford English Dictionary [16] is:

A large natural flow of water travelling along a channel to the sea, a lake, or another river.

The most obvious example of vagueness is 'large', though other aspects may also be vague such as the boundary between respective channels. But this isn't the only definition for a river; some may differ entirely, others may be more or less restrictive. In comparison, OpenCyc[3] is the open source version of Cyc, which is intended to be the largest and most complete general knowledge base in the world. The definitions of river and stream in OpenCyc are:

A *River* is a specialisation of *Stream*. Each instance of River is a natural stream of water, normally of a large volume.

---

[3] OpenCyc http://www.opencyc.org/

A *Stream* is a specialisation of *BodyOfWater*, *InanimateObject-Natural*, and *FlowPath*. Each instance of *Stream* is a natural body of water that flows when it is not frozen.

Again, these are vague and also do not include the restrictions of the water flowing into a particular feature. Yet at the same time, both definitions are perfectly valid within a given context to describe rivers.

The sorites paradox can be easily adapted to illustrate vagueness in geography [3,17], showing that an explicit boundary may not always exist between definitions, such as between rivers and streams. Geographical definitions are also dependant on the context in which they are made. For example, whilst UK rivers are defined usually as permanent flows, in Australia this is not necessarily the case, and thus temporal aspects enter the definition [18]. The application of UK based definitions in Australia could therefore fail to classify some rivers due to their temporal nature, whilst Australian based definitions may overly classify things as rivers when applied in the UK.

The principal approaches for handling vagueness at present are fuzzy logic and supervaluation theory. It is usually the case that the two are presented as opposing theories. However, this in part assumes that vagueness can only take one form, which as discussed in [19] is not true. Rather, there are instances where it is more appropriate to use fuzzy logic and instances where supervaluation theory is better.

The suitability of the two approaches to the proposed system were discussed in [6], where it was noted that supervaluation theory was more applicable as crisp boundaries were produced. This means that we use *precisifications* to represent user decisions and to set contexts.

5   Data Segmentation

In [6], an initial polygon representing the inland water network extending from the Hull estuary was segmented based upon linearity thresholds. This was done by first finding the medial axis of the polygon using a voronoi diagram based approach VRONI [20]. The medial axis of a polygon as first proposed in [21] is defined as the locus of the centre of all the maximal inscribed circles of the polygon. Here, a maximal inscribed circle is a circle that cannot be completely contained within any other inscribed circle in the polygon [22].

However, the medial axis is extremely sensitive to noise and variation along the edge of the input polygon. We want to be able to prune off arcs such that the remaining arcs still represent the topology of the polygon effectively, without disconnecting parts or removing arcs we wish to keep. The approach used to prune the medial axis skeleton here was contour portioning [23,24], which satisfies these requirements. The contours used here are manually defined; whilst an automatic approach is desirable (and work has been done in this area), it is beyond the scope of this project.

The results of using contour partitioning are shown in Fig. 2, where we see the remaining skeleton retains the topology whilst removing unnecessary arcs. This skeleton easily translates into a graph.

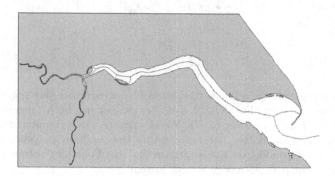

Fig. 2. The results of contour partitioning to reduce the medial axis of the Hull Esturary to a simplified skeleton whilst retaining topology of the shape

The next stage was to use this skeleton to determine linearity. In [6] this was done using a scale invariant approach looking at the variation in widths across stretches of the skeleton. From this we can determine linear lines in the skeleton.

To generate a polygon from this, we determine a left and right side for the skeleton, and combine this with the end points to create a simple polygon. For each side, we use the two boundary points closest to both end nodes, which we already know as these are the points that the maximal inscribed circle at each point touches the boundary.

Once these two points are selected, we find the shortest path between the two along the boundary. This is done by representing the boundary as a graph, and thus a path between is easy to calculate. If no path exists or the length of the path is too great in relation to the distance between the points, then we simply use a single edge with the points as end nodes. An example of this is shown in Fig. 3. This approach guarantees a unique polygon for each line that is simple. We can also use the technique on sets of lines to generate larger polygons.

The initial results of this segmentation stage is a series of segmented polygons, with a label attached representing whether the polygon is linear or non-linear. Further attributes could be used to generate further polygons and labels, such as different linearity measures or size and distance measurements. Some may require further segmentation of the data, whilst others can be performed without segmenting.

The initial results of marking all linear polygons is shown in Fig. 4. Although most parts we would like marked as linear are marked as such, there are some cases that are not. These may be rectified with alternative or refined definitions of linearity. For example, the mouth of the river does not seem linear, because although the width is not varying, the difference in the two banks is significant. Therefore a refined linearity definition may be that a polygon is required to also

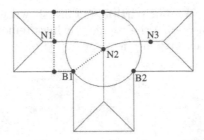

Fig. 3. An example of how a line is translated into a corresponding polygon. For this shape, we have taken the line *N1-N2*. For *N1* we just use both tangent points. For *N2* we choose *B1* as it is closer to *N1*. We then trace along the boundary using the shortest path, with the dotted line representing the resultant polygon. Had the line been *N1-N2-N3*, depending on the length around the boundary between *B1* and *B2* we may replace the path with edge *B1-B2* instead.

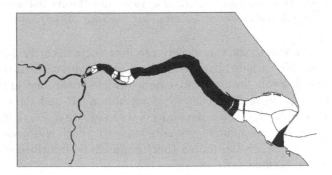

Fig. 4. The results of marking linear stretches, with the original skeleton once again shown. Black sections represent polygons marked as linear with respect to the width.

be linear in relation to its edges, in that the length of the edges should not vary too great from the length of the stretch.

However, there are always likely to be discrepancies in our data, because of variations in actual data in comparison to our abstract notion of a river as a constant line. So we would like a mechanism that can flag up such small discrepancies so that they can be filled in. A method for this was discussed in [6], where the discrepancies were referred to as gaps. To avoid confusion we have introduced the term *interstretch* to represent these features. So using a closeness threshold we can determine which polygons could be 'filled in' at a higher level to generate connected stretches.

6   Data Storage and Querying

Our initial system allows us to segment our data into a series of linear or non-linear polygons, as well as identify *interstretches*. However, we would like to reduce the amount of pre-computed features used, as the aim is to allow a user to generate their own definitions. For example, rather than explicitly calculating *interstretch*, we would like to be able to identify these parts based upon first order logic. So example definitions of stretch and *interstretch* are:

$$stretch[l](x) \leftrightarrow CON(x) \land P(x, WATER) \land linear[l](x) \quad (13)$$
$$\land \; \forall y(P(y, WATER) \land linear[l](y)$$
$$\land \; P(x, y) \rightarrow EQ(x, y))$$
$$interstretch[c, l](x, y, z) \leftrightarrow stretch[l](x) \land stretch[l](y) \quad (14)$$
$$\land \; P(z, WATER) \land EC(x, z) \land EC(y, z)$$
$$\land \; CON(z) \land \forall w(PP(w, z)$$
$$\rightarrow (close - to[c](w, x) \land close - to[c](w, y)))$$

Here, we use the form $p[v](x)$, where the predicate p is true for a given variable or *precisification* v for a given variable x, So, our previous definition of something being linear if the variation in widths is low translates to linear$[l](x)$, or x is linear for a given *precisification* l. Equation 13 defines stretch as a maximal self-connected region that is water and linear for a given *precisification*. Equation 14 defines an *interstretch* as a self-connected region of water that is connected to two stretches, such that all parts of the *interstretch* are close to the two stretches.

Now, instead of having *interstretch* as a primitive, we have a primitive *close-to*, representing the notion that they are close if the distance between is insignificant. As with linearity, this can be treated as a *precisification*. From these definitions, we wish to define water-channels to be maximal self-connected regions that are made up of stretches or *interstretches*. An initial attempt at representing this logically is:

$$waterchannel[c, l](x) \quad \leftrightarrow \quad CON(x) \land P(x, WATER) \land \forall w(PP(w, x) \quad (15)$$
$$\rightarrow \quad \exists s(stretch[l](s) \land P(w, s) \land TPP(s, x)) \lor$$
$$\exists d, e, f \; (interstretch[c, l](d, e, f) \land P(w, f)$$
$$\land \quad TPP(d, x) \land TPP(e, x) \land TPP(f, x)))$$

So a water-channel is a self-connected region of water such that all proper parts of the region are either part of a stretch or an *interstretch*, that is also proper parts of the channel.

This is not the only way such a query could be formed, and there may be further refinements required in order to capture exactly our intended definition. In order to represent a query such as water-channel, we require several stages of work. First, we need a data representation that allows more effective querying. We then need to consider how we can test for RCC relations. Finally, we then need a method of producing the union of resultant polygon sets to produce the final water-channel result.

Our aim at each stage is to find a balance between simplicity and computational complexity. The system is intended to use logic definitions that may not be known at this stage. So to accommodate for this, our design should be reasonably easy to understand and adapt, whilst remaining reasonably efficient.

6.1   Data Storage

The winged edge structure [25] and variations such as the half-edge winged structure [26] offer a more effective representation of polygons as opposed to simply storing the corner points. Our initial polygon data can be easily translated into such a structure. We can easily gain a list of all polygons, edges and points in Prolog.

6.2   Calculating the RCC Relations

We now move on to encoding RCC relations such that we can query the system to find the relations between polygons, thus allowing qualitative and quantitative queries. However, this move from an abstract level to the data level is computationally expensive. We therefore wish to reduce the calculations required at each stage in order to speed up the reasoning process. Previous work on an abstract level was done in [10], which illustrated the process could be broken down into a hierarchical tree, reducing the calculations required.

We first reduce down the potential relations that can occur between two polygons. A first approximation is to compare the bounding boxes of each polygon, hereby defined as the smallest rectangle that can entirely contain its polygon. This significantly reduces the initial calculations, and allows us to eliminate relations that are not possible. We do this using an approach similar to Allen's interval Algebra [27]. The algebra represents 13 different relations (hereby referred to as Allen relations) that can occur between two time intervals, as shown in Fig. 5.

If we treat the X-Axis and Y-Axis as separate dimensions, we can determine the Allen relations between two polygons in each axis. We then compare the resulting pair of Allen relations and determine what possible RCC relations these allow. Determining the Allen relations is straightforward; for a given axis we find

Fig. 5. A graphical representation of the 13 different Allen relations. With the exception of the final relation Equals, the other 12 are in fact 6 pairs of duals. So the first relation represents both white before black and black after white.

the minimum and maximum values of the two polygons and represent as two lines. We can then sort these numbers and determine what Allen relation they correspond to. We repeat for the other axis, so each operation is only working in a single dimension.

This results in a pair of Allen relations, which in turn represent a set of possible RCC relations, as shown in 6. In these examples, we have quickly determined that the first example shows two disconnected polygons, thus no further computation is required. With the other two examples we are left with a set of possible relations. However, we can use these reduced sets to determine the most effective approaches to take next, thus tailoring our deductions to each pair of polygons.

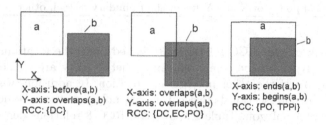

X-axis: before(a,b)
Y-axis: overlaps(a,b)
RCC: {DC}

X-axis: overlaps(a,b)
Y-axis: overlaps(a,b)
RCC: {DC,EC,PO}

X-axis: ends(a,b)
Y-axis: begins(a,b)
RCC: {PO, TPPi}

Fig. 6. Examples of how the bounding boxes of two polygons a and b may be related spatially, and what possible RCC relations these represent. We obtained the Allen relations for the X- and Y-axis', then compared these to see what the set of possible RCC relations are for the polygons.

Theoretically there are 169 different combinations, but in fact there are only 14 different possible combinations, listed in table 1. So we now have a method of reducing the possible RCC relations quickly; we can for example quickly determine polygons which are definitely disconnected.

Table 1. The possible relations as a result of comparing the Allen relations between the X- and Y-axis. Starred relations also have versions replacing TPP/NTPP with TPPi/NTTPi.

Possible RCC combinations from previous stage
DC
DC, EC
DC, EC, PO
DC, EC, PO, TPP *
EC, PO
EC, PO, TPP, NTPP *
EC, PO, TPP, TPPi, EQ
PO
PO, TPP *
PO, TPP, NTPP *

Table 2. The definitions of the RCC-8 relations used in the system. Only 6 are shown here, as TPPi and NTPPi are merely the inverses of TPP and NTPP, respectively.

RCC	Definition in system
DC(X,Y)	There is no intersection between X and Y, and no point of X is inside or on the boundary of Y (and vice versa)
EC(X,Y)	There exists a point that is on the boundary of both X and Y, but there are no points of X inside Y (or vice versa)
PO(X,Y)	There exists either at least one intersection between X and Y, and there are points that are inside one polygon but outside the other
TPP(X,Y)	All points of X are either inside or on the boundary of Y
NTPP(X,Y)	All points of X are inside Y
EQ(X,Y)	All points of X and Y lie on the boundary of each other

Our calculations for RCC relations are based upon the locations of the corners of the polygons to be compared, and whether they are inside, outside or on the boundary of the other polygon in question. In addition, we add to our set all points of intersection between the two polygons that are not already a corner point of a polygon. Table 2 shows the RCC-8 relations defined in terms of the tests that are required in order to make decisions as to the RCC relation between two regions. This is similar to [28], where the spatial domain was also restricted to polygons as opposed to arbitrary points due to the existence of efficient algorithms to handle polygons.

6.3   Intersections

The intersections of two polygons has been studied extensively, in an attempt to improve upon the brute force approach of comparing all lines against all others. More efficient methods are based upon the sweepline approach [29]. The aim of such algorithms is to reduce the comparisons between lines. For our approach, we use our Allen relations based approach to reduce the number of intersection tests.

Using the brute force algorithm as our basis, we order the polygons into two sets of lines. We then take a line in our first set and compare against the bounding box second set, since if an intersection exists the line must touch, intersect or be inside this bounding box. So using our Allen relations approach we can quickly test if the line falls inside the box, and thus eliminate lines that could not intersect the second polygon. For a line that satisfies this criteria, we wish to improve upon simply then comparing the line against all others. We once again use Allen relations, as lines can't intersect if they occur before/after each other in either axis. This once again eliminates many lines, leaving only a small set to be tested.

One final consideration is which of the polygons to use as our first set, as this choice can further speed up the process. Looking at the possible relations, we first see if the relation PP is possible. If so, we use the outer polygon, as our bounding box test would remove no lines if the inner polygon was used. If the relation PP is not possible, we use whichever polygon has fewer lines, as this will remove more intersections in the first part of the test.

So our intersection algorithm uses information previously calculated to speed up the calculation of all intersects whilst remaining simple to understand. Although further work is required to determine the actual efficiency of this approach in comparison to others, it has so far been successfully implemented within Prolog, where it has proven fast enough for the requirements of the project. The result of this stage is a set of points of intersection, which may include existing corner points of a polygon. These can be separated into existing and new points using set operations.

6.4 Points Inside

As with our intersection tests, we wish to reduce down the number of points we test to keep computation time down. Using the bounding boxes generated previously, we can again reduce down the possible points to be all those that are inside or touching the bounding box. This subset is then tested to find which points are inside using a standard test of extending a line horizontally from the point and then counting the number of intersections with the polygon; if the number of intersections is odd the point is inside and if it is even the point is outside. We can reduce the number of intersection tests by using Allen relations to eliminate lines that could not intersect the projected line. How to handle points that lie on the boundary is often an issue for such algorithms. However, we have previously found this set of boundary points in our intersection tests and so can use this set to remove points on the boundary, leaving only points explicitly inside.

6.5 Using the Results with RCC

The results of the previous stages give us a series of sets of points. We can therefore test for RCC relations using set operations on these points, as our previous definitions easily translate into set operations.

First, we find all the potential RCC relations using the Allen relations based approach mentioned earlier. From this stage we can make decisions based on which tests to do; for example if the results of this stage is the set [DC,EC], we know we only need to test for at least one intersection to determine if the answer is DC or EC. So for each of these sets of possible relations we can order the queries to be asked so that they are optimal. We can also find other relations that are implied; if DC is not a member of the list then we know that the two regions are connected, whereas if DC and EC have both been removed we know that at least some part of one region is part of the other.

By using Prolog, we are able to allow for variations of the query. So instead of simply being able to return the relation between two polygons, we can also ask such queries as "find all polygons that are connected to X" and "find all pairs of polygons that are externally connected".

6.6   Building New Regions Based on Queries

The aim of our system is to return regions that match particular queries from an ontology, so we require the system to be able to return sums of regions. For our water-channel example, we need to find all linear polygons, as well as all *interstretch* polygons that connect linear polygons together, and return the results as single connected regions.

We have previously defined self-connected as being the sum of connected regions. This is also applicable to our skeleton and the associated graph, whereby any subset of this graph can be considered self-connected if there is a path between all nodes in the subgraph generated from the subset. As illustrated in Fig. 3, the polygon generated for any given line is simple and self-connected. Thus using this technique on sets of lines is the equivalent of taking the union of all the polygons generated from all connected subsets of the set of lines. We can thus infer that the resulting polygon is self-connected if the skeleton used to generate it is a connected graph. To produce all linear polygons, we simply find the set of all linear lines and convert into a graph, then generate polygons for each maximal self-connected component, where maximal means that there does not exist an edge that is connected to our component that is not part of the component. For *interstretch*, we find the set of lines used to produce the polygons that satisfy our definition and repeat the process above (thus some non-linear polygons may have more than one *interstretch* proper part).

To generate our maximal self-connected polygons, we an approach similiar to a breadth-first search, marking neighbours of polygons as we find them. An example of this process is shown in Fig. 7.

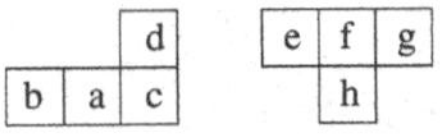

Fig. 7. An example of how self-connected regions are marked. Starting at *a* our breadth-first search returns the set *a,b,c,d*, and then finds polygons remain, so repeats to get *e,f,g,h*. These sets can then be spatially summed to return maximal self-connected regions.

So for our water-channel example, our criteria is that all polygons are either linear or an *interstretch* between linear polygons. We find this set, then using our breadth-first search type approach, travel through all connections until all have been visited. The result is sets of maximal self-connected regions.

We now wish to generate the sum of these polygons, to create our new polygon representing a water-channel. For this, we can use our winged edge structure coupled with our polygon generation approach. Firstly, if we have a set of polygons that are only ever EC, we can find the union by removing all edges that are incident to two or more polygons and traveling along the remaining edges, returning a polygon when we reach our start point (if further edges remain, these are holes and we simply repeat the process until there remains no unvisited edges). If we have overlapping polygons, then we can use our polygon generation approach to create a union by combining the sets of lines that make up our polygons and generating a new polygon that represents their union. We could simply use this approach for the union of all polygons, but this is slower than the union of existing polygons.

So our sum operation combines the previous operations; we find our set of candidate polygons, find maximal self-connected sets via our breadth-first search and then form the union either through union or additional polygon generation. Further operations such as spatial difference or intersection could also be developed, but are beyond the scope of this work. The results of running our water-channel query are shown in Fig. 8, where we see stretches have successfully been joined to form larger regions.

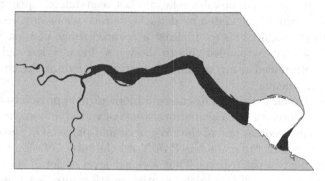

Fig. 8. The results of running the water-channel query. The linear stretches were segmented first, then a query representing interstretch was run. Finally, a water-channel was defined to be the self-connected sum of these two features, such that we find the set of polygons that are either linear or an interstretch, then used our traveling algorithm to find maximal self-connected sets.

7   Future Work

The next key stage of the research is into further logical definitions that can be used to represent inland water features, and thus construct an ontology that represents such features. This may require further primitive functions to be implemented in addition to the linearity and closeness tests present in the system. However, the aim is to keep such primitives to a minimum, as the system is intended to be as general as possible. Thus new features should be defined in logical definitions at the ontology level.

The system has been developed in Prolog and at present is designed to use first order logic based queries. However, a possible extension would be to integrate more closely with a language such as OWL, which can be inputted into Prolog [30,31]. By creating the ontology in OWL, we allow interaction with the semantic web, whilst retaining the segmentation level in Prolog allows us to reason with vague features and ground the ontology upon the data effectively. This is also proposed in [32], where it is shown that OWL cannot effectively handle RCC without modifying the rules of the language. However, such revisions may remove other favourable features of OWL, hence a hybrid system is more appropriate.

8 Related Work

The problem of combining qualitative and quantitative data has previously been discussed in [33]. Here, the combination of different levels of information are discussed, such that the intention is to bridge the gap between the primitive level of points, lines and polygons, and the object level describing the spatial relations and definitions of features.

Like the Allen relation based approach used in this paper, transitivity tables are formed representing the possible relations between different primitives. Thus, spatial relations can be calculated by deductive processes as opposed to computational geometric algorithms (or at least a reduced usage of such algorithms). In this paper, we have expanded this to show how intersection and point locations can be determined using similar approaches to reduce the computational geometry requirements.

As previously mentioned, [10] discussed a hierachical approach to determining RCC relations. Moreover, the calculations were converted to boolean terms, such that the problem becomes one of the closure of half-planes. On the other hand, in this paper decisions are made based on both the intersection and location of points with respect to the regions. Thus a richer level of detail is deducible.

Another approach to deducing the spatial relationships is to use constraint logic programming [34], as discussed particularly in [35]. Such an approach offers an interesting alternative, but is reliant on the efficiency of the constraint logic solver used, and as discussed in [35] further work is required to improve such an approach for effective implementation.

9 Conclusions

In this paper we have demonstrated a method of calculating and using RCC relations on segmented topographical data, thus allowing integration with an ontology grounded upon the data. This improves the handling of vagueness within geographical features, as we can make decisions on features based upon the context in which they are made, as opposed to using predefined regions.

We have shown that although the calculation of RCC relations is computationally expensive, we can still implement the relations effectively by using other knowledge representation approaches such as Allen's interval algebra. Further, Allen's relations were adapted to provide simple but effective methods of calculating the intersections and locations of points of polygons in relation to each other, although more efficient algorithms may exist. Further work is therefore required to determine the efficiency of the approaches discussed here, or whether a hybrid approach using deductive methods in conjunction with other computational geometric algorithms, thus providing the most efficient environment overall.

We have shown how previous queries used in the system could be written in first order logic instead of being specified in the code. Although these may require further clarification, this does highlight the possibility of defining features in first order logic. We have also shown how maximal self-connected regions satisfying such queries can be generated. Finally, we have shown where the work is intended to progress and how this will improve the handling of vagueness within geographical features.

Acknowledgments. I am grateful to our industrial partner Ordnance Survey, for their contribution towards PhD funding. I am also grateful to Brandon Bennett and Allan Third for their support. Finally, I am extremely grateful to the reviewers for their comments that helped me prepare the final text.

References

1. Fonseca, F., Egenhofer, M., Agouris, C., Cmara, C.: Using Ontologies for Integrated Geographic Information Systems. Transactions in Geographic Information Systems 6(3), 231–257 (2002)
2. Guarino, N., Welty, C.: Evaluating Ontological Decisions with Ontoclean. Communications Of The Acm 45(2), 61–65 (2002)
3. Varzi, A.C.: Vagueness in Geography. Philosophy & Geography 4(1), 49–65 (2001)
4. Smith, B., Mark, D.M.: Ontology and Geographic Kinds. In: Poiker, T., Chrisman, N. (eds.) Proceedings of the Tenth International Symposium on Spatial Data Handling, Burnaby, BC, pp. 308–320. Simon Fraser University (1998)
5. Jakulin, A., Mladenić, D.: Ontology Grounding. In: Proceedings of the 8th International multi-conference Information Society IS-2005, Ljubljana, Slovenia, pp. 170–173 (2005)
6. Mallenby, D.: Grounding a Geographic Ontology on Geographic Data. In: Amir, E., Lifschitz, V., Miller, R. (eds.) 8th International Symposium on Logical Formalizations of Commonsense Reasoning (Commonsense 2007), Stanford, USA, pp. 101–105 (2007)
7. Randell, D.A., Cui, Z., Cohn, A.G.: A Spatial Logic-Based on Regions and Connection. In: Principles Of Knowledge Representation And Reasoning: Proceedings Of The Third International Conference (Kr 92), pp. 165–176. Morgan Kaufmann Pub Inc., San Mateo (1992)
8. Harnad, S.: The Symbol Grounding Problem. Physica D 42(1-3), 335–346 (1990)
9. Taddeo, M., Floridi, L.: Solving the Symbol Grounding Problem: a Critical Review of Fifteen Years of Research. Journal Of Experimental & Theoretical Artificial Intelligence 17(4), 419–445 (2005)
10. Bennett, B., Isli, A., Cohn, A.G.: A System Handling RCC-8 Queries on 2D Regions Representable in the Closure Algebra Half -Planes. In: Moonis, A., Mira, J.M., de Pobil, A.P. (eds.) Methodology and Tools in Knowledge-Based Systems. LNCS, vol. 1415, pp. 281–290. Springer, Heidelberg (1998)
11. Giritli, M.: Who Can Connect in RCC? In: Günter, A., Kruse, R., Neumann, B. (eds.) KI 2003. LNCS (LNAI), vol. 2821, pp. 565–579. Springer, Heidelberg (2003)

12. Bennett, B.: The Application of Qualitative Spatial Reasoning to GIS. In: Abrahart, R. (ed.) Proceedings of the First International Conference on GeoComputation, vol. I, pp. 44–47 (1996)
13. Egenhofer, M.J.: Reasoning about Binary Topological Relations. In: Günther, O., Schek, H.-J. (eds.) SSD 1991. LNCS, vol. 525, pp. 144–160. Springer, Heidelberg (1991)
14. Egenhofer, M.J., Franzosa, R.D.: Point-Set Topological Spatial Relations. International Journal Of Geographical Information Systems 5(2), 161–174 (1991)
15. Bennett, B.: What is a Forest? on the Vagueness of Certain Geographic Concepts. Topoi-An International Review Of Philosophy 20(2), 189–201 (2001)
16. Simpson, J., Weiner, E. (eds.): Concise Oxford English Dictionary, 11th edn. Oxford University Press, Oxford (2004)
17. Varzi, A.C.: Vagueness, Logic, and Ontology. The Dialogue. Yearbooks for Philosophical Hermeneutics 1, 135–154 (2001)
18. Taylor, M.P., Stokes, R.: When is a River not a River? Consideration of the Legal Definition of a River for Geomorphologists Practising in New South Wales, Australia. Australian Geographer 36(2), 183–200 (2005)
19. Dubois, D., Esteva, F., Godo, L., Prade, H.: An information-Based Discussion of Vagueness. In: 10th IEEE International Conference On Fuzzy Systems, Vols 1-3 - Meeting The Grand Challenge: Machines That Serve People, pp. 781–784. IEEE PRESS, New York (2001)
20. Held, M.: Voroni: An Engineering Approach to the Reliable and Efficient Computation of Voronoi Diagrams of Points and Line Segments. Computational Geometry-Theory And Applications 18(2), 95–123 (2001)
21. Blum, H.: Biological Shape and Visual Science. Journal Of Theoretical Biology 38(2), 205–287 (1973)
22. Ge, Y.R., Fitzpatrick, J.M.: Extraction of Maximal Inscribed Disks from Discrete Euclidean Distance Maps. In: 1996 Ieee Computer Society Conference On Computer Vision And Pattern Recognition, Proceedings. Proceedings / Cvpr, Ieee Computer Society Conference On Computer Vision And Pattern Recognition, pp. 556–561. IEEE PRESS, New York (1996)
23. Bai, X., Latecki, L.J., Liu, W.Y.: Skeleton Pruning by Contour Partitioning. In: Kuba, A., Nyúl, L.G., Palágyi, K. (eds.) DGCI 2006. LNCS, vol. 4245, pp. 567–579. Springer, Heidelberg (2006)
24. Bai, X., Latecki, L.J., Liu, W.Y.: Skeleton Pruning by Contour Partitioning with Discrete Curve Evolution. IEEE Transactions On Pattern Analysis And Machine Intelligence 29(3), 449–462 (2007)
25. Baumgart, B.G.: Winged Edge Polyhedron Representation. Technical report, Stanford University, 891970 (1972)
26. Mantyla, M.: Introduction to Solid Modeling, vol. 60949. W. H. Freeman & Co, New York (1988)
27. Allen, J.F.: Maintaining Knowledge about Temporal Intervals. Communications Of The Acm 26(11), 832–843 (1983)
28. Haarslev, V., Lutz, C., Möller, R.: Foundations of Spatioterminological Teasoning with Description Logics. Principles of Knowledge Representation and Reasoning, 112–123 (1998)
29. Bentley, J.L., Ottmann, T.A.: Algorithms for Reporting and Counting Geometric Intersections. Ieee Transactions On Computers 28(9), 643–647 (1979)
30. Laera, L., Tamma, V., Bench-Capon, T., Semeraro, G.: Sweetprolog: A system to Integrate Ontologies and Rules. In: Antoniou, G., Boley, H. (eds.) RuleML 2004. LNCS, vol. 3323, pp. 188–193. Springer, Heidelberg (2004)

31. Wielemaker, J.: An optimised Semantic Web Query Language Implementation in Prolog. In: Gabbrielli, M., Gupta, G. (eds.) ICLP 2005. LNCS, vol. 3668, pp. 128–142. Springer, Heidelberg (2005)
32. Grütter, R., Bauer-Messmer, B.: Towards Spatial Reasoning in the Semantic Web: A Hybrid Knowledge Representation System Architecture. In: Fabrikant, S., Wachowicz, M. (eds.) The European Information Society: Leading the Way With Geo-information. Lecture Notes in Geoinformation and Cartography, vol. XVII, p. 486 (2007)
33. Abdelmoty, A., Williams, M., Paton, N.: Deduction and Deductive Database for Geographic Data Handling. In: Symposium on Large Spatial Databases, pp. 443–464 (1993)
34. Jaffar, J., Maher, M.J.: Constraint Logic Programming: A Survey. Journal of Logic Programming 19/20, 503–581 (1994)
35. Almendros-Jimenez, J.: Constraint Logic Programming Over Sets of Spatial Objects. In: Proceedings of the 2005 ACM SIGPLAN workshop on Curry and functional logic programming, Tallinn, Estonia, pp. 32–42. ACM Press, New York (2005)

Supporting Complex Thematic, Spatial and Temporal Queries over Semantic Web Data*

Matthew Perry[1], Amit P. Sheth[1], Farshad Hakimpour[2], and Prateek Jain[1]

[1] Kno.e.sis Center, Department of Computer Science and Engineering,
Wright State University, Dayton, OH, USA
[2] LSDIS Lab, Department of Computer Science, University of Georgia, Athens, GA, USA
{perry.66,amit.sheth,jain.18}@wright.edu,
fhakimpour@uga.edu

Abstract. Spatial and temporal data are critical components in many applications. This is especially true in analytical domains such as national security and criminal investigation. Often, the analytical process requires uncovering and analyzing complex thematic relationships between disparate people, places and events. Fundamentally new query operators based on the graph structure of Semantic Web data models, such as semantic associations, are proving useful for this purpose. However, these analysis mechanisms are primarily intended for thematic relationships. In this paper, we describe a framework built around the RDF metadata model for analysis of thematic, spatial and temporal relationships between named entities. We discuss modeling issues and present a set of semantic query operators. We also describe an efficient implementation in Oracle DBMS and demonstrate the scalability of our approach with a performance study using a large synthetic dataset from the national security domain.

1 Introduction

Analytical applications are increasingly exploiting complex thematic relationships between named entities as a powerful tool in the analysis process. Such "connecting the dots" applications are common in many domains, for example national security, drug discovery and medical informatics. Semantic Web data models, such as Resource Description Framework (RDF) [1], fit nicely with this analysis paradigm because relationships are modeled as first class objects. Fundamentally new analytical operators based on the graph structure of RDF have emerged (e.g., semantic associations [2] and subgraph discovery [3]) which allow querying for complex relationships between named entities where an ontology provides the context or domain semantics. We use the term *semantic analytics* to refer to this process of searching, analyzing and visualizing semantically meaningful connections between named entities. Many successful applications of semantic analytics can be seen in the literature (e.g., identifying conflict of interest [4], detecting patent infringement [5] and metabolic pathway discovery [6]).

---

* This work is partially funded by NSF-ITRIDM Award #0325464 & #0714441 entitled "SemDIS: Discovering Complex Relationships in the Semantic Web."

F. Fonseca, M.A. Rodríguez, and S. Levashkin (Eds.): GeoS 2007, LNCS 4853, pp. 228–246, 2007.
© Springer-Verlag Berlin Heidelberg 2007

While spatial and temporal data often play a crucial role in many analytical domains, research in semantic analytics has focused on thematic relationships. Current approaches do not adequately handle spatial and temporal data. Furthermore, traditional spatial and spatiotemporal data models used for GIS [7] excel at modeling and analyzing spatial and temporal relationships between geographic entities but tend to model the thematic aspects of a given domain as directly attached attributes of geospatial entities.

In a recent work [8], we have tried to overcome this limitation by modeling spatial, temporal and thematic data using ontologies and temporal RDF graphs [9]. An upper-level ontology is used to define the basic classes and relationships of the thematic and spatial domains. With this approach, thematic entities and relationships are represented as first class objects and are modeled separately from their spatial properties (basic spatial features, such as points and lines, termed *spatial entities*). Thematic entities and events are connected to spatial entities through *located_at* and *occurred_at* relationships modeled in the upper-level ontology. Deeper domain ontologies are integrated with this upper-level ontology through *rdfs:subClassOf* and *rdfs:subPropertyOf* statements. A unique aspect of this approach is that a 1-to-1 mapping between thematic and spatial entities is not enforced. Rather, a many-to-many mapping is achieved by utilizing indirect thematic connections (specified with domain ontologies) between entities. For example, using a military ontology, a soldier could be associated with the spatial properties of his residence through one set of relationships (*Soldier – lives_at – Residence – located_at – Spatial_Entity*) or with the locations of his training facilities using a different set of relationships (*Soldier – member_of – Military_Unit – trains_at – Base – located_at – Spatial_Entity*).

A variety of query operators are possible over this model which combine thematic relationships with spatial and temporal relationships, thus adding more expressive domain semantics to spatial and temporal queries. We argue that by incorporating more complex models and operators for thematic data, a GIS can be significantly more useful in applications which require complex thematic analysis in addition to spatial and temporal analysis.

Spatial and temporal data bring many unique challenges to semantic analytics applications. Thematic relationships can be explicitly stated in the RDF graph, but some spatial and temporal relationships (e.g., quantitative relationships like distance) are implicit and only evident after additional computation. RDF and RDF Schema (RDFS) inferencing rules [10, 11] are also affected as the temporal properties of asserted statements will have implications on the temporal properties of the corresponding inferred statements.

Example (biochemical threat detection): Suppose an intelligence analyst is assigned the task of monitoring the health of soldiers in order to detect possible exposure to a chemical or biological agent which may imply a biochemical attack. In this case, the analyst may search for relationships connecting a sick soldier to potential chemical or biological agents by matching the soldier's symptoms with known reactions to these agents. In addition, the analyst could further determine the likelihood of a particular chemical substance by querying for associations between the substance and enemy groups in the knowledgebase. For example, a member of the group may have worked at a facility which was reported to have produced the chemical. It is doubtful that such an analysis could produce definitive evidence of a biochemical attack, but incorporating spatial and temporal relationships could help in this regard. For instance, the analyst may want to limit the results to soldiers and enemies in close spatial proximity (e.g., find all soldiers with symptoms indicative of exposure to *chemical X* which fought in battles within 2 miles of sightings of any members of *enemy group Y*). We may pose the following SQL query involving the *spatial_eval* table function for such a search:

```
select a from table (spatial_eval ('(?a has_symptom ?b)
    (Chemical_X induces ?b)(?a fought_in ?c)', ?c,
    '(?d member_of Enemy_Group_Y)(?d spotted_at ?e)', ?e,
    'geo_distance(distance=2 units=mile)'));
```

With this query, we are using the *spatial_eval* operator to specify (1) a relationship between a soldier, a chemical agent and a battle location and (2) a relationship between members of an enemy organization and their known locations. We are then limiting the results based on the spatial proximity of the battles and enemy sightings. Additionally, we provide a *spatial_extent* operator which allows retrieving the spatial geometry associated with the spatial entities composing a thematic relationship and optionally filtering the results using a spatial predicate. For example, *find all soldiers participating in military events that take place within an input bounding box*. For temporal aspects, we provide an analogous *temporal_extent* operator which returns the temporal properties of a given relationship and allows optional filtering. For example, *return all soldiers exhibiting a given symptom during a specific time period*. We also provide a *temporal_eval* operator which can answer queries such as *find soldiers who exhibited symptoms after participating in a given military event*.

This paper focuses on providing a framework to support spatial and temporal analysis of RDF data. RDF is a World Wide Web Consortium (W3C) standard for representing ontologies and corresponding instance data. We address problems of both data storage and operator design and implementation. Specifically, the contributions of this paper are:

- A storage and indexing scheme for spatial and temporal RDF data
- An efficient treatment of temporal RDFS inferencing
- The definition and implementation of four spatial and temporal query operators
- A performance study using a large RDF dataset

The remainder of the paper is organized as follows. Section 2 discusses background information and related work regarding data modeling and querying. Section 3 further describes the set of spatial and temporal query operators. Section 4 describes the implementation of this framework in Oracle DBMS. An experimental evaluation of this implementation follows in Section 5, and Section 6 gives conclusions.

2 Background and Related Work

In this section, we discuss background information and related work with regards to data modeling and querying Semantic Web data.

RDF and Ontologies. RDF [1] has been adopted by the W3C as a standard for representing metadata on the Web. Resources in RDF are identified by Uniform Resource Identifiers (URIs) that provide globally-unique and resolvable identifiers for entities on the Web. These resources are described through participation in relationships. Relationships in RDF are called *Properties* and are binary relationships connecting resources to other resources or resources to *Literals*, i.e., literal values such as Strings or Numbers. These binary relationships are encoded as triples of the form *(Subject, Property, Object)*, which denotes that a resource – the *Subject* – has a *Property* whose value is the *Object*. These triples are referred to as *Statements*. RDF also allows for anonymous nodes called *Blank Nodes* which can be used as the *Subject* or *Object* of a statement. We call a set of triples an *RDF graph*, as RDF data can be represented as a directed, labeled graph with typed edges and nodes. In this model, a directed edge labeled with the *Property* name connects the *Subject* to the *Object*.

RDF Schema (RDFS) [10] provides a standard vocabulary for describing the classes and relationships used in RDF statements and consequently provides the capability to define ontologies. Ontologies serve to formally specify the semantics of RDF data so that a common interpretation of the data can be shared across multiple applications. RDFS allows us to define hierarchies of class and property types, and it allows us to define the domain and range of property types.

Additionally, a set of entailment rules are defined for RDF and RDFS [11]. These rules essentially specify that an additional triple can be added to the RDF graph if the graph contains triples of a specific pattern. Such rules describe, for example, the transitivity of the *rdfs:subClassOf* property.

Temporal RDF Graphs. In order to analyze the temporal properties of relationships in RDF graphs, we need a way to record the temporal properties of the statements in those graphs, and we must account for the effects of those temporal properties on RDFS inferencing rules. For this purpose, we adopt temporal RDF graphs defined in [9]. Temporal RDF graphs model absolute time and are defined as follows. Given a set of discrete, linearly ordered time points T, a temporal triple is an RDF triple with a temporal label $t \in T$. A statement's temporal label represents its valid time. The notation $(s, p, o) : [t]$ is used to denote a temporal triple. The expression $(s, p, o) : [t_1, t_2]$ is a notation for $\{(s, p, o) : [t] \mid t_1 \leq t \leq t_2\}$. A temporal RDF graph is a set of temporal triples. For example, consider a soldier $s1$ assigned to the 1st Armored Division ($1^{st}AD$) from April 3, 1942, until June 14, 1943, and then assigned to the 3rd Armored Division ($3^{rd}AD$) from June 15, 1943, until October 18, 1943. This would yield the following triples: $(s1, assigned\_to, 1^{st}AD) : [04:03:1942, 06:14:1943]$, $(s1, assigned\_to, 3^{rd}AD) : [06:15:1943, 10:18:1943]$. Any temporal ontology that defines a vocabulary of time units can be used to precisely specify the start and end points of time intervals.

As discussed in [9], we must account for the effects of temporal labels on RDFS inferencing rules. To incorporate inferencing into temporal RDF graphs, we must use a basic arithmetic of intervals to derive the temporal label for the inferred statements. For example, interval intersection would be needed for *rdfs:subClassOf* (e.g., $(x, rdfs:subClassOf, y) : [1, 4] \wedge (y, rdfs:subClassOf, z) : [3, 5] \rightarrow (x, rdfs:subClassOf, z) : [3, 4]$).

Related Work. We will first discuss our modeling approach using temporal RDF as it compares with other spatiotemporal models in the literature. For a recent survey, see [7]. Of the models discussed in the literature, the object-oriented and event-based models and the three domain model are most similar to our RDF-based approach. The three domain model, introduced by Yuan, is described in [12, 13]. This model represents semantics, space and time separately. To represent spatiotemporal information in this model, semantic objects are linked via temporal objects to spatial objects. This provides temporal information about the semantic (thematic) properties of a given spatial region. This is analogous to temporal *located_at* and *occurred_at* relationships in our model. The three domain model is quite similar to our approach in that it represents thematic entities as first class objects rather than attributes of geospatial objects. The key difference is that the three domain model relies on direct connections from thematic entities to spatial regions whereas our model allows indirect connections composed of sequences of thematic relationships, which is made possible by a richer modeling of the thematic domain. Additionally, relaxing the direct connection requirement better tolerates incompleteness of information – a necessity when handling Web data. In [14], the authors discuss a combination of the object-oriented and event-based modeling approaches for dynamic geospatial domains. They define an upper-level ontology similar to the one we present in [8]. They model the concept of a setting and a situate function which maps entities and events to settings. Settings can be spatial, temporal, or spatiotemporal. In contrast to our work, the authors focus on geospatial objects and events and model what we would consider a thematic entity (e.g., an airplane) as a geospatial entity. That is, the separation between the thematic and spatial domains is not as strongly emphasized. Our RDF-based modeling approach provides a means to assign spatial properties to those entities not directly connected to a spatial setting and allows deeper analysis of purely thematic relationships.

Many RDF query languages have been proposed in the literature. These include SQL-like languages (e.g., SPARQL [15]), functional languages (e.g., RQL [16]), rule-based languages (e.g., TRIPLE [17]) and graph traversal languages (e.g., RxPath [18]). Efficient implementations of these languages for persistent RDF data usually involve translation into a SQL query against an underlying RDBMS representation of the RDF data (e.g., Jena2 [19], RDFSuite [20]). As an alternative to defining a new query language, an approach for querying RDF data directly in SQL has been proposed [21]. This facilitates easy integration with other SQL queries against traditional relational data and saves the overhead of translating data from SQL to the RDF query language data format. Our implementation follows this approach and introduces new SQL functions for spatial and temporal querying of RDF data.

Work is somewhat limited with regards to incorporating spatial and temporal relationships into queries over Semantic Web data. Examples of querying geospatial RDF data are mostly seen in web applications and semantic geospatial web services [22, 23] in the spirit of the Geospatial Semantic Web [24]. In general, query processing proceeds by translating RDF representations of spatial features into geometric representations on the fly and then performing spatial calculations, and the focus is more on interoperability than efficient query processing. The SPIRIT spatial search engine [25] combines an ontology describing the geospatial domain with the searching and indexing capability of Oracle Spatial for the purposes of searching documents based on the spatial features associated with named places mentioned in the document. In contrast, our searching operators are intended for general purpose querying of ontological and spatial relationships. Querying for temporal data in RDF graphs is less complicated as RDF supports typed literals such as *xsd:date*, and corresponding query languages support filtering results based on literal values. However, this is far from supporting full temporal RDF as graphs discussed in this paper. In addition to formally defining temporal RDF graphs, Gutierrez et al. briefly discussed aspects of a query language for these graphs, but no implementation issues were mentioned [9]. Also, to the best of our knowledge, this paper is the first to investigate implementation of RDFS inferencing which incorporates the concept of valid time for RDF statements.

3   Query Operators

In this section, we introduce a set of spatial and temporal query operators for searching and analyzing spatial and temporal relationships between named entities in temporal RDF graphs. These operators are an adequate functional set in that they (1) allow precise specification of a thematic portion of the RDF graph (subgraph), (2) provide facilities to compute spatial and temporal properties of these subgraphs and (3) allow filtering and joins based on the computed spatial and temporal properties. The operators are implemented as SQL table functions. Table functions produce a set of rows as output which can be queried. They are used in SQL queries in the same manner as a database table name. For example, we may have the query `select x, y from table (table_func (...)) order by x`.

Graph Patterns. SPARQL-like graph patterns are the basic building block of these operators. Intuitively, a *graph pattern* is a set of RDF triples where the subjects, properties and/or objects may be replaced with variables. In general, a graph pattern query against an RDF graph *G* returns a set of mappings between the variables in the graph pattern and terms (URIs, Blank Nodes and Literals) in *G* such that substituting the mapped terms into the graph pattern results in a set of triples actually present in *G*. We refer to the set of triples resulting from a substitution as a *graph pattern instance*, and the result of a graph pattern query on a given RDF graph *G* is the set of variable bindings for all matching graph pattern instances in *G*. Fig. 1 illustrates these concepts for an example graph pattern query.

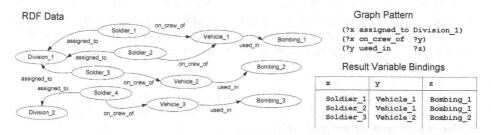

Fig. 1. Example graph pattern with resulting variable bindings

Spatial Query Operators. We define two spatial query operators for RDF graphs containing geospatial data: *spatial_extent* and *spatial_eval*. The following descriptions assume the existence of a class *Geometry* in the ontology which models spatial objects, and we use the term *spatial feature* to refer to an *SDO_GEOMETRY* object that would be stored in Oracle Spatial (i.e., the implementation of *Geometry*).

The first spatial operator, *spatial_extent*, is intended to retrieve the spatial feature of the *Geometry* connected to a thematic entity and optionally filter the results based on the properties of the spatial feature. The signature for the corresponding table function is shown below:

```
spatial_extent (graphPattern VARCHAR, spatialVar
  VARCHAR, ontology RDFModels, <geom SDO_GEOMETRY>,
  <spatialRelation VARCHAR>)
returns AnyDataSet;
```

The *graphPattern* parameter specifies the relationship between a thematic entity and a *Geometry*, for example (*Soldier*, *fought_in*, *Battle*) (*Battle*, *located_at*, *Geometry*). The *spatialVar* parameter identifies the variable in the graph pattern that corresponds to a *Geometry*, and *ontology* determines the ontology to search against. This function returns a table with rows containing columns for each variable in the graph pattern and one column for the spatial features. Each row contains the URI bound to each variable and the spatial feature corresponding to the *Geometry* bound to *spatialVar* (displayed as *well known text* format in Oracle). Two optional parameters, a spatial feature and a spatial relationship, can be used to filter the graph pattern instances. In this case, the table would only contain those graph pattern instances whose associated spatial features satisfy the specified spatial relation with the input spatial feature. We support the following spatial relationships: *touch*, *overlap*, *equal*, *inside*, *covered by*, *contains*, *covers*, *any interact* and *within distance*.

The second spatial operator, *spatial_eval*, acts as a spatial join between graph pattern instances. It is intended to allow for searching thematic entities based on their spatial relationships. The signature for the corresponding table function is shown below:

```
spatial_eval (graphPattern VARCHAR, spatialVar VARCHAR,
  graphPattern2 VARCHAR, spatialVar2 VARCHAR,
  spatialRelation VARCHAR, ontology RDFModels)
return AnyDataSet;
```

graphPattern and *spatialVar* specify the left hand side of the join operation, while *graphPattern2* and *spatialVar2* specify the right hand side. *spatialRelation* identifies the spatial join condition. This function returns a table containing a column for each variable in *graphPattern* and *graphPattern2* and a column for each associated spatial feature (*sf*1 and *sf*2). For each row in the resulting table, *sf*1 *spatialRelation* *sf*2 evaluates to *true*.

Temporal Query Operators. We define two temporal query operators for temporal RDF graphs: *temporal_extent* and *temporal_eval*. The basic idea behind the operators is that we compute a temporal interval for a graph pattern instance based on the temporal properties of the triples making up the graph pattern instance.

The first temporal operator, *temporal_extent*, is used to compute the temporal interval for a graph pattern instance and optionally filter the results based on the computed temporal interval. We support two basic intervals for a graph pattern instance: the interval during which the entire graph pattern instance is valid (*INTERSECT*) and the interval during which any part of the graph pattern is valid (*RANGE*). The signature for the corresponding table function is shown below.

```
temporal_extent (graphPattern VARCHAR, intervalType
   VARCHAR, ontology RDFModels, <start DATE>,
   <end DATE>, <temporalRel VARCHAR>)
   return AnyDataSet;
```

This function takes three parameters as input, specifically a graph pattern, a String value specifying the interval type (*INTERSECT* or *RANGE*), and a parameter specifying the ontology to search against. The table returned contains a column for each variable in the graph pattern and two *DATE* columns which specify the start and end of the time interval computed for the graph pattern instance. Three optional parameters, two *DATE* values to identify the boundaries of a time interval and a temporal relationship, can be used to filter the found graph pattern instances. In this case, assuming the *DATE* columns in the returned table are named *stDate* and *endDate*, each row in the result satisfies the condition [*stDate, endDate*] *temporalRel* [*start, end*]. We currently support seven temporal relationships: *before, after, during, overlap, during_inv, overlap_inv* and *any interact*.

The second temporal operator, *temporal_eval*, acts as a temporal join operator for graph pattern instances. The corresponding table function has the following signature:

```
temporal_eval (graphPattern VARCHAR, intervalType
   VARCHAR, graphPattern2 VARCHAR, intervalType2
   VARCHAR, temporalRel VARCHAR, ontology RDFModels)
   return AnyDataSet;
```

graphPattern and *intervalType* specify the left hand side of the join operation, while *graphPattern2* and *intervalType2* specify the right hand side. *temporalRel* identifies the join condition. This function returns a table containing a column for each variable in *graphPattern* and *graphPattern2* and four *DATE* columns (*start1, end1, start2, end2*) to indicate the time interval for each found graph pattern instance. For each row in the resulting table, [*start1, end1*] *temporalRel* [*start2, end2*] evaluates to *true*.

4 Implementation in Oracle

In this section, we describe the implementation of our spatial and temporal RDF query operators in Oracle DBMS. The implementation builds upon Oracle's existing support for RDF storage and inferencing and support for spatial object types and indexes. We create SQL table functions for each of the previously discussed query operators. Additional structures are created to allow for spatial and temporal indexing of the RDF data for efficient query evaluation. We should note that in general this approach is not limited to Oracle and could be implemented using any extensible DBMS that supports user-defined object types and functions.

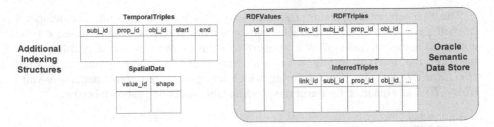

Fig. 2. Storage structures for RDF data. Existing tables of Oracle Semantic Data Store are shown on the right, and our additional tables for efficiently searching spatial and temporal data are shown on the left.

Existing Oracle Technologies. Oracle's Semantic Data Store [26] provides the capabilities to store, infer, and query semantic data, which can be plain RDF descriptions and RDFS based ontologies. To store RDF data, users create a model (ontology) to hold RDF triples. The triples are stored after normalization in two tables: an *RDFValues* table which stores RDF terms and a numeric id and an *RDFTriples* table which stores the ids of the subject, predicate and object of each statement. Users can optionally derive a set of inferred triples based on user-defined rules and/or RDFS semantics. These triples are materialized by creating a rules index and stored in a separate *InferredTriples* table. These storage structures are illustrated in Fig. 2. A SQL table function is provided that allows issuing graph pattern queries against both asserted and inferred RDF statements.

Oracle Spatial [27] provides facilities to store, query and index spatial features. It supports the object-relational model for representing spatial geometries. A native spatial data type, *SDO_GEOMETRY*, is defined for storing vector data. Database tables can contain one or more *SDO_GEOMETRY* columns. Oracle Spatial supports spatial indexing on *SDO_GEOMETRY* columns, and provides a variety of procedures, functions and operators for performing spatial analysis operations.

Data Representation. Our Framework supports spatial and temporal data serialized in RDF using an RDFS ontology discussed in [28]. This ontology models the concept of a *Geometry* Class and allows for recording coordinate system information and representing points, lines, and polygons. This model complies with the OGC simple feature specification [29]. Using this representation, spatial features are stored as instances of *Geometry* and are uniquely identified by their URI. Temporal labels are associated with statements using RDF reification, as suggested in [9]. Reification allows us to assert statements about RDF statements. Our framework supports time interval values serialized as instances of the Class *Interval* from this ontology. A property type, *temporal*, is defined to assert that a statement has a valid time which is represented as an *Interval* instance.

Indexing Approach. In order to ensure efficient execution of graph pattern queries involving spatial and temporal predicates, we must provide a means to index portions of the RDF graph based on spatial and temporal values. Basically, this is done by building a table mapping *Geometry* instance URIs to their *SDO_GEOMETRY* representation and by building a modified *RDFTriples* table which also stores the temporal intervals associated with the triple. In order to build these indexes, users first load the set of asserted RDF statements into Oracle Semantic Data Store and build an RDFS rules index.

Spatial Indexing Scheme. We provide the procedure *build_geo_index()* to construct a spatial index for a given ontology. This procedure first creates the table *SpatialData* (*value_id NUMBER, shape SDO_GEOMETRY*) for storing spatial features corresponding to instances of the class *Geometry* in the ontology. *value_id* is the id given to the URI of the *Geometry* instance in Oracle's *RDFValues* table, and *shape* stores the *SDO_GEOMETRY* representation of the *Geometry* instance (see Fig. 2). This table is filled by querying the ontology for each *Geometry* instance, iterating through the results and creating and inserting *SDO_GEOMETRY* objects into the spatial indexing table. Finally, to enable efficient searching with spatial predicates on this table, a spatial (R-Tree) index is created on the *shape* column.

Temporal Indexing Scheme. Our temporal indexing scheme is a bit more complicated, as it must account for temporal labels on statements inferred through RDFS semantics. However, we only need to handle a subset of the RDFS inference rules. This is the case because we are not interested in handling temporal evolution of the ontology schema. What we need to handle are temporal properties of instance data. Specifically, we need to account for temporal labels of inferred *rdf:type* statements and statements resulting from *rdfs:subPropertyOf* statements. *rdf:type* statements result from the following rules: (1) $(x, rdf:type, y) \wedge (y, rdfs:subClassOf, z) \rightarrow (x, rdf:type, z)$, and (2) $(x, p, y) \wedge (p, rdfs:domain, a) \wedge (p, rdfs:range, b) \rightarrow (x, rdf:type, a), (y, rdf:type, b)$. We infer instance statements from *rdfs:subPropertyOf* using the following rule: (1) $(x, p, y) \wedge (p, rdfs:subPropertyOf, q) \rightarrow (x, q, y)$. In each case, if we assume that schema level statements in the ontology are eternally true, the temporal label of an inferred instance statement s is the union of the time intervals of all statements which can be used to infer s.

We provide the procedure *build_temporal_index()* to create a temporal index for a given ontology. This procedure executes in three phases.

The first phase creates the temporary table *asserted_temporal_triples* (*subj_id NUMBER, prop_id NUMBER, obj_id NUMBER, start DATE, end DATE*). The ontology is then queried to retrieve all asserted temporal reifications. The subject, property, and object ids of each temporally reified statement and the start time and end time are inserted into this temporary table. The final step of this phase inserts statements without asserted temporal reifications into the *asserted_temporal_triples* table using *min_start_time* and *max_end_time* as the default start and end times. These values are specified during index creation. Additionally, all schema-level statements also receive these start and end values to denote that the ontology schema is always valid.

At this point, we have recorded the temporal values for each asserted statement, and the second and third phases perform the temporal inferencing process and create the final temporal triples table (see Fig. 2). In the procedure *TemporalInference* (shown below), we first create a second temporary table *redundant_triples* (*subj_id NUMBER, prop_id NUMBER, obj_id NUMBER, start DATE, end DATE*). Then, we iterate through the *asserted_temporal_triples* table and add any inferred statements to the *redundant_triples* table. In this step, the temporal label of the asserted statement is directly assigned to the corresponding inferred statements. This procedure results in possibly redundant and overlapping intervals for each statement, so a third phase iterates through this table and cleans up the time intervals for each statement. The cleanup phase first sorts *redundant_triples* by (*subj_id, prop_id, obj, start_id*) and then makes a single pass over the sorted set to merge overlapping intervals for statements with the same subject, property and object. The final result of this process is a table *TemporalTriples* (*subj_id NUMBER, prop_id NUMBER, obj_id NUMBER, start DATE, end DATE*) which contains the complete set of asserted and inferred temporal triples (see Fig. 2).

Procedure TemporalInference

```
 1:  create temporary table redundant_triples (subj_id, prop_id, obj_id, start, end)
 2:  for each row r ∈ asserted_temporal_triples do
 3:     if (r.prop = rdf:type) then
 4:        for each Class C ∈ SuperClasses(r.obj) do
 5:           insert row (r.subj, rdf:type, C, r.start_date, r.end_date) into redundant_triples
 6:        end for
 7:     else
 8:        for each property P ∈ SuperProperties(r.prop) do
 9:           insert row (r.subj, P, r.obj, r.start_date, r.end_date) into redundant_triples
10:        end for
11:        x ← domain(r.prop)
12:        for each Class C ∈ SuperClasses(x) ∪ {x} do
13:           insert row (r.subj, rdf:type, C, r.start_date, r.end_date) into redundant_triples
14:        end for
15:        y ← range(r.prop)
12:        for each Class C ∈ SuperClasses(y) ∪ {y} do
13:           insert row (r.obj, rdf:type, C, r.start_date, r.end_date) into redundant_triples
14:        end for
15:     end if
16:  end for
```

Operator Implementation. The SQL table functions were implemented using Oracle's *ODCITable* interface methods [30]. With this scheme, users implement a *start*(), *fetch*() and *close*() method for the table function. Generally, in the *start*() method, table function parameters are parsed, and a SQL query is prepared and executed against the underlying database tables. The *fetch*() method fetches a subset of rows from the prepared query and returns them. The *fetch*() method is invoked as many times as necessary by the kernel until all result rows are returned. The *close*() method performs cleanup operations after the last *fetch*() call.

Each of the table functions takes a graph pattern and ontology as input. In *start*(), the graph pattern is parsed and transformed into a self-join query against the *TemporalTriples* table corresponding to the input ontology. We will illustrate this process with the following example: (?a on_crew_of ?b)(?b used_in ?c).

First, URIs in the graph pattern are resolved to numeric ids through a lookup in the *RDFValues* table. Assume that in this case the ids of *member_of* and *used_in* are 1 and 2 respectively. Next, we perform a self join of the *TemporalTriples* table with two sets of conditions in the where clause: (1) we must restrict the rows of each table based on the ids of the URIs in the graph pattern and (2) we must create a join condition based on variable correspondences between different parts of the graph pattern. We must also join with the *RDFValues* table to resolve the ids of URIs bound to variables to actual URI Strings for return from the function. The graph pattern above results in the following query:

```
select  rv1.uri, rv2.uri, rv3.uri
from    TemporalTriples t1, TemporalTriples t2,
        RDFValues rv1, RDFValues rv2, RDFValues rv3
where   t1.prop_id = 1 and t2.prop_id = 2 and
        t1.obj_id = t2.subj_id and rv1.id = t1.subj_id
        and rv2.id = t1.obj_id and rv3.id = t2.obj_id;
```

Spatial operators are implemented by augmenting the base graph pattern query in *start()*. For the *spatial_extent* operator, we add an additional join with the *SpatialData* table to retrieve the *SDO_GEOMETRY* object corresponding to the *spatial_variable* parameter. In the case of optional result filtering, we need to modify the where clause so that we filter the spatial features from *SpatialData* according to the input spatial feature and spatial relation. This is done by adding the appropriate *sdo_relate* or *sdo_within_distance* predicate available in Oracle Spatial. For example, given the query `spatial_extent (..., sdo_geometry (...), 'geo_relate (inside)')`, we would modify the query as follows: `where ... and sdo_relate (SpatialData.shape, sdo_geometry (...), 'mask=inside') = 'true';`

For the *spatial_eval* operator, we implement what is essentially a nested loop join (NLJ) using the basic *spatial_extent* and filtered *spatial_extent* operators. We first construct and execute a basic *spatial_extent* query in the *start()* routine. Next, in the *fetch()* routine, we consume a row from the *spatial_extent* query and then construct and execute the appropriate filtered *spatial_extent* query using the second pair of graph pattern and spatial variable parameters and the spatial relation parameter. This is repeated until all rows in the outer *spatial_extent* query are consumed. This NLJ strategy is needed to avoid an awkward query plan on what would be a very large single base query.

The implementation of the temporal operators does not translate directly to a SQL query. We must do some extra processing of the base query results in *fetch()* to form a single time interval for each found graph pattern instance.

For the *temporal_extent* operator, we first augment the basic graph pattern query in *start()* to also select the start and end values for each temporal triple in the graph pattern instance. In the *fetch()* routine, to compute the final temporal interval for each graph pattern instance, we examine the start and end times for each triple and select the earliest start and latest end (*RANGE*) or the latest start and earliest end (*INTERSECT*). In each case, we ensure that the resulting time interval is valid (i.e., start time less than end time) before including it in the result. When the optional filtering parameters are specified, we must perform additional checking of the found graph patterns to ensure they satisfy the filter condition. In addition to these extra computations in *fetch()*, as an optimization, we augment the base query in *start()* with a series of predicates involving the start and end times of each statement in the graph pattern. This is done to filter the results as much as possible in the base query to reduce subsequent overhead in *fetch*(). To illustrate these additional predicates, consider the following *temporal_extent* query and corresponding base query:

```
select ...
from table(temporal_extent('(?x on_crew_of ?y)(?y
     used_in ?z)', 'range', 1942, 1944, 'during'));

select ...
from ..., TemporalTriples t1, TemporalTriples t2
where ... and t1.start > 1942 and t1.end < 1944
         and t2.start > 1942 and t2.end < 1944;
```

The implementation of the *temporal_eval* operator is similar to the implementation of *spatial_eval*. We first build a basic *temporal_extent* query involving the first pair of graph pattern and interval type parameters which is executed in the *start*() routine. Next, in *fetch*(), we consume a row from the basic *temporal_extent* query and execute an appropriate filtered *temporal_extent* query using the second pair of graph pattern and interval type parameters. This query uses the time interval from the current outer *temporal_extent* result and the inverse of the temporal relation parameter from the original *temporal_eval* query.

5 Experimental Evaluation

In this section, we describe the experimental evaluation of our spatial and temporal query operators. All experiments were conducted using Oracle 10g Release 2 running on a Red Hat Enterprise Linux machine with dual Xenon 3.0 GHz processors and 2 GB of main memory. The database used an 8 KB block size and was configured with an *sga_target* size of 512 MB and a *pga_aggregate_target* size of 512 MB. The times reported for each query are an average of 15 trials using a warm cache. Times were obtained by querying for *systimestamp* before and after query execution and computing the difference. Datasets and queries can be downloaded from http://knoesis.wright.edu/students/mperry/STData.html.

Dataset. Three synthetically generated datasets were used in our experiments. The datasets correspond to an ontology schema from the military domain that we created with the overall idea being to analyze historical entities and events of WWII. The ontology schema defined 15 class types and 9 property types. Each dataset was created in three phases. First we populated the thematic portion of the ontology. Second we added spatial information, and in the final step we generated temporal labels for the statements in the populated ontology.

To populate the thematic portion of the military ontology, we used the ontology population tool described in [31]. This tool inputs an ontology schema and relative probabilities for generating instances of each class and property type. Based on these probabilities, it generates instance data, which, in effect, simulates the population of the ontology. We generated three RDF datasets this way. The first contained 95,000 triples, the second contained 1.6 million triples and the third contained over 15 million triples (asserted and inferred statements). We integrated these military RDF graphs with the upper-level ontology described in [8] by adding a handful of *rdfs:subClassOf* statements to each RDF dataset.

To add spatial aspects to this dataset, we randomly assigned spatial features to each instance of *Geometry* in the ontology with uniform probability. We used year 2000 census block group boundary polygons from the US Census Bureau for the spatial features [32]. Differently-sized sets of contiguous US States were chosen in proportion with the ontology size. The total numbers of features for each dataset were 873; 9,352 and 83,236 for the small, medium and large ontology, respectively.

The final phase of dataset generation assigned temporal labels to statements in the ontology. Temporal intervals were randomly assigned to each asserted instance statement. Start times and end times for each interval were randomly selected with uniform probability from two overlapping date ranges. We ensured that each interval was valid (i.e., start time earlier than end time) before adding it to the dataset.

Experiments. Our experiments are designed to characterize the overall performance of our approach with respect to (1) ontology size and (2) graph pattern complexity. For testing, B-Tree indexes were created on each column of the *TemporalTriples* table and on the *value_id* column of the *SpatialData* table, and an R-Tree index was created on the *shape* column of *SpatialData*. We also created two additional B-Tree indexes (*prop_id, subj_id, obj_id, start, end*) and (*prop_id, obj_id, subj_id, start, end*) on the *TemporalTriples* table. For the 15 million triple dataset, the physical size of the *TemporalTriples* table was 642 MB, and the inferencing procedure took 1 hour and 31 min. to execute, which compared with 1 hour and 11 min. for Oracle RDFS rules index creation. The *SpatialData* table was 47 MB in size.

Query Execution Time. Table 1 summarizes the results of our experimentation with respect to ontology size.

Table 1. Experimental results for query execution time with respect to ontology size

Operator (Exp. #)	Graph Pattern Type		Queries	Avg. Result Size	Avg. Execution Time for each ontology (ms)		
	# Vars	# Triples			Small	Medium	Large
T-Ext (1)	4	3	4	N/A	394	390	385
(2)	3	3	5	221	22	32	48
S-Ext (3)	4	3	3	N/A	360	350	365
(4)	3	3	3	100	22	30	67
T-Filter(5)	4	3	4	312	157	345	714
S-Filter (6)	4	3	3	331	173	192	374
T-Eval(7)	2/2	2/2	3	129	414	411	437
	2/3	3/3	3	220	306	195	268
S-Eval (8)	2/2	2/1	3	244	343	467	485
	2/2	2/3	3	209	251	385	457

Experiments 1 through 4 were designed to test the general scalability of basic *temporal_extent* and *spatial_extent* queries. Experiments 1 and 3 measured the response time (i.e., time to return the first 1000 rows of results) for a very unselective query. Our unselective graph patterns consisted of 3 triples and 4 variables. For each triple in the pattern a constant URI was given for the *property*, and the *subject* and *object* were left as variables. We used 4 different graph patterns for *temporal_extent* with an *INTERSECT* type query in each case. For *spatial_extent*, 3 different graph patterns were used. In each case, the DBMS uses a nested loop joion (NLJ) strategy for evaluating the base query which results in response times which are essentially constant across each dataset as the execution time of a NLJ usually grows in proportion with the result set size. Experiments 2 and 4 are designed to measure scalability for a very selective graph pattern. For experiment 2, we used 5 different graph patterns consisting of 3 triples and 3 variables. For experiment 4, we used 3 different graph patterns with 3 triples and 3 variables. The graph patterns are of the same basic form as the previous experiment except we replace one of the variables in the *subject* or *object* position with a constant URI. This restricts the nodes in the resulting graph pattern instance instead of just the edges, providing a much more selective query. In each case, query execution time increases slightly as the ontology size increases, which is a consequence of scanning larger indexes during query evaluation and querying a larger *SpatialData* table.

In experiment 5, we measured the scalability of the *temporal_extent* operator using optional filtering with respect to dataset size. For these tests, we used very unselective graph patterns in combination with very selective temporal conditions. Note that this represents a worst case scenario for *temporal_extent*. Because we only store the temporal labels for single triples in the DB, we can only index these single triples. The temporal labels for graph pattern instances are constructed during query evaluation and therefore cannot be indexed. We must apply the temporal filter to each graph pattern instance as it is being constructed, which can potentially lead to very large intermediate result sets because in many cases we cannot exclude a graph pattern from the results until its time interval has been fully constructed. Our experiments show an increase in execution time which is roughly linear with respect to ontology size which reflects the growth of intermediate results processed during the query. Each query used the *INTERSECT* option and either a *before*, *after* or *during* temporal relation.

In experiment 6, we measured the performance of *spatial_extent* using the optional filtering capability as dataset size increases. As with experiment 5, we combined a low selectivity graph pattern with a highly-selective spatial predicate. We used three different queries. The first retrieved results which were within a short distance of a point; another retrieved results which were covered by an input polygon, and the final query retrieved results which intersected with an input polygon. The results show that *spatial_extent* with filter scales better than its temporal counterpart because we can effectively index the spatial features and quickly reduce the search space using the spatial index. The execution time increases because larger indexes must be scanned when evaluating the graph pattern.

Experiment 7 illustrates the scalability of selective *temporal_eval* queries. For this test, we used selective graph patterns for both the LHS and RHS input patterns. We varied the constant URIs in the graph pattern and the temporal condition so that the result set sizes were constant across each dataset. The results show that execution time is roughly constant across each dataset with variations resulting from slight

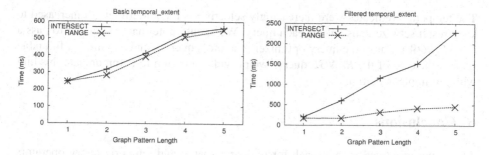

Fig. 3. Scalability of temporal operators with respect to graph pattern size

differences in the number of query restarts required in *fetch*() and the selectivity of the graph patterns used. Each query used the *INTERSECT* option and either a *before*, *after*, *during* or *any interact* temporal relation.

Experiment 8 characterizes the performance of selective *spatial_eval* queries as the dataset size increases. Again, we used selective graph patterns for both the LHS and RHS pattern and varied the constant URIs and spatial predicates so that result set size was consistent across each dataset. The results show that execution time grows slightly as ontology size increases, which is a result of scanning larger indexes and querying a larger spatial dataset.

Our next experiments were designed to test the scalability of the *temporal_extent* operator as the graph pattern size increased. We elected to present experimental results for only temporal queries due to space limitations, and, because temporal processing is less efficient than spatial processing in our scheme, these numbers should represent an upper bound. All queries in these tests were run against the 15 million triple dataset. The graph on the left side of Fig. 3 shows the response time (first 1000 rows) of basic temporal extent queries (*INTERSECT* vs. *RANGE*) for low selectivity graph patterns of increasing length. The times are the mean of 4 different queries for a given length. Each graph pattern has a constant URI in each predicate position and variables in each subject and object position. The results show that response time scales roughly linearly with graph pattern size. More processing time is required for *INTERSECT* because of extra join conditions needed to ensure valid time intervals. The graph on the right side of Fig. 3 shows the execution time for filtered *temporal_extent* queries using unselective graph patterns and selective temporal predicates. The idea behind this experiment was to bound the execution time for filtered *temporal_extent* queries. In some circumstances, our filtering optimization in the base query can only place weak conditions on the temporal properties of each triple in the result. For example, using *INTERSECT* and *during* [x, y], we can only enforce that each triple does not end *before x* or start *after y*. In contrast, using *RANGE* and *during* [x, y] we can enforce that each triple both starts *after x* and ends *before y*, which completely filters any unmatching graph patterns. The graph in Fig. 3 (right) shows the execution times for each scenario. Each value is the average of four different queries of that type. We can see that performance using the worst-case scenario scales much worse than the best case, but the growth is still roughly linear.

The temporal predicates were increasingly selective as the pattern length increased to keep result set size constant for each query. We should note that we needed to pass a *FIRST_ROWS* hint to the query optimizer to avoid a query plan containing a full table scan in the case of the *RANGE* query (we provide an option to communicate this hint with our implementation).

6 Conclusions

This paper discussed an approach for realizing spatial and temporal query operators for Semantic Web data. Our work was motivated by a lack of support for spatial and temporal relationship analysis in current semantic analytics tools. Spatial and temporal data is critical in many analytical applications and must be effectively utilized for semantic analytics to reach its full potential. Our approach built upon existing support for storage and querying of RDF data and spatial data in Oracle DBMS. A set of experiments using a synthetic RDF dataset of over 15 million triples showed that our implementation exhibited good scalability for a fairly large populated ontology. Basic *temporal_extent* and *spatial_extent* queries were quite fast in all circumstances. The worst performance was seen with filtered *temporal_extent* queries using low selectivity graph patterns with highly selective temporal predicates. However, the resulting execution times were quite manageable.

A possible limitation of this work is that Oracle Semantic Data Store does not support incremental maintenance of RDFS rules indexes. Consequently, our indexing scheme inherits this limitation. However, incremental maintenance of a materialized set of inferred triples upon updates of asserted triples is possible (e.g., [33]), and existing algorithms could be extended to incorporate temporal information.

In the future, we plan investigate this incremental maintenance issue and to perform further testing using other ontologies populated with both real and synthetic data. We also plan to investigate extensions of the SPARQL query language to support the types of operations discussed in this paper.

References

1. RDF, http://www.w3.org/RDF/
2. Anyanwu, K., Sheth, A.P.: ρ-Queries: Enabling Querying for Semantic Associations on the Semantic Web. In: 12th Int'l WWW Conf., Budapest, Hungary (2003)
3. Ramakrishnan, C., Milnor, W.H., Perry, M., Sheth, A.: Discovering Informative Connection Subgraphs in Multi-relational Graphs. SIGKDD Explorations 7(2), 56–63 (2005)
4. Aleman-Meza, B., et al.: Semantic Analytics on Social Networks: Experiences in Addressing the Problem of Conflict of Interest Detection. In: 15th Int'l WWW Conf., Edinburgh, Scotland (2006)
5. Mukherjea, S., Bamba, B.: BioPatentMiner: An Information Retrieval System for BioMedical Patents. In: 30th Int'l Conf. on VLDB, Toronto, Canada (2004)
6. Kochut, K., Janik, M.: SPARQLeR: Extended Sparql for Semantic Association Discovery. In: 4th European Semantic Web Conf., Innsbruck, Austria (2007)

7. Pelekis, N., et al.: Literature Review of Spatio-Temporal Database Models. The Knowledge Engineering Review 19(3), 235–274 (2004)
8. Perry, M., Hakimpour, F., Sheth, A.P.: Analyzing Theme, Space and Time: an Ontology-based Approach. In: 14th ACM Int'l Symposium on GISs, Arlington, VA (2006)
9. Gutierrez, C., Hurtado, C., Vaisman, A.: Temporal RDF. In: Gómez-Pérez, A., Euzenat, J. (eds.) ESWC 2005. LNCS, vol. 3532, pp. 93–107. Springer, Heidelberg (2005)
10. Brickley, D., Guha, R.V.: RDF Vocabulary Description Language 1.0: RDF Schema, W3C Recommendation (2004), http://www.w3.org/TR/rdf-schema/
11. Hayes, P.: RDF Semantics, http://www.w3.org/TR/rdf-mt/
12. Yuan, M.: Wildfire Conceptual Modeling for Building GIS Space-Time Models. In: GIS/LIS, Pheonix, AZ (1994)
13. Yuan, M.: Modeling Semantical, Temporal and Spatial Information in Geographic Information Systems. In: Craglia, M., Couclelis, H. (eds.) Geographic Information Research: Bridging the Atlantic, pp. 334–347. Taylor & Francis, Abington (2006)
14. Worboys, M., Hornsby, K.: From Objects to Events: GEM, the Geospatial Event Model. In: Egenhofer, M.J., Freksa, C., Miller, H.J. (eds.) GIScience 2004. LNCS, vol. 3234, pp. 327–344. Springer, Heidelberg (2004)
15. Prud'hommeaux, E., Seaborne, A.: SPARQL Query Language for RDF, http://www.w3.org/TR/rdf-sparql-query/
16. Karvounarakis, G., Alexaki, S., Christophides, V., Plexousakis, D., Scholl, M.: RQL: A Declarative Query Language for RDF. In: 11th Int'l WWW Conf., Honolulu, HI (2002)
17. Sintek, M., Decker, S.: TRIPLE - A Query, Inference, and Transformation Language for the Semantic Web. In: 1st Int'l Semantic Web Conf., Sardinia, Italy (2002)
18. Souzis, A.: RxPath Specification Proposal (2004), http://rx4rdf.liminalzone.org/RxPathSpec
19. Wilkinson, K., Sayers, C., Kuno, H., Reynolds, D.: Efficient RDF storage and retrieval in Jena2. In: VLDB Workshop on Semantic Web and Databases, Berlin, Germany (2003)
20. Alexaki, S., Christophides, V., Kaevounarakis, G., Plexousakis, D.: On Storing Voluminous RDF Descriptions: The Case of Web Portal Catalogs. In: 4th Int'l Workshop on the Web and Databases, Santa Barbara, CA (2001)
21. Chong, E.I., Das, S., Eadon, G., Srinivasan, J.: An Efficient SQL-based RDF Querying Scheme. In: 31st Int'l Conf. on VLDB, Trondheim, Norway (2005)
22. Kammersell, W., Dean, M.: Conceptual Search: Incorporating Geospatial Data into Semantic Queries. In: Terra Cognita - Directions to the Geospatial Semantic Web, Athens, GA (2006)
23. Tanasescu, V., et al.: A Semantic Web GIS based Emergency Management System. In: Int'l Workshop on Semantic Web for eGovernment, Budva, Montenegro (2006)
24. Egenhofer, M.J.: Toward the Semantic Geospatial Web. In: 10th ACM Int'l Symposium on Advances in Geographic Information Systems, McLean, VA (2002)
25. Jones, C.B., et al.: The SPIRIT Spatial Search Engine: Architecture, Ontologies, and Spatial Indexing. In: Egenhofer, M.J., Freksa, C., Miller, H.J. (eds.) GIScience 2004. LNCS, vol. 3234, pp. 125–139. Springer, Heidelberg (2004)
26. Oracle Spatial Resource Description Framework (RDF) 10g Release 2, http:// download-east.oracle.com/docs/cd/B19306_01/appdev.102/b19307/toc.htm
27. Oracle Spatial User's Guide and Reference 10g Release 2, http://download-east.oracle.com/docs/cd/B19306_01/appdev.102/b14255/toc.htm
28. Hakimpour, F., eAleman-Meza, B., Perry, M., Sheth, A.: Data Processing in Space, Time and Semantics Dimension. In: Terra Congita - Directions to the Geospatial Semantic Web, Athens, GA (2006)

29. Open GIS Consortium: Open GIS Simple Feature Specification for SQL (1999), http://portal.opengeospatial.org/files/
30. Oracle Database Data Cartridge Developer's Guide, 10g Release 2, http://download-east.oracle.com/docs/cd/B19306_01/appdev.102/b14289/toc.htm
31. Perry, M.: TOntoGen: A Synthetic Data Set Generator for Semantic Web Applications. AIS SIGSEMIS Bulletin. 2(2), 46–48 (2005)
32. U.S. Census 2000 Cartographic Boundary Files, http://www.census.gov/geo/www/cob/bg2000.html
33. Volz, R., Staab, S., Motik, B.: Incrementally maintaining materializations of ontologies stored in logic databases. Journal on Data Semantics 2, 1–34 (2005)

Semantic Similarity Applied to Generalization of Geospatial Data

Marco Moreno-Ibarra

PIIG Laboratory, CIC – National Polytechnic Institute, Mexico City, Mexico
marcomoreno@cic.ipn.mx

Abstract. The paper presents an approach to verifying the consistency of generalized geospatial data at a conceptual level. The principal stages of the proposed methodology are Analysis, Synthesis, and Verification. Analysis is focused on extracting the peculiarities of spatial relations by means of quantitative measures. Synthesis is used to generate a conceptual representation (ontology) that explicitly and qualitatively represents the relations between geospatial objects, resulting in tuples called herein semantic descriptions. Verification consists of a comparison between two semantic descriptions (description of source and generalized data): we measure the semantic distance (confusion) between ontology local concepts, generating three global concepts Equal, Unequal, and Equivalent. They measure the (in) consistency of generalized data: Equal and Equivalent – their consistency, while Unequal – an inconsistency. The method does not depend on coordinates, scales, units of mea-sure, cartographic projection, representation format, geometric primitives, and so on. The approach is applied and tested on the generalization of two topographic layers: rivers and elevation contour lines (case of study).

1 Introduction

The generalization is used to produce geographic data at coarser levels of detail, while retaining essential characteristics of underlying geographic information [1]. Therefore, generalization systems should assure the semantic consistency of generalized data. Traditionally, the problem of consistency is a numerical task based on measures that represent constraints at topological and geometrical levels [2]. These measures are difficult to interpret and adapt to different contexts. Thus, the present work is focused on using well studied quantitative measurements, but passing them at the conceptual operating level to facilitate the detection and interpretation of inconsistencies, which are commonly presented in generalization. To do this, we use a conceptual representation of topographic domain (ontology) those concepts (classes) are defined by numerical intervals, which in turn are the results of quantitative measurements of the geographic objects. In the case of study, the used concepts belong to two levels deep ontology fragment (ordered hierarchy). Each concept however can have more descendents at other ontological levels. In other words, a number of subclasses can be generated to conceptualize at deeper detail the numeric intervals. This allows measuring the distance between qualitative values instead of quantitative ones, that is to say, the distance between two ontology classes. In contrast

F. Fonseca, M.A. Rodríguez, and S. Levashkin (Eds.): GeoS 2007, LNCS 4853, pp. 247–255, 2007.

to numerical approaches the semantic distance facilitates the interpretation of the measurements and produces better results to user's satisfaction, because the concepts and their similarity can be easily understood and interpreted [3]. In addition, the detection and interpretation of inconsistencies is based only on the conceptual representation, which does not depend on coordinates, scales, units of measure, cartographic projection, representation format, geometric primitives, and so on.

The rest of the paper is organized as follows: Section 2 describes the conceptualization of topographic domain and ontology designed for the case of study. Section 3 presents the methodology. Section 4 exposes some results for the case of study. Section 5 sketches out our conclusions and future work.

2   Conceptualization of Topographic Domain

This work is based on a conceptualization of topographic domain, which describes the main properties and relations of geographic objects (elevation contour lines and rivers). To conceptualize the domain we are use the following documents: Environmental Data Coding Specification (EDCS) [4], WordNet [5], and documents of the National Institute of Statistics, Geography and Informatics (INEGI) [6][7].

2.1   Relations Between Rivers and Elevation Contour Lines

Relations between rivers and elevation contour lines depend on the flow direction. In consistent data, a river can cross just once an elevation contour line; more than one crossing represents an inconsistency. From the point of view of cartographic representation, in the best case, the river should *pass by* a maximum convexity of the contour line (Fig 1a). Sometimes the river *passes by* a convexity of the elevation contour line (Fig 1b, c and d) or *passes by* its concavity (Fig. 1f, g and h); these cases represent inconsistencies. In other cases, the river passes by straight part of the elevation contour line (Fig. 1e and j).

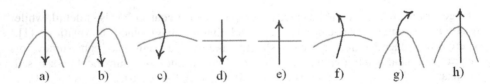

Fig. 1. Different cases of the relations between rivers and elevation contour lines

Thus, six different relations are to be considered in the following: *pass by maximum convexity, pass by almost maximum convexity, pass by convexity, pass by straight, pass by concavity, pass by almost maximum concavity,* and *pass by maximum concavity.*

2.2   Ontology

We define the terms of ontology to describe the relations between elevation contour lines and rivers (Fig 2). We include the relation between elevation contour lines and

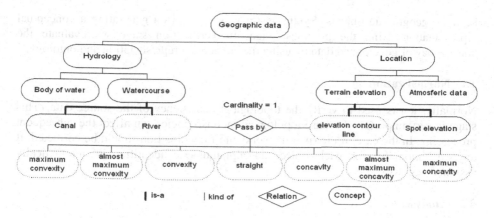

Fig. 2. Fragment of ontology to describe the relations between rivers and elevation contour lines

rivers (*pass by*) as a concept of ontology. Additionally this relation is specified in order to enrich the expressiveness of ontology by using other concepts, which describe the general term (*pass by*) at deeper level of details.

3 Methodology

The methodology consists of five stages: *Normalization, Processing, Analysis, Synthesis,* and *Verification* (Fig. 3). In normalization stage we verify the consistency of the geographic data prior to performing the following stages. In processing stage the data are automatically generalized. In analysis stage we automatically extract the relations

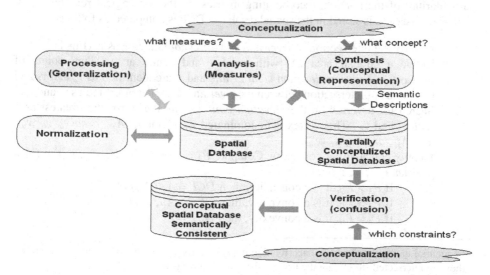

Fig. 3. Methodological framework

between geographic objects. Synthesis stage is focused on generating a conceptual representation using the previous stage. In verification stage we evaluate the consistency of generalized data by using the conceptual representation and ontology.

3.1 Normalization

Normalization is used to verify the topological consistency of the source data.†This stage is very important, because topological inconsistencies can affect the subsequent processing. In the case of the river networks the flow direction should be corrected, if any. Additionally, for each object an alphanumeric identifier is automatically assigned, e.g. Rio_1.

3.2 Analysis

Analysis stage is based on measures of geospatial data. A measurement is a computing procedure for evaluating characteristics (features, attributes, etc.) of geographical objects [8]. A measurement is a numerical value assigned to an observation that reflects the magnitude or amount of a characteristic. As result of this stage we obtain *quantitative descriptors (DC)*.

Quantitative descriptors of pass by relation between rivers and elevation contour lines. Herein we are focused on the measure of *pass by* relation between rivers and elevation contour lines. As result of this measure we obtain a value denominated *quantitative descriptor of pass by relation between rivers and elevations contour lines (DCP)*. Additionally, *Number-of-Relation (NR)* property is calculated in this stage[1]. To better describe the relation, we follow the premise that "*topology matters, while metric refines*" [9]. By using the *9-intersection model by* Egenhofer we can not identify all the cases of *pass by* relation (Section 2.1). Instead, we use a measure to extract the particularities of the relation, incorporating metrics for the topological relations as in [10]. Each measure is stored in the spatial database. DCP is computed as follows:

1. Identify the intersection of *river* and *elevation contour line* as v_p (Fig 4).
2. Define a circular area (*A*) with radio *r* and center at v_p. Two points of intersection between *elevation contour line* and *A* are identified as v_a and v_b and two points of intersection between the *river* and *A* are denominated as v_c and v_d.
3. Search for a vertex on the *elevation contour line* that form the greatest area with v_a and v_b . This vertex is denominated v_m. Compute $A_P = area(v_a, v_b, v_p)^2$ and $A_{M=} area(v_a, v_b, v_m)$.
4. Identify concavity or convexity. Compute $P_A = perimeter(v_a, v_b, v_p, v_c)^3$ y $P_D = perimeter(v_a, v_b, v_p, v_d)$
   If $P_A > P_D$, it is a concavity, then $DCP = A_P / A_M$ (1)
   If $P_A > P_D$, it is a convexity, then $DCP = A_P / A_M$ (-1)
   If $P_A = P_D$, it is a convexity, then $DCP = 0$

---

[1] Sequential number assigned to each relation between a river and elevation contour whenever they are intersected (following the flow of the river), starting at 1.

[2] *area(a,b,c)* computes the triangular area composed of 3 points.

[3] *perimeter (a,b,c,d)* computes the perimeter of figure composed of 4 points.

The interpretation of the values of DCP is (Fig. 5):

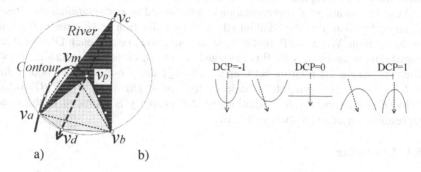

Fig. 4. Components of the river and elevation tour line to compute *DCP*

Fig. 5. Interpretation of *DCP*

- *DCP* = 1 means that the river *passes by* the point of maximum convexity of elevation contour line.
- 1 > *DCP* > 0 means that the river *passes by* a convexity of the elevation contour line.
- *DCP* = 0 means that the river *passes by* a straight part of an elevation contour line (concavity and convexity do not exist).
- 0 > *DCP* > -1 means that the *river passes by* a concavity of the elevation contour line.
- *DCP* = -1 means that the river *passes by* the point of maximum concavity of an elevation contour line.

3.3 Synthesis

Synthesis stage is used to generate a conceptual representation that explicitly describes the relations between geographic objects. Thus, a *Partially Conceptualized Spatial Database* (*PCDB*) is obtained. PCDB is composed of *tuples* denominated

Table 1. Criteria to define the classes/relations between rivers and elevation contour lines

Class	Criteria
<PASS-BY-MAXIMUM-CONVEXITY>	$[0.95 < DCP < 1]$
<PASS BY ALMOST MAXIMUM CONVEXITY>	$[0.85 < DCP < 0.94]$
<PASS BY CONVEXITY>	$[0.35 < DCP < 0.84]$
<PASS BY STRAIGHT>	$[-0.34 < DCP < 0.34]$
<PASS BY CONCAVITY>	$[-0.84 < DCP < -0.35]$
<PASS BY ALMOST MAXIMUM CONCAVITY>	$[-0.94 < DCP < -0.85]$
<PASS BY MAXIMUM CONCAVIYTY>	$[-1 < DCP < -0.95]$

semantic descriptions (SD). SD has the form {O$_i$, **R**, O$_j$}, where O$_i$, y O$_j$ are the identifiers of geographic objects and **R** represents the relation between the objects O$_i$ and O$_j$. This conceptual representation can be stored in any *relational database.* SD is generated by mapping the DCP into the conceptualization and expresses the semantics of the relation. We use different criteria to define the class of each DCP (Table 1) and constraints in order to a DCP is belonged to certain class. For instance, suppose that after measuring the relation between the *River_1* and *Contour_8* we obtain that *DCP = 0.99.* Then, we generate a tuple of the form: *{River_1, PASS BY MAXIMUM CONVEXITY, Contour_8}.* Additionally, *NR* property is included in the conceptual representation as an attribute in PCDB.

3.4 Processing

Processing stage consists of an automatic generalization system based on the generalization operators proposed by McMaster and Shea [11]. The system is used to generalize the rivers and elevation contour lines. The user defines the parameters of generalization operators to modify the scale from 1:50,000 to 1:250,000. Here the INEGI specifications [6][7] are also used.

3.5 Verification

This stage is used to verify the consistency of generalized data. We compare the *semantic descriptions of the source data (SD$_F$)* with the *semantic descriptions of the generalized data (SD$_G$).* Specifically this stage is based on the *semantic invariants,* which are relations that do not change after the generalization. The invariants depend on the consistency of the geospatial data and therefore on their semantic content.

In order to find the semantic invariants the hierarchical structure of ontology classes is used. The method to evaluate the consistency is based on *confusion* (the confusion conf(r, s) in using qualitative value r instead of the intended or correct value s). The concept of confusion allows defining the closeness to which an object fulfills a predicate as well as deriving other operations and properties among hierarchical values [12].

Thus, we obtain a *semantic distance* between concepts. This distance allows us to define new concepts (*equal, equivalent, unequal*) that represent the difference between the conceptualizations and captures the change of the relations after generalization. The consistent relations (*equal, equivalent*) are stored in a *Spatial Database Semantically Consistent (SDSC).*

Cases to verify the consistency. The consistency is evaluated by using confusion over a fragment of ontology. The fragment of ontology has the form of *ordered hierarchy* [12] (*pass by* relation in Fig.2). Confusion (conf (r, s)) is computed by the relative distance from the concept r to the concept s (the number of steps needed to jump from r to s in the ordering) divided by the (*cardinality* – 1) of the father (*pass by*) [12]. We consider three different cases to evaluate the consistency that depend on the confusion value:

- *Equal,* conf(r, s) = 0, the concepts are equal. This means that the relation is consistent and the semantics is preserved after the generalization.

- *Unequal,* $0 < \text{conf}(r, s) \leq 1$, the concepts are unequal. This means that the relation is inconsistent and the semantics is not preserved after the generalization.
- *Equivalent,* some cases are considered to be consistent. Define a threshold (u). If $u < \text{conf}(r, s) < 1$, we consider that the relation is unequal. If $0 < \text{conf}(r, s) \leq u < 1$, we consider that the relation is equivalent. In a certain sense we can say that here the semantics is preserved.

By using this methodology, other errors can be identified. These errors are produced by the operators of line simplification. The identification of these errors is based on the premise "*an elevation contour line and river must cross once*".

4 Results

The system is implemented in Arc/Info 8.1.2, using Arc Macro Language (AML). The geospatial data are provided by INEGI. Some results for the case of study to verify the consistency are presented in this section. We put threshold $u = 1/6$.

Fig. 6 shows two semantic descriptions of the same relation prior (left) and after generalization (right). Applying confusion we have conf(r, s) = conf(*PASS BY MAXIMUM CONVEXITY, PASS BY MAXIMUM CONVEXITY*) = 0; in this case the relation is *equal*. This means that the semantics is preserved in the generalized data and they are consistent.

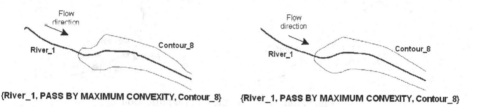

{River_1, PASS BY MAXIMUM CONVEXITY, Contour_8}   {River_1, PASS BY MAXIMUM CONVEXITY, Contour_8}

Fig. 6. Identifying a relation *Equal*

Fig. 7 depicts two semantic descriptions of the same relation prior (left) and after generalization (right). Applying confusion we have conf(r, s) = conf(*PASS BY MAXIMUM CONVEXITY, PASS BY ALMOST MAXIMUM CONVEXITY*) = 1/6 that is less than u. Thus, this relation is *equivalent* and therefore the relation in the generalized data is consistent.

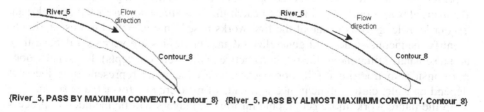

{River_5, PASS BY MAXIMUM CONVEXITY, Contour_8}   {River_5, PASS BY ALMOST MAXIMUM CONVEXITY, Contour_8}

Fig. 7. Identifying a relation *Equivalent*

Fig. 8 shows two semantic descriptions of the same relation prior (left) and after generalization (right). Applying confusion we have conf(*r, s*) = conf(*PASS BY ALMOST MAXIMUM CONVEXITY, PASS BY STRAIGHT*) = 3/6 that is more than u; this means that this relation is unequal and represents an inconsistency.

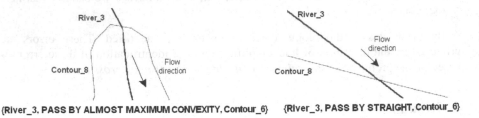

{River_3, PASS BY ALMOST MAXIMUM CONVEXITY, Contour_6}   {River_3, PASS BY STRAIGHT, Contour_6}

Fig. 8. Identifying a relation *Unequal*

Using the method proposed in Section 3.4 some relations are checked out in the generalized data (Fig. 9), finding that a river crosses the elevation contour line more than once and represents inconsistency. To identify these relations, we use the *NR* property: for consistent relations *NR* = 1, while for inconsistent – NR > 1.

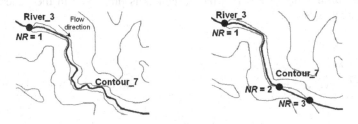

Fig. 9. Relations after the line simplification

5 Conclusions and Future Work

In this paper we presented an approach to conceptually verify the consistency of generalized data. The method is based on a conceptual representation of spatial relations (ontology); it is generated by analyzing the quantitative metrics of topological relations between rivers and elevation contour lines. The concepts represent the interpretation of the geospatial data and the meaning of spatial relations. By using this approach, we attempt to catch the *semantic content* of the spatial data. To our knowledge, this is one of the first works based on conceptual representation to identify the inconsistencies of generalized data. Ontologies are very useful since they add a semantic component (the relations between different concepts) that usually does not consider in traditional GIS approaches. The conceptual representation does not depend on scale, cartographic projection, units of measure and format, and so on.

In the future work, more geographic information layers will be included to process the identification of the inconsistencies of generalized data as well as different kind of relations between objects from those layers will be considered.

Acknowledgements

Thank you to Dr. Serguei Levachkine for his noble support and ideas to enrich this work. The author of this paper wishes to thank the IPN, CIC, SIP, and CONACYT for their support.

References

1. Weibel, R.: A Typology of Constraints to Line Simplification. In: Kraak, M.J., Molenaar, M. (eds.) Advances in GIS Research II. 7th International Symposium on Spatial Data Handling, UK, pp. 533–546. Taylor & Francis, Abington (1996)
2. Beard, M.K.: Constraints on Rule Formation. In: Buttenfield, B.P., McMaster, R.B. (eds.) Map Generalization: Making Rules for Knowledge Representation, London, U.K, pp. 121–135. Longman (1991)
3. Schwering, A., Raubal, M.: Measuring Semantic Similarity between Geospatial Conceptual Regions. In: Rodríguez, M.A., Cruz, I., Levashkin, S., Egenhofer, M. (eds.) GeoS 2005. LNCS, vol. 3799, pp. 90–106. Springer, Heidelberg (2005)
4. ISO 18025 (2005), http://standards.iso.org/ittf/PubliclyAvailableStandards/
5. Wordnet: A Lexical Database for the English Language, http://wordnet.princeton.edu/
6. INEGI: Base de datos geográficos. Diccionario de datos topográficos (escala 1:50,000), Instituto Nacional de Estadística Geografía e Informática (INEGI) (1995)
7. INEGI: Base de datos geográficos. Diccionario de datos topográficos (escala 1:250,000), Instituto Nacional de Estadística Geografía e Informática (INEGI) (1996)
8. Project Agent, http://agent.ign.fr/
9. Egenhofer, M., Mark, D.: Naive Geography. In: Kuhn, W., Frank, A.U. (eds.) COSIT 1995. LNCS, vol. 988, pp. 1–15. Springer, Heidelberg (1995)
10. Nedas, K., Egenhofer, M., Wilmsen, D.: Metric Details of Topological Line-Line Relations. International Journal of Geographical Information Science 21(1), 21–48 (2007)
11. McMaster, R.B., Shea, K.S.: Generalization in Digital Cartography. In: Association of American Geographers (1992)
12. Levachkine, S., Guzman-Arenas, A.: Hierarchy as a New Data Type for Qualitative Variables. Expert Systems with Applications 32(3), 899–910 (2007)

Towards Semantics for Map Styles

Neeharika Adabala

Microsoft Research India
"Scientia", 196/36, 2nd Main, Sadashivnagar
Bangalore 560080
neeha@microsoft.com

Abstract. Study of semantics in the context of Geographic Information Systems (GISs) usually focuses on association of meaning with spatial data that constitute the input to these systems. The goal is to create new data models that enable richer interaction with GISs. In this paper we widen the perspective of such studies and explore the implications of associating semantics with map styles that are present implicitly in the output of GISs. Traditionally in computer science style of rendering/presenting data has been viewed as extraneous information that does not add semantic value to data. In GIS however, several styles of rendering maps are possible, and these styles are often motivated by functionality. Thus a map style has a strong association with meaning. In this paper we explore the methods of associating semantics with the style of a map. We discuss the various levels at which semantic associations with map styles can be established, and how this lead to creation of new map rendering styles that exhibit coherence in visualization of spatial information.

1   Introduction

Maps, diagrammatic representations of spatial information, have existed from the beginning of civilization. They have evolved over time and several variations in map appearance have come into existence [1]. We focus on map appearance in this work.

There are two key aspects to maps making namely, collection/selection of spatial data and visualization of the data as a map. Much work has been done in the context of enabling rich semantics based interactions with data, and it is possible to make a rich collections/selections of data for a map. However, visualization of the data, namely making maps, still involves working at a primitive level (non-semantic level) of choosing fonts, colors, thickness of features, choosing icons and symbols, etc. Enabling semantics based definition of map styles is essential for the following reasons:

- facilitating easy development of new map styles,
- achieving the full potential of semantically rich data like Geographical Markup Language (GML), and
- exploiting the developments of technology in areas of computer graphics and image synthesis

we explain each of these in detail below.

F. Fonseca, M.A. Rodríguez, and S. Levashkin (Eds.): GeoS 2007, LNCS 4853, pp. 256–267, 2007.

Facilitate easy development of new map styles: Map styles have semantic significance, therefore it is possible to describe them with subjective terms like decorative, classic, fine detailed, quaint, simple, eye-catching, playful children's illustration, richly colored, bold, hand colored, outlined, clear, beautiful, woodcut,etc. It should be feasible to interact with a GIS at a semantic level during map display, for example, we should be able to request for a classic, clear, detailed rendering of a map. In the current GIS frameworks it is feasible only to display maps in a few pre-created fixed styles, or to work at a primitive level of defining the value for each of the entities to be displayed in the map, which is tedious.

Achieve the full potential of semantically rich data like GML: The ability to define map styles at a semantic level is especially relevant in the context of the current trend of creating semantically rich data representations with GML [2] and CityGML [3]. The full potential of these information rich data formats is not reached if they are always displayed in a few fixed predesigned map styles without the ability to customize the map to emphasize the information of interest to the user.

Exploit the developments of technology in areas of computer graphics and image synthesis: The advances in computer graphics and image synthesis have made it feasible to generate various rich rendering styles on the computer by developing suitable rendering algorithms. In the current framework of GIS, programs developed in the context of one style of rendering cannot be easily reused in the context of another style at runtime. Creating a semantic framework for defining styles will enable more effective use of code developed in computer graphics and image synthesis.

In this paper we present a study on association of semantics with map style. The semantics associated with *map style* has the property of defining the visual significance of a symbols in the context of a map, as opposed to conventional semantics that associates the symbol with the linguistic or physical concept it represents. We investigate and identify the semantics that are explicitly and implicitly associated with entities presented in a map. We describe various levels at which semantic associations can be made, and relate them to studies is semiotics and syntactics. The topic of associating semantics with style has not been investigated formally before, it is vast and has several ramifications into perception, cognition, abstraction, etc, which cannot be addressed in a single paper. In this paper we introduce a formal study in this area, and we discuss a preliminary implementation where we have incorporated a very rudimentary style semantics in the context of building representation in a GIS, and have generated some alternative map styles.

The rest of this paper is organized as follows, section 2 describes related work. In section 3 we consider possible methods of representing the semantics associated with style, and discuss issues in the context of granularity of semantic associations. We also examine denotational and connotational semantics and the criterion for distinguishing them in the context of map styles. Section 4 describes with examples the semantics that can be associated with lettering, structural and color aspects of map style. Section 5 discusses the implications of the suggestions presented in this paper. We give conclusions and directions for future work in section 6.

2   Related Work

Related work includes studies on: map appearance, design of maps, semiotics and syntactics in the context of maps, GML, rendering of maps and tools to aid map design. Details are given below.

There are several studies on the appearance of maps, which are summarized in books like "The Look of Maps" by Robinson [4], "How maps work? Representation, Visualization and Design" by MacEachren [5], "Thematic Maps" by Slocum et al. [6], and "Mapping It Out: Expository Cartography for the Humanities and Social Sciences" by Monmonier [7].

Other related work include the books that focus on mapdesign, we mention two recent books in this field one by authors Krygier and Wood [8] and the other by Brewer [9]. These books capture the expertise of the authors in designing maps, they focus on defining the style of entities that are represented in a map with emphasis on the final appearance of the map. They do not discuss the semantic implications of the style selected. The books by Tufte [10,11] are useful in understanding the semantics behind some of the best practices in design that have be followed down the centuries.

The book by MacEachren [5] contains detailed discussions on semiotics and syntactics in the context of visual elements in a map, which are very relevant in the context of the work presented in this paper.

The GML [2] and City GML [3] are approaches that enable association of semantics with data which is the input to a GIS, in this work we investigate techniques to associate semantics with map styles that are part of the output of GISs.

Techniques in computer graphics that address the problem of rendering maps in various styles include [12,13] are also relevant to this study. They prove that it is possible to create maps in various styles using computer graphics. However, once a particular style is programmed, it is not straightforward to use the same program to create a new map style. The program has to be rewritten for map styles that are significantly different. Also there is no automation to create new map styles, the programmer establishes the semantic associations mentally and redesigns the program. The work presented in [14] contains a rudimentary attempt to associate semantics with a map style. Two styles of rendering the maps are developed in it, namely "modern tourist" style, and "19th century panoramic map" style were implemented in this work.

In the context of map design tools, ColorBrewer [15] can be viewed as an excellent work that demonstrates the power of associating semantics with color schemes of a map. The approach to design/select colors for maps using ColorBrewer operates at a semantic level. It enables the selection of a group of colors that are "sequential", "diverging" or "qualitative". Any result it generates is associated with style semantics of "color blind friendly", "photocopy friendly", "LCD projector friendly", "laptop friendly", "CRT friendly" and "printer friendly". The elegance of the solution presented in ColorBrewer is obvious to any map designer who has spent several hours trying to come up with a color scheme by sequentially selecting individual colors.

3   Method of Associating Semantics with Map Style

In this section we look at how semantics associated with a map, which is embedded in an image, can be separated from an image and represented in a form that can be processed by a GIS. We know that the appearance of maps is dependent on several factors like, accuracy of information represented, scale of representation, details represented, technology used to create maps, functionality of maps, colors used, design decisions made by cartographers regarding appearance of the map, just to name a few. Choices of these factors dictate the map style. While several map styles can be created, only a subset of the possible map styles are effective. A map style can be considered effective if it is optimal for performing the task it was designed for, in case of special purpose maps. In case of general purpose maps, a map style can be considered effective if it optimizes the information presented to the user without reducing legibility. In both cases the optimization should be achieved without ruining the map's appearance.

A map style can be defined by specifying the values of the factors that influence the maps appearance. Some of the factors are objective and quantitative, these include factors like accuracy, details of information presented, and scale of map, while others are subjective. The objective factors tend to have unambiguous values that are easy to specify. The subjective factors are usually design decisions that the artists make, and are associated with more than one semantic implication.

Semantics is associated with map style by creating a database that associates semantic significance to any definition that can occur in a style file. As mentioned previously, semantics associated with *style* has the property of defining the visual significance of symbols in the context of a map, as opposed to conventional semantics that associates a symbol with the linguistic or physical concept it represents. This approach to associating semantics leads to two issues:

 – Granularity of semantic associations, and
 – Representation of denotational and connotational semantics

These aspects are discussed in detail in the following subsections.

3.1   Granularity of Semantic Associations

Maps are a collection of symbols, these symbols have individual semantics and they also can have collective semantic implications on the style of a map. When we associate semantics with a map style, we can associate it at the granularity of individual symbols represented in the maps, groups/collections of entities

Fig. 1. Example map to demonstrate granularity of semantic associations. The buildings are associated with "realistic" and "detailed", while the ground is associated with "schematic" while the map itself is associated with "schematic", "detailed" and "modern".

or with the whole map. We explain this concept by considering and example map presented in figure 1. The buildings in this map are rendered realistically with textures from photographs, therefore, the semantics associated with them indicates that the style of rendering is "realistic" and "detailed". In the same figure the style of rendering of the ground map is a functional road map style and therefore can be associated with the semantics of "schematic". The style of the lettering on the map can be associated with the semantics of "modern" and "no embellishment". The overall style of the map when considering both the ground plane, lettering and buildings can be considered as "schematic", "detailed" and "modern". The semantics that is associated with groups of entities present in a map due to their interaction is known as syntactic semantics [5].

In the previous example we notice that we often associate two or more semantic labels with a single entity, as in the case of the lettering. This happens frequently in semantics associations with entities and is known as polysemic [5]. When an entity is polysemic it is possible to distinguish between denotational and connotational semantics and we discuss this aspect in detail in the following subsection.

3.2   Denotational and Connotational Semantics

The semantics associated with an entity may be explicit or implicit. The semantics that is directly perceived in the appearance of an entity is called denotational semantics and the semantics that is more deeply embedded and results from prior cartographic knowledge in the context of the entity is the connotational semantics. For example, in the case of the lettering "no embellishments" is a semantic property that is directly perceived, and is therefore a denotational semantics, while "modern" is a semantics associated with the lettering by observing several maps and imbibing the knowledge that such lettering appears in modern maps.

It is useful to make these distinctions in the kind of semantics as it enables one to aggregate the semantic implications meaningfully while developing new map styles.

4   Example Entities, Styles, and Associated Semantics

This paper has so far discussed why there is a need for associating semantics with style, and how this is done by associating semantic labels with entities present in the map. In this section we describe with few examples the association of semantics with map styles. We group all entities that are visualized in a map into lettering, map structure and color elements following the approach used by Robinson [4], and we examine each of these groups of entities in detail in the following three subsection.

4.1   Lettering

Historically lettering has always occupied an important position in maps. The style of lettering is sometimes so unique in antique maps, that it is possible to identify the cartographer who made the map based on the appearance of lettering alone.

North America
NORTH   AMERICA
NORTH AMERICA
North America
North America
North America

Fig. 2. Example lettering. The semantics that can be associated includes "serious", "regular", "clear", "all capital", "antique style", "embellished", "hand written", "artistic", "comic".

Lettering has three main appearance parameters associated with it, namely font style, size and color. We defer the discussion on color to the final subsection of this section. Figure 2 presents some example font styles of lettering that may be present in a map. It is straightforward to associate the semantic tags "serious", "regular", "old", "modern", "not embellished", "clear", "all capital", "antique style", "embellished", "hand written", "artistic", "comic", etc. suitably with these font styles. The tags "all capital", "clear", "not embellished", "hand written", "bold", "italic" and "embellished" are denotational semantic tags. The remaining tags can only be associated with the font styles based on prior knowledge of seeing other maps, therefore they are considered as connotational semantics.

The size of the font dictates the relative importance of the information presented in the font. Also, the semantics of "bold" and "italics" come with an associated relative importance. Therefore, in the context lettering in map styles a quantitative semantics parameter called "importance" is defined. It captures the syntactics of relative importance in a group of lettering present in a map style based on their size and font. The quantitative semantics label has the form "importance level 1", "importance level 2", "importance level 3", etc.

4.2   Map Structure

In this sections we describe semantics associations that can be established with all entities of a map that result in the overall map structure and appearance, these include:

1. Scale
2. Two-dimensional image on terrain
3. Three-dimensional map elements
4. Artist style/Technique of map making
5. Projection

The appearance of entities 2 and 3 can be influenced by choice of color, discussion on this aspect is deferred to the next subsection.

Scale. The scale applied to a map may be uniform for all entities or it can be separate for each entity. When it is uniform and is applied to online maps it is often referred to as zoom level. An example map in which all elements are not scaled by the same factor is shown in figure 3. The denotational semantics associated with a scale factor that is non-uniform for all elements is "non-uniform scale" while the connotational semantics is "comic scale", "non-scientific", while that of a constant scale for all elements denotes "uniform scale" and connotes "Scientific" or "accurate". Scale can also be associated with "detailed" or "non-detailed" scale values.

Scaling is also associated with a quantitative syntactics that gives a value of relative importance based on the variation in scale applied to similar data entities that are displayed in a single map. The form of this association is similar to the syntactics captured in "importance level" described in the context of lettering.

Two-dimensional image on terrain - This refers to the 2D-texture draped on the terrain. Some of the style semantics that can be associated are "schematic", "abstract" like road map style or "realistic", "aerial photograph", "detailed", "bird-eye view", "top view","flat shaded", "elevation included", "relative elevation shading", "absolute elevation shading", etc. In this context also some of the labels are denotational and others are connotational.

Three-dimensional map elements - These refer to entities like: buildings, trees, terrain, and other elements including - traffic lights, sign posts, street lights, etc.

Fig. 3. An example map where the semantics associated with scale is "comic scale". There is no correlation between the building scale and the scale applied to the road map appearing in on the 2D plane or the trees scale.

Fig. 4. Example abstractions of textures of a building. The semantics that can be associated is quantitative and the associated semantics labels cab be "abstraction level1", "abstraction level2" and "abstraction level3".

The parameters available to varying in the case of these entities are texture, geometry and color. We discuss color in the next subsection. An example of varying the texture of buildings is presented in figure 4 that shows various degrees of abstraction of representation of a building facade texture. Here a quantitative syntactics is associated with abstraction, for example "abstraction level1", "abstraction level2" and "abstraction level3". All quantitative semantic definitions occur in relation to other elements in a map, therefore we refer to them as syntactic definitions. We can also associate the semantics "detailed", "photograph", "realistic" or "photo-realistic" with the textures.

Apart from the abstraction of textures and geometry, one can vary the appearance of a map by varying the number of these elements displayed on the map. For example, all the buildings in an area may be represented and the semantics associated with such a map style is "complete", "detailed", "general purpose" when only a subset of the entities are selected then the map style is associated with terms that define the basis of selection of the entities like "tourist", "educational", "historic", etc. This is an example of how the functional aspect of a map is incorporated into semantics of map style.

Technique - Refers to the aspect of appearance of a map that does not directly constitute part of the spatial information presented in the map, rather the semantics of technique is inferred based on the appearance of the map. A typical example of such a semantic information relates to the technology used to create the map, this information is always implicity present, and often prominently visible in a map style. Some of the possible semantic labels based on this aspect are "Pen and ink", "Water color", "Woodcut", "Cartoon" etc. Figure 5 includes maps that can be associated with the semantic style labels "woodcut" and "pen and ink". The technique semantics labels are examples of low granularity of semantic association. The technique semantics label can also indicate the name of the cartographer who's style is captured in the creation of the map.

Projection - Various types of projections have evolved for drawing maps. We can apply the same projection for the whole map or vary the projection for various types of elements present in the map or sub-regions of the map. The semantics associated with projection can include "comic" which is demonstrated in the figure 3 or it can be associated with other more technical semantics like, "Mercator", "equal area", "direction preserving", "cylindrical", "conical", etc.

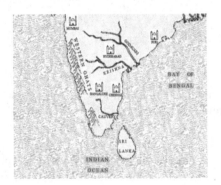 

Fig. 5. Example maps with associated technique semantics "woodcut" and "pen and ink"

4.3   Color

Color is perhaps the most important element present in a map that influences our perception of a map. The design of color for maps has been extensively studies and impressive tools like the ColorBrewer [15] have been developed to aid selections of colors for maps.

The appearance of lettering, 2D map on terrain and 3D map entities can be influenced by choice of color. The colors can be applied as flat shading or styles of color application like "water color", "hand colored outline" can be employed. In computer aided map synthesis the appearance of these color application styles can be simulated with programs. An effect of light interaction/illumination is often incorporated into the coloring this can emphasize the underlying structure as in the case of relief maps, the semantic labels that are applicable in this case include "expressive", or "photo-realistic" to indicate the utilization of photo-realistic rendering algorithms to assign colors to the map.

Color can be used to encode information or it can be used to enhance the realistic appearance based on this the semantic labels "informative" or "realistic" can be associated with color.

The groups of colors associated with "informative" semantics style label can also be associated with the labels "sequential", "diverging", "qualitative", "color blind friendly", "photocopy friendly", "LCD projector friendly", "laptop friendly", "CRT friendly" and "printer friendly" which are the syntactics associations defined for groups of colors in ColorBrewer [15].

Apart from this there are other subjective semantics that can be associated with color schemes like "vibrant", "warm", "bright", "rich", "dark", "light", "pastel", etc.

5   Discussion

In the previous section we have discussed with examples some semantic labels that can be associated with various aspects of appearance/style of maps. In this section we discuss the impact that the availability of semantics, in addition to the map style information, has on creation or rendering of maps in GISs.

In the current implementations of a GIS it is unaware of the kind of map it renders. It is only aware that it has completed display of the spatial information that was requested by the user. It creates possible variations in the appearance of the map style based on a user's selection of parameters. The system itself has no knowledge regarding the style of map it has displayed whether it was a "woodcut map", "tourist map" or "road map", etc. Nor does it associate any semantic qualities with the individual entities it renders into the map.

With the association of semantics it is possible to build a render that is aware of the semantic content of the map. The render can be designed to apply some fundamental logical inferences on the semantics of maps style and derive variations that create effective maps. Figure 6 presents two variations of map appearances that can be derived by choosing consistent semantics for various entities in the map. The significance of selecting consistent representation of entities is brought out by comparing the appearance of these maps with the example map presented in figure 1.

This paper does not describe the details of incorporating the semantics information into the rendering module of a GIS. In our preliminary implementations we have mainly considered a monosemic semantic association with each style of an entity. A case by case approach is required to resolve polysemic semantic associations for style. The semantics associated with figure 1 is an example where hard coded logic is required to resolve the final appearance of a map to be "schematic". Here a rule that states that the presence of at least one entity that is rendered in a map with "schematic" as its map style semantics, will result in

Fig. 6. Example maps with associated technique semantics "realistic" and "schematic"

the final map style being "schematic" helps to ensure that the right semantic association is made with the map.

6 Conclusions and Future Work

In this paper we have described the need for associating semantics with style. We have significantly departed from the traditional viewpoint that visualization styles are extraneous details that have to be separated from data for semantics studies. We proposed a method that associates semantic labels with specifications that occur in a style file.

This paper only introduces the idea of semantics for map style and presents some preliminary examples in this context and illustrates some possible benefits. However the problem of associating semantics with map styles is a large problem and we have not investigated all the ramification of it. Several issues arise once one recognizes the need to create a semantics for map styles, including creating systematic approaches to specify semantics, establishing standards for specifying the semantics, redesigning GIS rendering modules to support semantics. Also, there is a need to develop techniques to formalize the process of resolving polysemic semantics in the context of style. The implications of associating semantics with maps styles in the context of online map information systems [16,17] is also a problem that can be investigated.

References

1. Geography, and MapsDivision. Geography and Map Reading Room (2007), http://www.loc.gov/rr/geogmap/
2. OGC - Open Geospatial Consortium, Inc.: GML - the Geography Markup Language (2007), http://www.opengis.net/gml
3. Special Interest Group 3D (SIG 3D). CityGML - Exchange and Storage of Virtual 3D City Models (2007), http://www.citygml.org/
4. Robinson, A.H.: The Look of Map, Madison. University of Wisconsin Press (1952)

5. MacEachren, A.M.: How Maps Work: Representation, Visualization, and Design. The Guilford Press (1995)
6. Slocum, T.A., McMaster, R.B., Kessler, F.C., Howard, H.H.: Thematic Cartography and Geographic Visualization, 2nd edn. Prentice Hall, Englewood Cliffs (2003)
7. Monmonier, M.: Mapping It Out: Expository Cartography for the Humanities and Social Sciences. University Of Chicago Press, Chicago (1993)
8. Krygier, J., Wood, D.: Making Maps: A Visual Guide to Map Design for GIS. The Guilford Press, New York (2005)
9. Brewer, C.A.: Designing Better Maps: A Guide for GIS Users. Esri Press (2005)
10. Tufte, E.R.: Visual Explanations: Images and Quantities, Evidence and Narrative. Graphics Press (1997)
11. Tufte, E.R.: Envisioning Information. Graphics Press (1990)
12. Adabala, N., Toyama, K.: Semantics-guided procedural rendering for woodcut maps. In: SIGGRAPH 2005: ACM SIGGRAPH 2005 Posters, ACM Press, New York (2005)
13. Adabala, N., Toyama, K.: Customizable panoramic maps. In: SIGGRAPH 2006: ACM SIGGRAPH 2006 Sketches, vol. 131, ACM Press, New York (2006)
14. Adabala, N., Varma, M., Toyama, K.: Computer aided generation of stylized maps. Computer Animation and Virtual Worlds 18(2), 133–140 (2007)
15. Brewer, C.A., Harrower, M.: ColorBrewer - Selecting good color schemes for maps (2007), http://www.colorbrewer.org/
16. Microsoft Virtual earth 3d - online map service (2007), http://maps.live.com/
17. Google. Google earth - broadband, 3d application (2007), http://earth.google.com/

DAGIS: A Geospatial Semantic Web Services Discovery and Selection Framework

Ashraful Alam, Ganesh Subbiah, Latifur Khan, and Bhavani Thuraisingham

Department of Computer Science,
University of Texas at Dallas, Dallas, TX 75083
{malam,ganesh.subbiah,lkhan,bhavani.thuraisingham}@utdallas.edu

Abstract. The traditional Web services architecture uses a keyword based search to match a query to one or more service providers. However, a world-to-word matching to discover a service provider is too simplistic for geospatial data and fails to capture matches that advertise their functionality using domain-dependent terminology. In this paper, we present DAGIS (Discovering Annotated Geospatial Information Services) – a semantic Web services based framework for geospatial domain that has graphical interface to query and discover services. It handles the semantic heterogeneities involved in the discovery phase and we propose algorithms for selecting the best service through QoS (Quality of Service) based semantic matching. The framework is capable of performing dynamic compositions on the fly through a back chaining algorithm. The framework is evaluated by solving queries posed by users in various geospatial decision making scenarios.

1 Introduction

Geospatial data plays a pivotal role in value-added content exchange between software agents or amongst people. The ability to provide additional dimensions to otherwise monotonic information has led to an enormous increase in the use of geospatial services. A rather underrated aspect behind such an escalation is the fact that spatially-aware data is more amenable to human cognition than strictly textual information. A far more appreciated aspect is that the integration of diverse data types with geospatial sources has yielded practical business and research benefits. Medical data overlapped with digital maps provides wealth of information in forecasting epidemics; population research centers can trace genealogical data over a region to discover social trends and so forth. This growing interest and activity level in the geospatial domain is further edified by more than 232 million hits on Google [TM] for the keyword 'geospatial.' Geospatial data is characterized by multitude data formats and data models and integration of this valuable data is crucial for the businesses and applications on the World Wide Web. But lack of a common unified framework for discovery, collection, and dissemination of geospatial data is characterized by the coherent heterogeneities present at both the syntactic and the semantic level.

Web Services driven Service Oriented Architecture model provides a mechanism to handle the syntactic heterogeneities to an extent for geospatial data sources. The

F. Fonseca, M.A. Rodríguez, and S. Levashkin (Eds.): GeoS 2007, LNCS 4853, pp. 268–277, 2007.
© Springer-Verlag Berlin Heidelberg 2007

current geospatial standards recommended by OGC- a flagship consortium that specifies standards for describing the geospatial data and services are founded on these principles of providing geospatial data interoperability. On the other side, emergence of semantic web and its associated technologies which aims to transform the web data sources into intelligent knowledge repositories that will use web agents to reason and infer information in more sophisticated manner. Semantic Web technologies provide strikingly similar standards for better interoperability of data and services with less human intervention for the World Wide Web. This prompted the researches from both the communities towards the vision of geospatial semantic web for realizing semantic interoperability of geospatial data. The recent OGC geospatial semantic web interoperability experiments are a major step towards this vision.

It's argued by researches Semantic interoperability is an important goal but hard to pin down due to lack of common accepted formal specifications. Kuhn [13] establishes that Service Signatures needs to be semantically annotated to achieve semantic interoperability but the challenge of annotating proper semantics for web services description and automatic discovery is imminent.

In this paper, we propose DAGIS – Discovery of Annotated Geospatial Information Services framework for building geospatial semantic web services using the OWL-S Service ontology coupled with the geospatial domain specific ontology for automatic discovery, dynamic composition and invocation. The algorithms developed for this framework enables semantic matching of functional and non-functional services during each phase of Service Orientation. In addition, our approach makes use of [2] since its hybrid mechanism seems to produce better results.

There has been major work done on geospatial data interoperability. Vokovski et al. [7] and Goodchild et al. [8] address various interoperability issues related to spatial data processing of vectors and graphics, semantics, heterogeneous databases and representation. OGC identified that the key to solve interoperability issues are through the interface of software components where data and its operations are inseparable. This resulted in syntactic specification for geospatial data exchange through Geography Markup Language [9]. Operations on features in GML are implemented through web services [1]. Web Feature Service (WFS), Web Map Service (WMS), Web Coverage Service (WCS) are the core standards for Web services being developed by OGC to allow distributed geo-processing systems to provide complex services.

The rest of the paper is organized as follows. Section 2 presents the DAGIS architecture, its automatic discovery mechanism, dynamic composition algorithms and the invocation mechanism. Section 3 presents QoS based service selection. Finally, section 4 presents complex queries.

2   DAGIS Framework

Integration of geospatial and non-geospatial information tasks involves separate data sources and service providers. Executing the tasks with minimal human intervention is the motivation behind our proposed architecture. The implementation of the architecture -- called DAGIS -- focuses on devising improved query mechanisms through automated reasoning using a domain specific ontology. We have built DAGIS as a prototype application that is useful for finding information for local businesses over a geographical region. We have identified the major phases in developing this framework. These phases are discussed in the following sections.

DAGIS provides an immediate advantage over other web 2.0 and GIS based map solutions. The latter products have limitations when the following types of queries are encountered: "Find Movie Theaters between Richardson, TX and Irving, TX". This geospatial query is commonly posed by users looking for local information around the geographical regions of interest. Current solutions do not recognize the semantics of the geospatial operator "between" in this query. We posed this query on Google Maps and observed that it is oblivious to the presence of such operators.

2.1 DAGIS System Architecture

DAGIS system architecture is described in this section. Functionality of each of the components is addressed through a running example. We distinguish the layers that constitute an end-to-end query execution and result display. The major layers are the presentation layer, semantic middleware layer and the ontology data layer. DAGIS Framework has major components at each of this layer.

DAGIS Query Browser Portlet: In the presentation side, the DAGIS query browser portal gets the user query. We have developed a Java™ portlet that provides the required interface for the query.

DAGIS Agent: DAGIS agent, placed at the semantic middleware layer, fetches the query parameters from the user. We can deploy multiple DAGIS agents in this layer. In our current application we describe the behavior of a single DAGIS agent. This agent communicates with the DAGIS Matchmaker using OWL-S (formerly known as DAML-S) [5] service ontology language. It automatically constructs an OWL-S query for the given user query.

DAGIS Matchmaker: DAGIS Matchmaker is the component that performs semantic matching between the submitted queries and the semantic web service providers present in the registry. It performs both functional and non functional based selection and service discovery.

DAGIS Composer: DAGIS Composer dynamically builds service chain to solve the user query when there is no single service provider available to match user query requirements. This dynamic composition is done automatically and the composed service URI is returned back to the Matchmaker.

OWL-S Registry: The semantic web services are stored in this registry, which acts like a catalogue of useful services.

WSDL Registry: The WSDL registry is any standard UDDI or public web services registry such as www.x-methods.net and www.salcentral.com.

WSDL2OWLS Converter: This converter converts the WSDL service description file to OWL-S file. The XSLT conversions are currently done manually, but in the future there would be full fledged automatic conversion package.

Figure 1 shows how the aforementioned components fit into the DAGIS framework. Initially a user requests for service through a query browser (i.e., portlet). DAGIS agent receives the query and forwards it to a matchmaker. The matchmaker inquires the OWL-S registry to determine a match. The matchmaker is responsible for talking to the domain ontologies through a common OWL-S API and performing the semantic interpretation of the terms. Figure 1 also shows the separation of layers based on their functional requirements. The presentation layer allows the client to actually input the query. Then we have the middleware layer that allows interchangeable components to provide meta-service related functionality such as service search and reasoning. It is important that the middleware layer is not tied to a special platform or architecture. It should be abstracted in a way so that other layers do not have dependency on the underlying details of the middleware components. This abstract also encourages extensibility by swapping in and out modules to fit one's needs. The third layer consists of the ontologies including the service and domain ontologies. We describe the workflow of the DAGIS architecture in more details in the following sections.

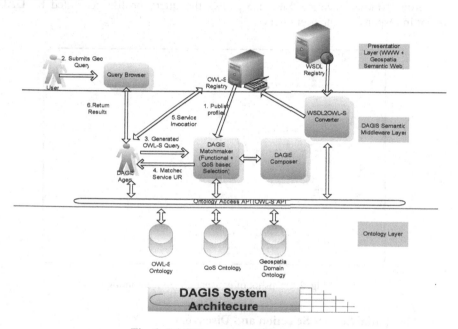

Fig. 1. DAGIS system architecture

2.2 Geospatial Ontology Development Phase

In our work, we have developed geospatial service ontology to describe concepts used by geospatial web services. The concepts defined in our ontology were developed in accordance with OGC Web Services Specification Architecture. The QoS ontology developed is described along with QoS selection process. Figure 2 shows the snapshot of our geospatial ontology developed for DAGIS. The businesses are categorized under the geocoder results class. City, Latitude, Map, State, Zip code are also subclasses of this class. The different kinds of geospatial web services are categorized

under the main class OGCSemanticWebServices. These subclasses are Feature
Handling Services and Mapping Services. Web Feature Service like Gazetteer Service
is part of Feature Handling Service. Coverage Portrayal Service, Feature Portrayal
Service, Web Map Services are subclasses of Mapping Services.

2.3 Automatic Semantic Query Profile Generation

After the user submits the query, it is disambiguated using our developed ontology;
subsequently, an OWL-S service profile is automatically generated. In the next step
the query profile is used by the DAGIS Agent for service discovery and selection of
the service providers that will solve this query.

The DAGIS Agent uses this semantic profile for selecting the appropriate service
provider from the matchmaker agent. The following figure shows a snapshot of the
profile for a simple query: 'Find Movie Theaters within 30 miles of zip code 75080'.
The profile of this OWL-S file has input ZipCode, distance 30 miles and output
required is movie theaters. Figure 3 shows the query profile generated by DAGIS
agent in response to the user query.

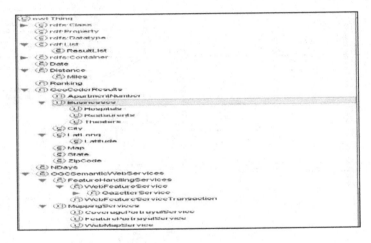

Fig. 2. Snapshot of geospatial service ontology

2.4 Geospatial Service Selection and Discovery

The service selection based on the functional and non-functional requirements of the
generated query profile is used by the DAGIS Matchmaker agent for selecting the
appropriate service providers. The Matchmaker in our framework does capability
based reasoning using the Pellet OWL-DL reasoner. Our implementation of the
Matchmaker for this framework is developed by extending the OWL-S MX
Matchmaker [2]. It is Java™-based and uses Pellet for logic based filtering. It also
uses loss-of-information, extended Jacquard, and Jensen-Shannon information
divergence based similarity metrics for complementary approximate matching. We
extend this hybrid matchmaker to handle service selection based on QoS. There are
different degrees of matches based on the similarity. The similarity criteria form a

```
        <profile:Profile rdf:about="#QueryProfile">
        <profile:hasInput>
        <process:Input rdf:ID="ZipCode">
        <process:parameterType
rdf:datatype="http://www.w3.org/2001/XMLSchema#anyURI">http://127.
0.0.1/Ontology/OGCServiceontology.owl#ZipCode</process:parameterTyp
e>
        </process:Input>
        </profile:hasInput>
        <profile:hasOutput>
        <process:Output rdf:ID="Movie Theaters">
        <process:parameterType
rdf:datatype="http://www.w3.org/2001/XMLSchema#anyURI"
        >http://127.0.0.1/Ontology/OGCServiceontology.owl#MovieTheaters</
process:parameterType>
        </process:Output>
        </profile:hasOutput>
        </profile>
```

Fig. 3. Generated query profile

lattice based on how relaxed the similarity is. EXACT match is least relaxed and FAIL is most relaxed.

3 QoS Based Service Selection

The QoS based automatic service selection plays a crucial role in the matchmaking process when there is more than one registered service provider providing similar functionalities. In our proposed system, trust calculations are established through capability based matching of the QoS parameters. QoS parameters are the nonfunctional attributes that aid in the dynamic service discovery and selection. This facilitates the dynamic computation of the trust for the service provider and selection can be made for a suitably trusted service by the client.

Our architecture is based on agent-based trust framework where the different QoS parameters characterized under various dimensions for describing the quality are captured in the client profile and the providers' profiles. The proposed geospatial services ordering metric (GSOM) for QoS evaluation and for establishing trust is described in the following section.

3.1 QoS Ontology

Our QoS ontology is developed in line with the upper and middle ontologies as described in [11]. This facilitates modular development and can easily be extended for our geospatial domain concepts defined in the geospatial ontology. The main concepts in the QoS ontology are:

- Quality: Representing the measurable nonfunctional concept of a service.
- QAttribute: The value of a quality concept is determined by the type of QAttributes that constitute that concept.

- QMeasurement: This described the measurement of quality which can be subjective or objective
- QRelationship: For describing relationship between two or more quality concepts.

During the service discovery phase, the query profile of the user is submitted to the matchmaker for determining the functional matches from the set of published services. The Matchmaker returns a set of functionally similar services if the query to be solved involves single service provider; otherwise, it returns a dynamically composed service. To incorporate the QoS based selection, we add a step to this service discovery process. The new algorithm operates as follows.

1. Service providers publish profiles to Matchmaker
2. User submits query and corresponding semantic query profile is generated
3. Find semantically similar services for the queBry using the functional parameters – that is, the input and output parameters
4. If there is no such service from step 3, dynamically compose complex service using the services registered using DAGIS composer algorithm
5. Sort the functionally similar semantic services using the GQoS Algorithm
6. Return the URI of the best service from step 5 to user

We will describe the approach developed by us for performing the step 5 of the above service discovery algorithm. The QoS selection differs when we have a dynamic composition. In that case, it involves computing the aggregate QoS values of the services dynamically, which is one of our contributions in this paper.

3.2 QoS Selection Algorithm

Interaction Model: The environment is comprised of registered service providers S_1, S_2 ... S_j, users U_1, U_2 ... U_i, matchmakers M_1, M_2 ... M_k. In our interaction model we assume only one matchmaker. We employ special monitoring services that get user reports on QoS relevance feedback called trust monitors TM_1, $TM_{2 ...} TM_l$.

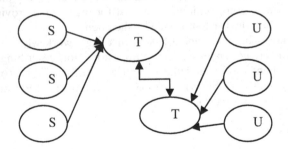

Fig. 4. Interaction model

Service providers publish their QoS values $(sq_1, p1)$, $(sq_2, p2)$ ...where (sq_i, p_i) are vector pairs of concepts and their values. Users provide QoS requirements for every query as (uq_1, r_1), (uq_2, r_2) , ... where (uq_i ,r_i) are vector pairs of concepts and user required values (see Figure 4). During feedback loop, users submit their feedbacks as

U_{1j}, U_{2j}, U_{3j} where j is index for the service provider j selected during each query iteration process.

In the first phase, for each registered service provider j in the functional match set F of the query Q, a G_{val} is evaluated using the advertised QoS parameters. The QoS similarity matching algorithm is illustrated in Figure 5. All the service providers are initially set with $G_{val} = 0$ and the target concept matches between query and service provider concept are set to 3. In step 1 for every service Sj a functional set F is returned from the Matchmaker. The aggregated difference in the user expected and provided values is stored in diff, which was initially set to 0. For every quality concept q_i in Vector uq, if there is a concept match (exact, subsumes etc.) with a concept in sq_j, ConceptMatch is incremented. The diff is updated for this match. In step 7 we check if there are at least target number of matches for meeting the user requirement; then we compute the G_{val} as average diff in step 8. Step 9 ensures that as G_{val} is updated through propagation algorithm (discussed next), when it goes above the threshold T, service s_j is considered to be untrustworthy and removed from set F. Step 11 returns the F in ascending order of G_{val}.

In the second phase we use the user feedback to update the advertised GQoS parameters of the selected service S_i as follows. For every query Q posed by U_i, C_{ij} is the conformance value vector submitted by U_i for S_j to TM_l. The satisfaction of the user on each QoS parameter he had specified is measured qualitatively through Cij on a fuzzy scale. This is used to get the weighted expectation vector (Uij * Cij) of a user. The feedback vector is used to update the P_i of Service S_i in step 4 in QoS propagation algorithm (Not reported here). In our model, user reports are considered to be credible only for authenticated users of the system, who log on to the system for service discovery. We assume that the service providers that publish their service descriptions to the matchmaker do not cancel their registration during the interaction for at least a certain number of iterations. The current model sets a hard number on the lower bound of the provider availability period to determine untrustworthy providers. The period is defined in terms of the number of iterations a provider was available for the Matchmaker. Right now this number is 10, but in the future we will maintain logs of the interactions to capture these cancellation scenarios also.

4 Complex Queries Using DAGIS

The scenario described in section 2.1 is a relatively simple one that involves selection of a single service provider. Real world scenarios often involve complex queries that necessitate dynamic composition of different service providers. To explain the complexities further, we restate the example from section 2.1. Consider the following query *"Find movie theaters within 30 miles of Richardson?"*.

We use the DAGIS visual interface to drive the user query, thereby bypassing the need to parse natural language based queries. Based on the client query profile a search is performed in service registry to discover matching OWL-S profiles. Since there is no service that takes city as input and returns movie theaters within a certain radius, the matchmaker resorts to decomposing the query into multiple atomic processes using DAGIS decomposer algorithm. Decomposing the query into two atomic parts results in a successful Web service execution since there is a profile that

```
User Query List UQ = {(uq₁, r1), (uq₂, r2) …. (uqₙ, rn)}
TargetMatch = 3
G_val = 0 for all services
findSimilarityMatch()
1. ∀S_j in Functional Match Set F
2. diff = 0.0
3. ∀q_i:q_i=quality concept in uq
4. If q_i matches with a concept in sq_j
5.        conceptmatch = conceptmatch +1
6. diff += |p_j-r_i|
7. If concept match >= TargetMatch
8.         G_val =   diff/conceptmatch
9. If G_val > T
10.     remove S_j from F.
11. Return F sorted by ascending order   of G_val scores.
```

Fig. 5. QoS similarity match algorithm

outputs zip-codes given a city and there is a second profile that outputs movie theaters given a zip-code. The Compose Sequencer component constructs the composite service.

4.1 DAGIS Composition and Sequencing Algorithm

The composer and sequencer algorithms in this section are based on the Recursive Back Chaining algorithm proposed in [12]. To construct the service chain, our algorithm is recursively called for each likely service available in the service registry. A service is selected only if its output is equivalent to desired output of the requesting client. We also have a sequencer algorithm that provides composite process chaining for non-atomic processes. This algorithm uses a trivial bind function to create a mapping between input and output parameters of two processes (a hash map can be used to represent the mapping data structure in the actual implementation).

4.2 Service Invocation

In this phase, the DAGIS Agent has the selected service provider's OWL-S URI from the discovery process and invokes the service provider. In this scenario, the selected service has an Atomic Process – GetTheaterProcess. As the service provider agent also uses the same domain ontology as the DAGIS Agent for semantic annotations of its services. This is the major benefit of sharing the semantic concepts using a unified ontology framework. The DAGIS agent does the invocation of the service through OWL-S grounding. The OWL-S grounding then uses WSDL grounding to invoke the Web Service using AXIS in our framework. The OWL-S API used in this system provides the execution engine and monitoring environment to monitor the process execution and for exception handling.

References

1. Sondheim, M., Gardels, K., Buehler, K.: GIS Interoperability. In: Longley, P., Goodchild, M., Maguire, D., Rhind, R. (eds.) Geographical Information Systems 1 Principles and Technical Issues, John Wiley & Sons, New York (1999)
2. Klusch, M., Fries, B., Sycara, K.: Automated Semantic Web Service Discovery with OWLS-MX. In: AAMAS. Proceedings of 5th International Conference on Autonomous Agents and Multi-Agent Systems, Hakodate, Japan, ACM Press, New York (2006)
3. Srinivasan, N., Paolucci, M., Sycara, K.: An Efficient Algorithm for OWL-S Based Semantic Search in UDDI. In: Cardoso, J., Sheth, A.P. (eds.) SWSWPC 2004. LNCS, vol. 3387, pp. 96–110. Springer, Heidelberg (2005)
4. Martin, D., Paolucci, M., McIlraith, S., Burstein, M., McDermott, D., McGuinness, D., Parsia, B., Payne, T., Sabou, M., Solanki, M., Srinivasan, N., Sycara, K.: Bringing Semantics to Web Services: The OWL-S Approach. In: Cardoso, J., Sheth, A.P. (eds.) SWSWPC 2004. LNCS, vol. 3387, pp. 26–42. Springer, Berlin (2005)
5. OWL-S: Semantic Markup for Web Services. W3C member submission (2004), available: http://www.w3.org/Submission /OWL-S/
6. Egenhofer, M.: Toward the Semantic Geospatial Web. In: Voisard, A., Chen, S.-C. (eds.) ACM-GIS, McLean (2002)
7. Včkovski, A., Brassel, K.E., Schek, H.-J. (eds.): Interoperating Geographic Information Systems. In: Včkovski, A., Brassel, K.E., Schek, H.-J. (eds.) INTEROP 1999. LNCS, vol. 1580, Springer, Berlin (1999)
8. Goodchild, M.F., Egenhofer, M., Feeas, R., Kottman, C.: Interoperating Geographic Information Systems. Proceedings of Interop 1997, Santa Barbara, CA, Norwell, MA. Kluwer, Dordrecht (1998)
9. Geography Markup Language (GML) version 3.1.1 Specification, available: http://opengis.net/gml/
10. Sirin, E., Bijan Parsia, B.: The OWL-S Java API. In: McIlraith, S.A., Plexousakis, D., van Harmelen, F. (eds.) ISWC 2004. LNCS, vol. 3298, Springer, Heidelberg (2004)
11. Li, L., Horrock, I.: A Software Framework for Matchmaking Based on the Semantic Web Technology. In: Proceedings of 12th Int Conference on the World Wide Web, Workshop on E-Services and the Semantic Web (2003)
12. Renier Gibotti, F., Gilberto, R.: GeoDiscover – a Specialized Search Engine to Discover Geospatial Data in the Web. In: 7th Brazilian Symposium on GeoInformatics (2005)
13. Kuhn, W.: Geospatial Semantics: Why, of What, and How? In: Spaccapietra, S., Zimányi, E. (eds.) Journal on Data Semantics III. LNCS, vol. 3534, pp. 1–24. Springer, Heidelberg (2005)

The Gravity Data Ontology: Laying the Foundation for Workflow-Driven Ontologies

Ann Q. Gates[1], G. Randy Keller[2], Leonardo Salayandia[1],
Paulo Pinheiro da Silva[1], and Flor Salcedo[3]

[1] University of Texas at El Paso, El Paso TX 79902, USA
agates@utep.edu, leonardo@miners.utep.edu, agates@utep.edu
[2] University of Oklahoma, Norman OK 73019, USA
grkeller@ou.edu
[3] Rockwell Collins, Cedar Rapids, IA 52498, USA
fisalced@rockwellcollins.com

Abstract. Ontologies can be tailored in ways that can facilitate the description of workflows by specifying how concepts representing services are used to access and create concepts representing data and products. Early work on the development of such ontologies, and reported in this paper, has resulted in the construction of a gravity data ontology. The relationships that are defined in the ontology capture inputs and outputs of methods, e.g., derived data and products, as well as other associations that are related to workflow computation. This paper presents the basis for a computation-driven ontology that evolved into the workflow-driven ontology approach. In addition, the paper describes the process used to construct an ontology for gravity data using the computation-driven approach, and it presents a gravity ontology that documents the processes and methods associated with gravity data and related products.

1 Introduction

Numerous institutions and organizations around the country have collected geospatial data, algorithms, and processes for manipulating and integrating these data with other diverse data sets, generating results that are useable by them, other scientists, or the general public. The goal of the work presented in this paper is to move from an environment in which a scientist relies on a professional network and manual processes to complete their work to one in which a scientist uses an automated system to accomplish tasks. An approach for realizing this goal is to capture knowledge through an ontology and then leverage the knowledge to support the design and execution of scientific workflows that compose software services to compute a particular result or generate a product.

There are several challenges that scientists face when creating any ontology: defining the scope of knowledge capture, determining the level of abstraction used to describe concepts and relationships, and identifying useful concepts and relationships. Clearly, creation of an ontology should be a continuing process that requires revision and refinement.

F. Fonseca, M.A. Rodríguez, and S. Levashkin (Eds.): GeoS 2007, LNCS 4853, pp. 278–287, 2007.

This paper presents an overview of a computation-driven ontology. The main contributions of the paper are to provide the rationale for establishing the key concepts in the computation-driven ontology and to document the process used to create a computation-driven ontology for gravity data. The paper also presents an overview of the effort to develop tools that assist a scientist during the process of creating and validating an ontology and generating abstract workflows. These workflows denote how a result is achieved by presenting the composition of methods (software services or algorithms) including the flow of data and control among the methods.

2   Basis for Computation-Driven Ontologies

The basis for the concept of a computation-driven ontology was inspired by a February 2004 Seismology Ontology workshop held at Scripps Institution in San Diego. The attendees of the workshop included experts in the areas of seismology and information technology.

While the initial focus of the workshop was on creating a discipline-based ontology, i.e., an ontology focused on capturing knowledge about a particular discipline, it ended with a categorization and a set of relationships that were based on a general workflow that describes a common task performed by seismologists. After struggling with identifying the concepts that should be captured in a seismology ontology and motivated by a desire to identify concepts and relationships that would be useful to the community, the workshop participants defined concepts of interest by constructing the workflow shown in Figure 1. For the scientists, the workflow captured the steps for completing the task of creating a P-wave velocity model and the necessary concepts that are involved in completing such a task. After completing the workflow, the seismologists next partitioned the diagram into three categories: "Data," "Method," and "Product," where *Data* denotes input to or output from a *Method*, *Method* is a software service or algorithm, and a *Product* is an artifact.

A summary of observations from the workshop includes the following:

1. *The benefits of using a workflow to drive creation of a specialized ontology*- If one considers how a desired product or result is generated, a discipline expert

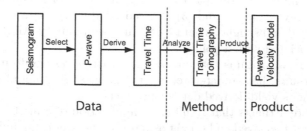

Fig. 1. A workflow created at the 2004 Seismology Ontology Workshop

can identify the data, derivation algorithms, transformation algorithms, and other data processing algorithms involved as well as the relationships between them.

2. *The benefits of using a workflow to determine missing concepts or relationships-* It's important to note that the workflow given in Figure 1 is not complete. The step from *P-Wave* to *Travel Time* requires a transformation method that is not depicted in the diagram. The ability to view a workflow based on concepts captured in an ontology can assist in the iterative process of refining an ontology.

3. *The importance of using abstraction in the ontology-construction process-* Related to the second observation, this promotes the need to focus on a particular product or result at a high-level while neglecting other aspects. Moving from a high-level abstraction to detail allows one to manage the complexity in defining an ontology. For example, one can specify that *P-Wave derives Travel Time* and in subsequent iterations specify the method by which this is done.

4. *The importance of having ontologies that are created by scientists and for scientists-* While technology is critical for the development of cyberinfrastructure, the tools that scientists use to define and manage ontologies and workflows must be scientist-friendly and relevant to them.

3   Overview of the Computation-Driven Ontology

The observations that were made at the 2004 Seismology Workshop led to the definition of a specialized ontology called a *computation-driven ontology*, an ontology that encodes discipline-specific knowledge in the form of concepts and relationships supporting visualizations that depict how data is derived or results are obtained, e.g., in the form of a workflow. It is important to note that a computation-driven ontology casts concepts from a discipline-specific ontology into pre-defined concepts and relationships.

As a proof-of-concept, Salcedo and Keller [10] applied the approach to develop a gravity-data ontology. The top-level categories of the ontology are described as they apply to the gravity domain:

- *Data* define three types of concepts: (1) Field Observations, the purest form of gravity data; (2) Principal Facts, i.e., latitude, longitude, elevation and observed gravity values; and (3) Derived (Reduced) Data, i.e., values that are perceived and sought as data by the user community. All three types are values associated with a point.
- *Methods* are algorithms that are applied to the various forms of data to produce results that are interpretable from a geologic point of view. Results from methods yield derived data or products.
- *Products* are artifacts that result from application of a method. These artifacts are not perceived and sought as data by the user community. Examples include maps, models, or images.

Table 1 summarizes the main relationships that are defined for a computation-driven ontology. The table gives the inverse relationships and indicates whether the relationship supports transitivity, i.e., if a is related to b and b is related to c, then a is related to c.

Table 1. A summary of relationships for a computation-driven ontology

Tuple	Inverse	Trans.	Description
$[c1, isInputTo, c2]$	getsInputFrom	No	c1 is a Data or Product concept with raw numerical values; c1 is input into Method c2
$[c1, isOutputOf, c2]$	outputs	No	c1 is a Data or Product concept; c2 is a Method concept
$[c1, isDerivedFrom, c2]$	isConvertedTo	Yes	c1 is a Data or Product concept; c2 is a Data or Product concept; c1 has been created through a transformation of c2; c1's existence depends upon the existence of c2
$[c1, includes, c2]$	isIncludedIn	Yes	Method c1 includes Method c2 as a helper Method
$[c1, uses, c2]$	isUsedFor	Yes	c1 is a Method concept; c2 is a Data or Product concept; a Method uses a Product or Data when neither one is direct input into the Method

Consider the following statement: the adjusted gravity reading in milligals is derived from the raw gravity reading via the equation:

$$AGR = (RGR * CC) + DC + TC$$

where AGR is the adjusted gravity reading, RGR is raw gravity reading, CC is calibration constant for the gravity meter, DC is drift correction, and TC is tidal correction. From this text, we identify a method MAGR that computes AGR, and we identify the following relationships:

- $[RGR, isInputTo, MAGR]$
- $[CC, isInputTo, MAGR]$
- $[DC, isInputTo, MAGR]$
- $[TC, isInputTo, MAGR]$
- $[AGR, isOutputOf, MAGR]$

In the initial iteration of the ontology, one could state: [AGR, isDerived-From, RGR], if the equation was not available or not considered because that level of detail was being abstracted. The next example shows the application of the include relationship, and makes an argument for incorporating it in a computation-driven ontology. Consider the text: Gridding methods include interpolation methods. This could be denoted as: $[M_{Grid}, includes, M_{Inter}]$. There are a number of interpolation algorithms that could be used with a gridding algorithm, and the includes relationship is used to capture this notion. To illustrate the uses relationship, consider the following statement: a Regional Gravity Map (RGM) is used to determine whether to use a Directional Filter Method because the user must visualize the anomaly values to decide whether to use this filter. This denotes a manual process and should be considered when deriving a workflow description. The relationship would be expressed as: $[M_{Filter}, uses, RGM]$.

4   Constructing a Computation-Driven Ontology

Ontology 101: A Guide to Creating Your First Ontology [6] presents guidelines for creating an ontology, which are applicable to a computation-driven ontology. In particular, use case modeling is an effective approach for driving the creation of any ontology.

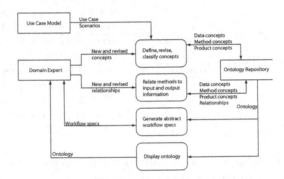

Fig. 2. Flow of information when constructing a computation-based ontology

The computation-driven approach places the primary focus on methods and data that generate results of interest to the scientist as well as on workflow-based relationships. Figure 2 presents a data flow diagram that depicts the processes or steps for defining a computation-driven ontology. The square in the diagram represents a source or sink, the rounded boxes depict transformation of information, and the open rectangle a store. As depicted in the figure, creation of an ontology is a continuing process, and it includes the use of an abstract workflow (as depicted in Figure 3). The processes are described next.

Identify concepts. Use cases allow one to scope the knowledge capture and identify useful concepts. In use-case modeling, the scientist identifies the primary uses of the ontology. Identifying use cases is complementary to developing workflows as an initial approach for specifying appropriate concepts. The discipline expert should consider the following questions: What types of data are available or can be derived? What existing algorithms, tools, or steps are used to generate data? What results are important to me or the community?

To illustrate the benefit of use cases, consider the following use cases in the gravity domain: "determine the Complete Bouguer Anomaly for points in a gravity data set," and "create a free-air anomaly map." Given the use case as a starting point, the scientist would identify related algorithms for generating the desired data or product. For example, starting with the concept *Complete Bouguer Anomaly* and knowing that "Variations in Simple or Complete Bouguer Anomaly values are the major input into interpretations of the geological features present in the area of a geophysical study" would lead to the following concepts (types in parenthesis): *Simple Bouguer Anomaly (Derived Data)*, *Complete Bouguer Anomaly (Derived Data)*, *Interpretation Method (Method)*. The following statement, "Calculation of the Complete Bouguer Anomaly uses the Free Air Correction value," leads to the following concepts: *Calculate Complete Bouguer Anomaly (Method)* and *Free Air Correction Value (Derived Data)*. The following statement, "Observed Gravity Data is input to the Calculate Free Air Anomaly method where it has modifications performed on it and this produces a Free Air Anomaly," leads to the following concepts: *Observed Gravity Data (Processed Data)*, *Calculate Free Air Anomaly (Method)*, and *Free Air Anomaly (Processed Data)*.

To elucidate the process of using a workflow to drive elicitation of concepts, consider that a discipline expert identifies *Anomaly Map* as an important result. Geospatial-mapping software, such as GMT (Generic Mapping Tools) [13] and denoted in the figure as *Mapping*, takes *Anomaly Values*, grids them, and contours them to generate an *Anomaly Map*. *Anomaly Values* are the result of raw gravity data reduction (e.g., [3]), which can be obtained through a series of steps programmed in Excel (e.g., [4]). In this example, *Anomaly Map* would be classified as *Product* and *Anomaly Values* would be classified as *Derived Data*. *Mapping* and *Excel Reduction* are classified as *Method*. Figure 3 presents two views for specifying this workflow. In the first depiction, methods are shown on the right side of the diagram, data and products are shown on the left. The relationships are marked above the arrows. In the second, the text in bold denotes the desired output. Questions regarding "how the output is generated" results in the specification of the next step. This continues until the base or initial concept is reached, i.e., *Raw Gravity Data*. The darkened arrows denote the outputs from methods and the text within parenthesis denote the inputs to the methods. Defining a simple workflow as shown in Figure 3 can be useful for defining concepts as well as refining concepts. For example, if the discipline expert had not included the *Excel Reduction* method and instead used the relationship *Anomaly Values isDerivedFrom RawGravityData* in the first diagram of Figure 3, then the expert would recognize that the ontology is underspecified; he or she would specify the method *Excel Reduction* during refinement.

Identify relationships. The discipline expert also identifies the relationships between concepts. All *Derived Data* and *Product* concepts should be associated with at least one *Method* class, and all *Method* classes should have input and output relationships.

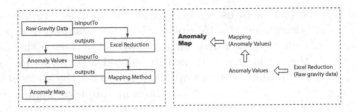

Fig. 3. Two views for illustrating the steps towards generating an abstract workflow specification for an Anomaly Map

The gravity data ontology is represented in the Ontology Web Language (OWL) [12], and the concepts described in this paper are referenced as classes in OWL. As a result, the class hierarchies are grounded in the OWL class *Thing*. During construction of the gravity data ontology, super class *Product* was divided into subclasses *Gravity map* and *Gravity model*, and subclasses *Anomaly Map* and *Contour Map* were defined under *Gravity Map*.

As described earlier, creation of an ontology is a continuing process that requires revision and refinement. For example, refinement of the ontology resulted in refining the *Interpretation* concept to include subclasses *Modeling* and *Mapmaking*. A similar refinement process occurred in which concepts *Complete Bouguer Anomaly* and *Free Air Anomaly* were classified as *Corrected Gravity Data*. Figure 4 shows a portion of the gravity data ontology that was created with experts in the field of geophysics using Protégé Ontology Editor and Knowledge-Base Framework tool, Version 3.1 Beta Full. Because of space constraints, the graphical depiction does not show relationships or annotations associated with each concept. See http://trust.utep.edu/ciminer/collaborations/ for documentation of the ontology.

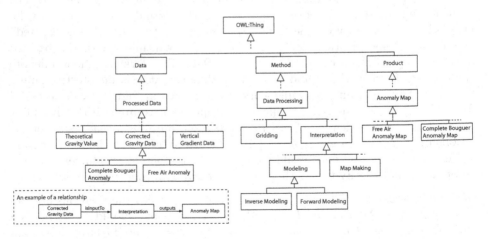

Fig. 4. A portion of the Gravity Data Computation-Driven Ontology

5 Tool Development Efforts

The experience of creating a workflow-driven ontology for gravity data provided a number of insights. The scientist involved in defining the gravity data ontology found it more amenable to work on an Excel worksheet to initially store the concepts and relationships prior to specifying them in a formal ontology language such as OWL [12] and with the aid of an ontology editor such as Protégé. Moving toward a scientist-friendly approach to specification of ontologies has become a focus of the research. Indeed, the computation-driven ontology has evolved in the workflow-driven ontology (WDO) approach [8,9].

The WDO-It! tool provides a graphical-user interface that is consistent with the concept classification requirements of workflow-driven ontologies and that can guide the scientist to elicit concepts and relationships from which abstract workflows can be generated. In addition, the WDO-It! tool provides workflow-generation functionality that allows the scientist to select a target data concept, and to generate graphical representations of abstract workflows that derive the selected data concept. The workflow generation functionality of WDO-It! is based on the Jena Ontology API [14] that supports inference engines that can interpret and reason about ontologies specified in OWL. The graphical representation of abstract workflows generated by WDO-It! serve as scientist-friendly devices that can be used towards the refinement and validation of the ontology, and that can be leveraged by scientists and technologists towards the development of executable workflow specifications. The authors are in the process of validating the usability of WDO-It!.

In addition, the capture of provenance information [7] provides the scientist with the ability to annotate data and method concepts with source metadata. For example, metadata regarding *Raw Gravity Data* could include information about the instrument used to collect the data, accuracy estimates, and the individual or entity that recorded the readings, while metadata regarding *Gravity Data Reduction Method* could include information about the specific implementation of the method and its constraints. As a result, once workflows are constructed from these data and method concepts, more complex data concepts or products could be automatically annotated with provenance information that includes the source data, methods, and workflow process used to generate them. Probe-It! is a prototype tool that provides the visualization of provenance data of an executing workflow. Assuming that the executable workflow is composed from provenance-annotated concepts, Probe-It! traces the provenance and constructs provenance proofs on the fly as a workflow is executed.

6 Related Work

There are numerous published ontologies. This section summarizes three: the Gene Ontology (GO), the Transparent Access to Multiple Biological Information Sources (TAMBIS), and the Semantic Web for Earth and Environmental Terminology (SWEET) ontologies.

GO [11] provides a controlled vocabulary to capture gene information. In the GO ontology, a function describes methods, and the process ontology describes a series of steps similar to a workflow. TAMBIS [2] is a bioinformatics ontology whose design is based on description logics in order to allow dynamic creation and reasoning about the concepts. TAMBIS is organized into multi layer divisions. For example, a structure can be separated into its physical and abstract representations. The ontologyalso has separate concept divisions for biological processes and biological functions. Similar to TAMBIS, the computation-driven ontology approach adopts the separation of concerns with respect to concepts.

The SWEET ontologies [15] were developed to capture knowledge about Earth System science. There are two main types of ontologies in SWEET: facet and unifier ontologies. Facet ontologies deal with a particular area of Earth System science, e.g., earth realm, non-living substances, living substances, physical processes; unifier ontologies were created to piece together and create relationships that exist among the facet ontologies.

7   Conclusions

The computation-driven ontology was devised to support scientists' ability to capture discipline-specific knowledge that supports their research. Such an ontology focuses on the capture of processes as well as data and reduces the dependence on a technologist to construct an ontology. Computation-driven ontologies are distinguished from discipline-based ontologies that capture basic knowledge about a discipline by capturing concepts and relationships that are tied to how results are generated. In particular, all defined methods are tied to the inputs, outputs, and other computation-associated relationships required to generate a result from a specified method. The gravity data ontology is the first comprehensive ontology that was developed using this approach.

The work reported in this paper has transitioned to the development of a prototype WDO API [8] to facilitate the integration and reuse of WDOs by the WDO-It! tool and other WDO-related tools that are being prototyped. The WDO API is built on top of the Jena2 Ontology API [14] that provides functionality to access OWL ontologies through Java programming. The WDO API offers specific methods that facilitate the development of WDOs, as well as functionality to create abstract workflow specifications. The WDO-It! tool provides a GUI to assist scientists to create new WDOs. Work is in progress to extend domain ontologies into WDOs and to transform abstract workflows to executable workflows. Future work will examine the use of WSDL-S [1]and OWL-S [5] to refine abstract workflows into executable workflows implemented as web service compositions. Both WSDL-S and OWL-S are specifications targeted specifically to enhance web service technology with semantic information.

Acknowledgements. The work described in this paper was partially funded by the NSF GEON project EAR-0225670 and the NSF CREST project HRD-0734825.

References

1. Akkiraju, R., et al.: Web Service Semantics - WSDL-S. World Wide Web Consortium (W3C) recommendation, WSDL-S (November 2005),
   http://www.w3.org/Submission
2. Baker, P.G., Goble, C.A., Bechhofer, S., Paton, N.W., Stevens, R., Brass, A.: An Ontology for Bioinformatics Applications. Bioinformatics 15(6), 510–520 (1999)
3. Hinze, W.J., Aiken, C., Brozena, J., Coakley, B., Dater, D., Flanagan, G., Forsberg, R., Hildenbrand, T., Keller, G.R., Kellogg, J., Kucks, R., Li, X., Mainville, A., Morin, R., Pilkington, M., Plouff, D., Ravat, D., Roman, D., Urrutia-Fucugauchi, J., Véronneau, M., Webring, M., Winester, D.: New standards for reducing gravity data: The North American gravity database. GEOPHYSICS 70, 325–332 (2005)
4. Holom, D.I., Oldow, J.S.: Gravity reduction spreadsheet to calculate the Bouguer anomaly using standardized methods and constants. Geosphere 3(2), 86–90 (2007)
5. Martin, D., et al.: OWL-S: Semantic Markup of Web Services. World Wide Web Consortium (W3C) recommendation (November 2004),
   http://www.w3.org/Submission/OWL-S/
6. Noy, N.F., McGuinness, D.: Ontology Development 101: A Guide to Creating Your First Ontology. Stanford Knowledge Systems Laboratory Technical Report KSL-01-05 (March 2001)
7. Pinheiro da Silva, P., et al.: Knowledge Provenance Infrastructure. IEEE Data Engineering Bulletin 26(4), 26–32 (2003)
8. Salayandia, L., Pinheiro da Silva, P., Gates, A., Salcedo F.: Workflow-Driven Ontologies: An Earth Sciences Case Study. In: Proceedings e-Science 2006, Amsterdam, Netherlands (December 2006)
9. Salayandia, L., Pinheiro da Silva, P., Gates, A., Rebellon, A.: Domain-Level Workflows for Scientific Applications. In: Proceedings 6th OOPSLA Workshop on Domain-Specific Modeling (October 2006)
10. Salcedo, F.: A Method for Designing Computation-Driven Ontologies in the Geosciences. Master's Thesis, University of Texas at El Paso (May 2006)
11. Smith, B., Williams, J., Schulze-Kremer, S.: The Ontology of the Gene Ontology. In: Proceedings AMIA Symp. 2003, pp. 609–613 (2003)
12. Smith, M. K., Welty C., McGuiness, D.L.: OWL Web Ontology Language Guide. World Wide Web Consortium (W3C) recommendation (February 2004),
    http://www.w3.org/TR/owl-guide/
13. Wessel, P., Smith, W.H.F.: New version of Generic Mapping Tools released. EOS. Transactions American Geophysical Union 76, 329 (1995)
14. Jena: Jena2 Ontology API (July 2006),
    http://jena.sourceforge.net/ontology/index.html
15. SWEET: Guide to SWEET Ontologies (June 2007),
    http://sweet.jpl.nasa.gov/guide.doc

Author Index

Lecture Notes in Computer Science

Sublibrary 3: Information Systems and Application, incl. Internet/Web and HCI

For information about Vols. 1– 4443
please contact your bookseller or Springer

Vol. 4658: T. Enokido, L. Barolli, M. Takizawa (Eds.), Network-Based Information Systems. XIII, 544 pages. 2007.

Vol. 4656: M.A. Wimmer, J. Scholl, Å. Grönlund (Eds.), Electronic Government. XIV, 450 pages. 2007.

Vol. 4655: G. Psaila, R. Wagner (Eds.), E-Commerce and Web Technologies. VII, 229 pages. 2007.

Vol. 4654: I.-Y. Song, J. Eder, T.M. Nguyen (Eds.), Data Warehousing and Knowledge Discovery. XVI, 482 pages. 2007.

Vol. 4653: R. Wagner, N. Revell, G. Pernul (Eds.), Database and Expert Systems Applications. XXII, 907 pages. 2007.

Vol. 4636: G. Antoniou, U. Aßmann, C. Baroglio, S. Decker, N. Henze, P.-L. Patranjan, R. Tolksdorf (Eds.), Reasoning Web. IX, 345 pages. 2007.

Vol. 4611: J. Indulska, J. Ma, L.T. Yang, T. Ungerer, J. Cao (Eds.), Ubiquitous Intelligence and Computing. XXIII, 1257 pages. 2007.

Vol. 4607: L. Baresi, P. Fraternali, G.-J. Houben (Eds.), Web Engineering. XVI, 576 pages. 2007.

Vol. 4606: A. Pras, M. van Sinderen (Eds.), Dependable and Adaptable Networks and Services. XIV, 149 pages. 2007.

Vol. 4605: D. Papadias, D. Zhang, G. Kollios (Eds.), Advances in Spatial and Temporal Databases. X, 479 pages. 2007.

Vol. 4602: S. Barker, G.-J. Ahn (Eds.), Data and Applications Security XXI. X, 291 pages. 2007.

Vol. 4601: S. Spaccapietra, P. Atzeni, F. Fages, M.-S. Hacid, M. Kifer, J. Mylopoulos, B. Pernici, P. Shvaiko, J. Trujillo, I. Zaihrayeu (Eds.), Journal on Data Semantics IX. XV, 197 pages. 2007.

Vol. 4592: Z. Kedad, N. Lammari, E. Métais, F. Meziane, Y. Rezgui (Eds.), Natural Language Processing and Information Systems. XIV, 442 pages. 2007.

Vol. 4587: R. Cooper, J. Kennedy (Eds.), Data Management. XIII, 259 pages. 2007.

Vol. 4577: N. Sebe, Y. Liu, Y.-t. Zhuang, T.S. Huang (Eds.), Multimedia Content Analysis and Mining. XIII, 513 pages. 2007.

Vol. 4568: T. Ishida, S. R. Fussell, P. T. J. M. Vossen (Eds.), Intercultural Collaboration. XIII, 395 pages. 2007.

Vol. 4566: M.J. Dainoff (Ed.), Ergonomics and Health Aspects of Work with Computers. XVIII, 390 pages. 2007.

Vol. 4564: D. Schuler (Ed.), Online Communities and Social Computing. XVII, 520 pages. 2007.

Vol. 4563: R. Shumaker (Ed.), Virtual Reality. XXII, 762 pages. 2007.

Vol. 4561: V.G. Duffy (Ed.), Digital Human Modeling. XXIII, 1068 pages. 2007.

Vol. 4560: N. Aykin (Ed.), Usability and Internationalization, Part II. XVIII, 576 pages. 2007.

Vol. 4559: N. Aykin (Ed.), Usability and Internationalization, Part I. XVIII, 661 pages. 2007.

Vol. 4558: M.J. Smith, G. Salvendy (Eds.), Human Interface and the Management of Information, Part II. XXIII, 1162 pages. 2007.

Vol. 4557: M.J. Smith, G. Salvendy (Eds.), Human Interface and the Management of Information, Part I. XXII, 1030 pages. 2007.

Vol. 4541: T. Okadome, T. Yamazaki, M. Makhtari (Eds.), Pervasive Computing for Quality of Life Enhancement. IX, 248 pages. 2007.

Vol. 4537: K.C.-C. Chang, W. Wang, L. Chen, C.A. Ellis, C.-H. Hsu, A.C. Tsoi, H. Wang (Eds.), Advances in Web and Network Technologies, and Information Management. XXIII, 707 pages. 2007.

Vol. 4531: J. Indulska, K. Raymond (Eds.), Distributed Applications and Interoperable Systems. XI, 337 pages. 2007.

Vol. 4526: M. Malek, M. Reitenspieß, A. van Moorsel (Eds.), Service Availability. X, 155 pages. 2007.

Vol. 4524: M. Marchiori, J.Z. Pan, C.d.S. Marie (Eds.), Web Reasoning and Rule Systems. XI, 382 pages. 2007.

Vol. 4519: E. Franconi, M. Kifer, W. May (Eds.), The Semantic Web: Research and Applications. XVIII, 830 pages. 2007.

Vol. 4518: N. Fuhr, M. Lalmas, A. Trotman (Eds.), Comparative Evaluation of XML Information Retrieval Systems. XII, 554 pages. 2007.

Vol. 4508: M.-Y. Kao, X.-Y. Li (Eds.), Algorithmic Aspects in Information and Management. VIII, 428 pages. 2007.

Vol. 4506: D. Zeng, I. Gotham, K. Komatsu, C. Lynch, M. Thurmond, D. Madigan, B. Lober, J. Kvach, H. Chen (Eds.), Intelligence and Security Informatics: Biosurveillance. XI, 234 pages. 2007.

Vol. 4505: G. Dong, X. Lin, W. Wang, Y. Yang, J.X. Yu (Eds.), Advances in Data and Web Management. XXII, 896 pages. 2007.

Vol. 4504: J. Huang, R. Kowalczyk, Z. Maamar, D. Martin, I. Müller, S. Stoutenburg, K.P. Sycara (Eds.), Service-Oriented Computing: Agents, Semantics, and Engineering. X, 175 pages. 2007.

Vol. 4500: N.A. Streitz, A.D. Kameas, I. Mavrommati (Eds.), The Disappearing Computer. XVIII, 304 pages. 2007.

Vol. 4495: J. Krogstie, A. Opdahl, G. Sindre (Eds.), Advanced Information Systems Engineering. XVI, 606 pages. 2007.

Vol. 4480: A. LaMarca, M. Langheinrich, K.N. Truong (Eds.), Pervasive Computing. XIII, 369 pages. 2007.

Vol. 4473: D. Draheim, G. Weber (Eds.), Trends in Enterprise Application Architecture. X, 355 pages. 2007.

Vol. 4471: P. Cesar, K. Chorianopoulos, J.F. Jensen (Eds.), Interactive TV: A Shared Experience. XIII, 236 pages. 2007.

Vol. 4469: K.-c. Hui, Z. Pan, R.C.-k. Chung, C.C.L. Wang, X. Jin, S. Göbel, E.C.-L. Li (Eds.), Technologies for E-Learning and Digital Entertainment. XVIII, 974 pages. 2007.